FRONTIERS IN ENVIRONMENTAL GEOGRAPHY

ABOUT THE EDITORS

Dr. Onkar Singh (*b.* 1941), Professor of Environment and Head, Department of Geo-graphy, Banaras Hindu University, Varanasi, is currently engaged with a few research projects, including one under Indo-Dutch collaboration, related to environmental degradation and management. Besides several research papers he has authored a monograph on India's Urban Environment. He has been to Germany and Hungary where he delivered lectures in important institutions on environmental issues in India.

Dr. Prithvish Nag (*b.* 1951) is M.Sc. (Gold Medalist) and Ph.D. from Banaras Hindu University. He is the Chairman of the International Commission on Population Cartography. He has been closely associated with international activities, such as a Member of the International Working Group on the Thematic Maps of the World, and a Member of the Editorial Board of the International Journal of the Geographical Information Systems. Dr. Nag is the author and editor of several books and about 60 research papers. He is a holder of the Commonwealth Geographical Borsary awarded by the Royal Geographical Society on its 150th anniversary. Recently he has been UNDP Consultant in Oman. He is now attached to the National Atlas and Thematic Mapping Organisation, Calcutta as a Deputy Director.

Dr. V.K. Kumra (*b.* 1954), Reader in Geography, Banaras Hindu University, took his M.Sc. (1975) and Ph.D. (1978) from B.H.U. He has published a number of books and research papers on environmental issues. Dr. Kumra has been recipient of Commonwealth Academic Staff Fellowship during 1989-90 at the University of Cambridge, U.K.

Dr. Jagadish Singh (*b.* 1952) is a Senior Lecturer in Geography, Banaras Hindu University. Dr. Singh has to his credit about two dozen research papers and four books including a monograph on Upper Damodar Valley. His major fields of interest are Settlement Geography and Environmental Studies. Recently he has been to Japan and U.K. in connection with Conferences on Environmental Aspects.

FRONTIERS
IN
ENVIRONMENTAL GEOGRAPHY

Edited by

ONKAR SINGH
Professor & Head, Department of Geography
Banaras Hindu University
Varanasi

PRITHVISH NAG
Deputy Director
National Atlas and Thematic Mapping Organisation
Calcutta

V.K. KUMRA
Reader in Geography
Banaras Hindu University
Varanasi

JAGADISH SINGH
Lecturer in Geography
Banaras Hindu University
Varanasi

CONCEPT PUBLISHING COMPANY, NEW DELHI-110059

ISBN 81-7022-462-4

First Published 1993

Published and Printed by
Ashok Kumar Mittal
Concept Publishing Company,
A/15-16, Commercial Block, Mohan Garden,
NEW DELHI 110059 (India)

Lasertypeset by
Microtech Advance Printing Systems (Pvt.) Ltd.,
H-13, Bali Nagar,
NEW DELHI-110015

FOREWORD

It was during the Platinum Jubilee Session of the Indian Science Congress, Poona University, 1988, that I had the privilege of delivering the Presidential Address of the Geology and Geography Section on *Frontiers in Environment*. It is indeed heartening to see the trail leading on to a book on *Frontiers in Environmental Geography*, jointly edited by Prof. Onkar Singh, Dr. Prithvish Nag, Dr. V.K. Kumra, and Dr. Jagadish Singh. All the editors are geographers with sound grasp of environmental geography.

Environmental concern has travelled a long way in the two decades since the U.N. Stockholm Conference on Human Environment in 1972, to Earth Summit in Rio de Janerio in June 1992. Greater awareness and concern has been generated on environmental issues. However, the world remains confronted with even greater environmental problems and challenges. The Rio Earth Summit is therefore of vital concern to us all. It is expected to be the largest assembly of world leaders in history, and could well be the most important for environment and mankind. The goal of the U.N. sponsored summit is challenging, that is, to chart a course that will halt degradation of earth's air, land, and water, and protect the multitudes of animals and plants threatened with extinction. The U.N. Conference on Environment and Development (UNCED) proposes to bring out several landmark documents, including the Earth Charter, that would constitute a set of principles designed to ensure a programme of action, called Agenda 21, and treaties aimed at curbing climatic change and limiting the loss of biodiversity. However, if such agreements are not backed by political will and people's support, they will end as mere cosmetic approaches to environmental problems. The central theme at Rio is sustainable development in which economic growth on longer results in net destruction of natural resources. There is no unanimity on the approaches and time-schedule. The European Community wants to reduce CO_2 emissions to 1990 levels by the year 2000. America's refusal to agree to this programme has already stymised the latest round of climate change negotiations. The changes advocated in the drafts of documents like Agenda 21, such as commitments to mass transits and high efficiency energy systems could ultimately improve the standards of living. A recent survey contends that by cutting down CO_2 emissions by 70 per cent over the next 40 years would cost the U.S. economy $ 2.7 trillion, but would trim fuel and utility bills by $ 5 trillion. Reducing waste and pollution will make fundamental changes to national economies, observes Maurice Strong. Germany is already endeavouring to move to a trashless society. The economic consequences of not making the changes will have to be borne later. Environmental planning and management would require all the wisdom and foresight. The real competitive disadvantages could come to nations failing to protect the environment. Already Japan is developing a 100-year plan to capitalize on environmental concerns by designing "Green Technologies", and the European nations are also trying to move in the same direction. It is obvious that those who do not plan ahead on the present and emergent environmental issues, and do not pursue environmentally responsible development are going to be left behind in dust and smoke.

Like other countries, India has no choice but to plan ahead on environmental issues. India carries the burden of a large growing population, along with the problems of poverty, unemployment, haphazard development, erosion of resources, and environmental degradation. There is little room for complacency. For balanced integrated development and environmental protection, the main tasks would be integrated land, soil, water, and forest management, pollution control, development of non-polluting energy resources, waste-utilization through recycling, conservation of biological diversity, slum removal and development of human settlements for healthy living, environmental awareness and education, population control and health-care and the will to rehabilitate and optimize the environment with sustainable development.

There has been much environmental change since man first appeared on this earth. Human dominance has ushered a new era of extinction, primarily by usurping natural habitats. Pollutants in environment wreak unforeseen results, among them a global warming trend. Yet with the potential of self-destruction, through war and destruction of environment, the present moment of technology's course has delivered tools of knowledge never before in hand. Among then are earth-sensing satellites, and computers, which are now available for the first time to watch the global mechanism of life at work. We must try to comprehend the state of environment, and mend the losses of disruption of nature's systems, the self-renewing cycles that rejuvenated land, water and air. When we temper with these systems, we threaten the basis of life itself.

To safeguard India's environment, the administrators, the politicians and the people must be well-informed on environmental issues. It is such important issues that have been highlighted in the present volume, which will be of use to the academics, the administrators and the voluntary agencies, who are involved in the solution of environmental problems.

What kind of world shall the coming generations inherit? The answer largely depends on the actions taken now by the governments and the individuals all over the world. It is of prime importance that sincere efforts are made towards conservation of environment and development of a conscience for sanctity of life. In this respect, India has long traditions of environmental ethics that extend to all living and non-living things. The resources exist to start building a world that is in ecological and human equilibrium. This will happen only when an informed and dedicated people want it to happen. If the man's greatest advances occurred in the 20th century, — the greatest challenges lie in the 21st century. The future generations will praise our foresight, if we cooperate in global spirit to save the environment of our planet. The hope rests on the actions taken by the present generation towards wise stewardship of the Planet Earth for a better quality of life. May our endeavours continue in this direction.

In concluding the foreword to this book, I share the hope that it will be widely read by those who feel concern for the well-being of the environment.

22nd April, 1992
Varanasi

S.L. KAYASTHA
Formerly, Professor & Head
Department of Geography
Banaras Hindu University, Varanasi

PREFACE

The study of environment has been the core of geography. Geographers from time to time have attempted to define this discipline in relation to the environment. Some were more inclined towards the physical aspects, while others were inclined to give more emphasis on the human aspects. Further, when the study of geography and environment are so close, is the frontier in environmental research different than geography? What are the frontier areas in environmental geography and where is the focus if there is a sub-discipline at all ? In such a situation how can there be a sub-discipline or thrust area on *environmental geography?*

There is no end to the above questions. But, such introspections compel us to have a survey of current trends in research in geography. The environmental issues in the frontier areas of the subject denote that certainly there has been a shift in approaches and research methodologies in the last decade. But there is no indication that the feedback from such researches has influenced the core concepts in geography. The man-environment relationships have different meaning when a population group unconnected with nuclear radiation or chemical fallouts is found to be concerned though the event happened or it is likely to happen thousands of kilometres away. Thus the concept of distance of space is fast changing. Further, the scientific language and vocabulary has changed in the frontier areas of researches. Geography is fast getting intermingled with other disciplines regarding researches on environmental issues. This has become essential because the study of environment requires such concerted efforts. With current initiatives, such study has been revitalized and there is a case for promoting a sub-discipline of environmental geography. The human factors have been high on the agenda. Behavioural approaches have modified the concept of space — the foundation of geographical thought. Based on the survey, the definition closer to Hartshorne can be worked out. The environmental geography is concerned to provide with accurate, orderly and rational description and interpretation of the variable environmental parameters over space in relation to human behaviour, priorities and expectations.

To define environmental geography and consequently redefine the discipline of geography has been the main thrust of this volume. The examples, conceptual or case studies, have a common goal towards this objective. The spectrum of environmental issues is so wide that the thirty two case studies are not sufficient to have a total gammut. However, within the limited scope, each representative case is supposed to highlight the emerging frontiers in this discipline.

The introductory chapter by P. Nag brings out the issues and challenges in the study of environment in the last decade. The issues shifted from urban to rural are more behavioural in nature. The study of disasters, crisis and catastrophes will influence the core of this discipline, i.e. the man-environment relationship. Smita Sengupta's paper on the environment and development underlines the problems in the third world situation in general. The issues are related to development, exploitation of resources, global climatic changes, bio-technological revolution, pollution, industrial fallouts, international co-operation, policy, legislation and possible approaches. This overview provides a basis for consideration about emerging issues in the environmental studies at regional, national global levels. The study of Kumaon Hills by D.S. Singh and A.K. Singh highlights similar issues in a regional perspective. How different development activities have resulted in deteriorating the environment? The most significant part of their contribution has been the linkages of human and natural factors in shaping catastrophes. Another case study of Damodar Valley Region by Swapna Basu also points out the fact that once the environment is disturbed for the sake of development, the nature may not heal the effects of the silver bullets of technology in the environment. It may contribute towards crisis-like situation. Pollution, waterlogging and creation of wastelands may become permanent features. Further, the contribution of Kishan S. Rajpurohit and Daksha C. Barai clearly indicates that relief measures connected with environmental degradation are partial, temporary and *ad hoc.* Relief in any form is no prescription for environmental regeneration.

How far agricultural practices contribute to environmental degradation is a debatable issue. The work by R.K. Rai on Meghalaya signifies that shifting cultivation does not cope up the in-

Subramaniam and P. Sanjeeva Rao in their study on climatic models in the semi-arid district of Andhra Pradesh, make an assessment of the Kharif rice production using monsoon index and yield moisture index. The impact of different intensities of drought on rice crop were discussed.

The study of the urban environment has a perennial interest for geographers. M.D. Nandeshwar, S.A. Abbasi, P.C. Nipaney and R. Soni made an assessment of the pressure created by the environmental pollution as a consequence of the urban development. The detail study for this purpose was taken for the water pollution in Bombay. This contribution also examines the prevailing environmental standards and their relevance in the light of recent research findings. The attention for studying the environmental Pollution in a small town like Ghazipur is a healthy trend. Onkar Singh and K.K. Singh have rightly mentioned the increasing menace of environmental aspects, such as housing condition, water, air and noise pollution. In the same continuation, B.B. Dhar has studied the impact of noise pollution. Noise abatement technology in heavy industries including mining has been suggested. Another study of air pollution by Onkar Singh highlights the impacts in a non-industrial environment like Varanasi city. The spatial pattern of the air pollution elements, such as suspended particulate matter (SPM), Sulphur dioxide (SO_2) and Oxides of nitrogen (NO_x) have been mapped in order to set on integrated picture.

The study of the association of crime with the environment do make an important point. How environmental conditions encourage criminal activities? These environmental conditions are not only natural but man-made as well. The case of Agra district, as studied by R. Prasad, should not however be considered as an isolated case. The man-environment factors have combined in different ways to harbour undesirable activities.

Modern technologies are now considered quite useful in chalking out strategies for the environmental management. P.S. Roy and B.K. Ranganath have attempted to redefine such strategies in the forest management using remote sensing data. The digital data can be manipulated in different ways in order to identify various features of the environment, including forests. However, remote sensing has become an essential part of the information system required for the environmental assessment and management. P. Nag and H. Prasad have pointed out how different information sources can be integrated with the technological options now knocking the door steps of the geographers. These options can no longer be ignored considering the issues and challenges in the emerging frontiers of environmental geography. On the other hand, the environment is also to be assessed at the lowest level, particularly when it relates to human health. R.S. Singh's study of the biocide induced environment is a valuable contribution in this direction. The last contribution is related to depleting vascular cryptogams in Kumaun Himalaya. Wherein P.C. Pande, A.K. Singh and H.C. Pande have described the deteriorating ferns and fern-allies population from environmental point of view.

Obviously, the above arguments make a case for assessing the development in the frontier areas of environmental geography. With this idea in mind, we the editors, specialising in different fields of environmental geography, decided to bring out a volume consisting of studies on different environmental issues and from various parts of the country. The authors from other disciplines were requested to make their contributions close to our requirements. We also attempted to get contributions from multidisciplinary teams of authors, involving geographers. This approach would have helped us in understanding the linkages of environmental geography with other disciplines. We were only partially successful on this account.

The organisation of this volume took about two years. It was time consuming process to get in touch with the potential authors through correspondence and personal contacts. Nevertheless, the authors were kind enough to spare their valuable time in honouring our requests. The editors met several times in Varanasi and Calcutta to give a final shape to this volume. Needless to say that we found Shri Naurang Rai Mittal and Shri Ashok Kumar Mittal of the Concept Publishing Company willing to accept our request to undertake the publication of this volume. We are grateful to them.

ONKAR SINGH
PRITHVISH NAG
V.K. KUMRA
JAGADISH SINGH

CONTENTS

LIST OF FIGURES

LIST OF CONTRIBUTORS

Shri Abbasi, S.A., Salim Ali School of Ecology, Pondicherry University, Pondicherry-605001

Prof. Barai, Daksha C., Department of Geography, Bangalore University, Jpana Bharati, Bangalore-560056

Dr. Basu, Swapna, Department of Geography, Calcutta University, Calcutta.

Shri Chaudhary, G.C., Department of Geology, Banaras Hindu University, Varanasi-221005

Dr. Das, Debashis, Department of Geography, Visva Bharati University, Shantiniketan.

Prof. De, N.K., Department of Geography, Burdwan University, Burdwan.

Dr. (Mrs.) Desai, Anjana P., Department of Geography, Gujarat University, Ahmedabad-380009

Dr. (Mrs.) Desai, Mamta, Netaji Institute for Asian Studies, Calcutta.

Prof. Dhar, B.B., Department of Mining Engineering, Institute of Technology, Banaras Hindu University, Varanasi-221005

Prof. Indra Pal, Department of Geography, Rajasthan University, Jaipur

Prof. Gupte, S.C., School of Environmental Sciences, Department of Geography, University of Poona, Pune-411007

Dr. Kale, V.S., School of Environmental Sciences, Department of Geography, University of Poona, Pune-411007

Dr. (Mrs.) Khanna, Sulbha, Department of Geography, Gujarat University, Ahmedabad-380009.

Dr. Kumra, V.K., Department of Geography, Banaras Hindu University, Varanasi-221005.

Prof. Mitra, Champa, Department of Geography, Burdwan University, Burdwan.

Dr. Nag, P., National Atlas and Thematic Mapping Organisation, Calcutta.

Dr. Nandeshwar, M.D., Surface Water Division, Centre for Water Resources Development and Management, Kumaramangalam, Calicut-673571.

Shri Nipaney, P.C., Salim Ali School of Ecology, Pondicherry University, Pondicherry-605001.

Dr. Pande, H.C., Department of Botany, Kumaun University Campus, Almora-263601.

Dr. Pande, P.C., Department of Botany, Kumaun University Campus, Almora-263601.

Dr. Pande, R.K., Department of Geography, Kumaun University, Campus, Almora-263601.

Dr. Pawar, N.J., School of Environmental Sciences, Department of Geography, University of Poona, Pune-411007.

Dr. Phadke, V.S., Department of Geography, University of Bombay, Vidyanagari, Bombay-400098.

Shri Phondekar, D.G., Department of Geography, University of Bombay, Vidyanagari, Bombay-400098.

Shri Pirazizy, A.A., Department of Geography, University of Delhi-110007.

Dr. Prasad, Haushila, Department of Geography, Kenyatta University, Nairobi.

Dr. Prasad, Rajendra, National Atlas and Thematic Mapping Organization, Calcutta.

Prof. Rai, Lallan, Institute of Agricultural Sciences, Banaras Hindu University, Varanasi-221005.

Prof. Rai, R.K., School of Environmental Sciences, North Eastern Hill University, Shillong-793004.

Dr. Rajpurohit, Kishan S., Department of Geography, Bangalore University, Jnana Bharati, Bangalore-560056.

Shri Ranganath, B.K., Regional Remote Sensing Service Centre, Bangalore.

Shri Rao, P. Sanjeeva, Rubber Research Institute of India, Rubber Board, Kottayam-686009.

Dr. Roy, P.S., Forestry and Ecology Division, Indian Institute of Remote Sensing, Dehradun.

Dr. Sarkar, H.K., National Atlas and Thematic Mapping Organization, Calcutta.

Dr. Sengupta, Smita, Department of Geography, Calcutta University Calcutta.

Shri Singh, A.K., Engineering Geology Cell, Almora.

Dr. Singh, Anil Kumar, Department of Geography, Kumaun University Campus, Almora-263601

Shri Singh, D.S., Engineering Geology Cell, Almora.

Dr. Singh, Jagadish, Department of Geography, Banaras Hindu University, Varanasi-221005.

Dr. Singh, K.K., Department of Geography, Banaras Hindu University, Varanasi-221005.

Prof. Singh, Onkar, Head, Department of Geography, Banaras Hindu University, Varanasi-221005.

Dr. Singh, R.B., Department of Geography, University of Delhi, Delhi-110007.

Dr. Singh, R.S., Bharat Kala Bhawan, Banaras Hindu University, Varanasi-221005.

Dr. Singh, S.K., Department of Geography, Kumaun University Campus, Almora-263601.

Shri Soni, R., Salim Ali School of Ecology, Pondicherry University, Pondicherry-605001.

Prof. Subramaniam, A.R., Department of Meteorology and Oceanography, Andhra University, Waltair-530003.

Dr. Tripathi, B.D., Pollution Ecology Research Lab, Centre for Advance Study in Botany, Banaras Hindu University, Varanasi-221005.

Km. Verma, Santosh, Department of Geography, Kumaun University Campus, Almora-263601.

1

EMERGING FRONTIERS IN ENVIRONMENTAL GEOGRAPHY

PRITHVISH NAG

In 1985 an edited volume was brought out on *Geography and Environment : Issues and Challenges.* The main thrust of this book was to re-emphasise the significance of the study of environment in geography. Geographers have been studying the environment traditionally. This book was one of its kind in India (Singh et. al., 1986). Attempt was to provide some conceptual basis to study environment in geography. Since then there has been a flood of literature in environmental studies in geography (Singh, 1990). One gets perplexed looking at the range of issues taken up for the environmental studies. It includes the traditional study of man-environment relationship to the introduction of G.I.S. and remote sensing technologies. The obvious questions are: What is the focus of environmental studies in geography ? How can a geographer contribute more in the interpretation of the environmental related issues ?

Concern for Environment

The study of environment has almost become synonymous to the study of crisis. It is not always the frequency of mishaps which has increased, but also more and more information on such events is now available. The news-papers, electronic media and fast means of communications have helped in disseminating the information which may be time specific. In any case, events may have localized impact but can be a matter of regional, national or global concern. The concern for the environment has no bounds. Due to this reason, environmental issues have become a political tool. Individuals, institutions, industries and governments have taken the advantages of these issues for their own aims and goals.

At the international scientific level, there are several bodies, agencies, commissions and groups who are directly or indirectly concerned with the environment. IUCN, WMO, UNEP, IGU and their affiliated bodies are a few examples. Further, the International Council for Scientific Unions (ICSU) sponsored an International Geosphere Biosphere Programme (IGBP) for studying environment, but it is considered as a *hard science* approach. In order to include human aspects, a parallel research programme entitled *The Human Dimension to Global Change* has been mooted in collaboration with the United Nations University (UNU), International Social Science Council (ISSC) and the International Federation of Institutes of Advanced Sciences (IFIAS). Hence, the human aspects of environmental studies are becoming relevant in the world today. In the geographical approach to the environmental studies, this aspect has been considered very important since the evolution of the subject.

Human Dimensions of Environmental Studies

In the last decade a new set of events and related consciousness have to certain extent demanded a fresh explanation of man-environment relationship. These events are related to nuclear fallout in the power plant of the Three Mile Island due to human error; nuclear leakage in Chernobyl plant, toxic release at Love Canal and Bhopal, and situations like collisions, explosions, breakdowns, collapses and crisis. There are several more examples of the technological disasters. These disasters become news immediately, though the casualties involve lesser in number of people than other disasters, such as floods, road accidents and coastal cyclones (Nag, 1991). The reason is the acceptability of natural disasters. The dangers and risks are known. People are aware of the consequences. On the other hand, the people have not resigned to the technological disasters. As a result, the efected persons may be limited but such events become a matter of global concern. The spatial dimensions of human concern are a new phenomenon in the environmental studies.

Two important volumes have been brought out in past 15 years which establishes the role of human dimensions in environmental studies. The first volume edited by Gilbert F. White (1974) on *Natural Hazards: Local, National, Global* is an international survey of the process of human adjustment to hazardous situations. How do people adjust ? What are the risks involved ? What are the precautions taken ? Further, in a recent volume edited by John I. Clarke, Peter Curson, S.L. Kayastha and P. Nag (1989) on *Population Disasters,* the population-environment relationship has been high-lighted again, but thc issucs arc slightly different. This study includes disasters of human causation and epidemics. The issues are related to the effects on fertility, mortality, migration, age and sex selective damages, social disturbance, AIDS, toxic shock and syndrome.

The disaster syndrome continues. Disasters, hazards, catastrophes, crises, risks and the like are some of the few words which are now frequently being used in popular, scientific and geographical writings. All these events are considered significant if the damages to population are substantial. But, with the replacement of natural disasters by toxic, technological and nuclear disasters, it is the fear, psychosis, possibility of permanent and genetic damages which have become the focus of man-environment relationship. The issues are related to evacuees reactions, unending misseries, radiological accidents, toxic emergencies, and invisible contaminants. The technological roster is endless.

Spatial Dimensions of Environmental Studies

The study of hazards and disasters have in a way modified the spatial aspects of the study of the environment. People concerned with a particular disaster or event can be in a part of the other continent, thousands kilometres away. The International Geographical Union (IGU) Commission on Population Geography mooted a study on population and scale (Kosinski & Webb, 1976). The attempt was to investigate the population issues with scale. Attempts have also been made to find out whether environmental issues can be related to scale (Nag, 1991). What are the spatial dimensions of environmental issues ? Do we have region specific environmental issues ? Conversely, whether such issues help in forming a region ? These questions pertain to very basic issues in geography where one would like to refer back to Kant, Humbolt and Hartshorne (Hartshorne, 1961 and Harvey, 1979). The man-environment relationships have to be reinvestigated in geography in the light of current environmental issues. With hazards, disasters and crises, the concept of region becomes more nebulous. Continuation of such events in temporal and spatial frame further complicates the basic issucs in geography. Furthermore catastrophes due to combination of the events and due to mismanagements of a hazard may create a region somewhat semi-permanent in nature, like droughts in Ethiopia or war like situation in West Asia, Afghanistan or Gulf. In addition, the political considerations have enlarged or even reduced the spatial dimensions of environmental issues.

The occurrence of hazards, their frequencies and intensities will certainly have implications in the spatial understanding. The earlier studies in natural hazards are worth mentioning here (White, 1974). Natural hazards are those elements of the physical environment, harmful to man and caused by forces extraneous to him (Burton & Kates, 1964). Such hazards are accepted as a part of the environment. They generally have some spatial or regional bearings. But, the natural hazards are being replaced by more damaging technological hazards. Every day a new section of population and additional land is exposed to risks. Even the most suitable areas for human settlements are not free of hazards. There is a threat to the civilizations with the inclusion of land, population and settlements in the advancing risk-zones (Nag, 1983). The risks for natural hazards are known and experienced, and thereby the spatial dimensions of such hazards can be estimated. But, regarding technological hazards the risk are undefined and unexperienced (Erikson, 1990).

The environmental perception studies in geography has re-emphasized the behaviour components in man environmental relationship. Sonnenfeld identifies a hierarchy of environments, each of which defines a spatial relation of environment with man (Nag, 1983). The broadest level is the *geographical environment,* constituting both proximal and distal elements of man's universe. This reduce in stags to the *operating* environment, which is that environment impinging on man with which he is likely to be involved in one way or other, to the *perceptual environment* of which he is aware, and finally to the *behavioural environment* which elicits a behavioural response from the individual (Sonnenfeld, 1978). With unknown risks, the perceptual environment become obscure. Such studies are likely to face new challenges with the increase in unexpected, unknown and less experienced situation relating to environment.

The Emerging Frontier

The man-environment relationship is changing and due to this fact the basic concepts and postulates in geography are likely to be reviewed or revised in near future. The research frontier has definitely come out of the preconceived shell. The concept of space is changing. The human association to environment is finding new equations. New issues have replaced the old ones. The environmental geography is spearheading new concepts which may ultimately influence the core of geography. What is the focus and where is the frontier? What is the emerging relationship between them? What will be the frontier if we are trying to find a focus in environmental geography in particular and this discipline of geography in general ?

On the other hand, how will the expanding frontier influence in the search for the focus ? Furthermore, how the issues at frontier take shape at different levels up to the core or focus of this discipline? These questions are becoming relevant in fast changing scenario in environmental geography due to emerging population issues, changing concept of space, quick, accurate and detailed information generating systems and expanding research frontiers in the sister disciplines.

This contribution is aimed to have an assessment of the research initiatives in the frontiers of geography and in other disciplines. The conventional topics related to agriculture, development, landslides, migration and ecological balance have been taken but the issues are different, e.g. agricultural practices with population pressure, or ecological adoption of the migrants. There has been a shift in approach in studying the rural environment. The environmental issues in rural India can not be neglected any longer. Poverty, energy, water supply, sanitation and modernization related issues *per se* are essential part of the development strategy. Further, the pollution studies are not only confined to big cities. The environmental problems in such cities are already over emphasized. The attention has shifted to medium size urban centres, such as Ghazipur, a dusty town in eastern Uttar Pradesh. The specific urban pollution and related aspects do have relationships with the findings in other disciplines, such as water and noise pollution.

The region specific environmental issues have made lasting impact in the frontiers of geographical studies. But the themes and the methods have changed. Studies have become more and more complex. These studies may be related to deforestation or exploitation of resources. The complexities of environmental studies have increased with the more and more understanding of the chemical parameters.

It is not only in pollution studies, where the analysis is now being done with the help of BOD, carbon monoxide, etc., the study of plants or protection of cultural heritages can not be completed without chemical analysis.

The frontiers of environmental studies have not only expanded in understanding or analysis of the environment, but also to process voluminous information which is now pouring in continuously. The flow of data, their processing facilities and spatial superimposition have made our tool sharper, which facilities in understanding the complexities as well as to establish rapport with other disciplines. Remote sensing, GIS technology and chemical and pollution analysis have joined together to push the frontier in environmental geography in all direction.

Conclusion

The study of crisis, hazards, disasters and catastrophes have introduced new issues for consideration, analysis and interpretation. These issues have called for the changes in the basic tenets of geography. The core concept of *man-environment relationship* in geography have to be redefined with the new initiatives in researches in frontiers areas in the sub-discipline of environmental geography. People have reacted in different ways to an environmental crisis situations. Population may not directly be affected by a crisis, but they are concerned with a possible crisis or a crisis-like situation. The unknown and unexperienced crisis, such as technological hazards, have not only been able to make a wider impact but the understanding of spatial relationship with the environment has undergone change. The electronic media, fast communication, and dependable sources of information alongwith technological facilities for processing have raised issued which have push the frontier of geography closer to other disciplines. The feedback from the sister disciplines have become essential in order to have an idea of the frontier in environmental geography.

REFERENCES

Burten, Ian and R.W. Kates (1964), The perception of natural hazards in resource management, *Natural Resource Journal*, vol. 3, No. 412, pp. 412-41.

Clarke, John I, Peter Curson, S.L. Kayastha, P. Nag (eds) (1989), *Population and Dis-*

aster, Besil Blackwell, Oxford.

Erikson, Kari (1990), Toxic reckoning: Business Faces a new kind of fear, *Harvard Business Review,* January-February, pp. 118-26.

Hartshorne, Richard (1961), *Perspective on the Nature of Geography,* Rand McNally & Co., Chicago.

Harvey, David (1979), *Explanation in Geography,* Edward Arnold.

Kosinski, L.A. & J.W. Webb (eds) (1976), *Population at Microscale,* IGU Commission on Population Geography and New Zealand Geography Society, Christ Church.

Nag, P. (1983) Perception of the environment: Issues and Challenges, *National Geographical Journal of India,* Vol. XXIX, pp. 217-31.

Nag. P. (1991), Population and Natural disasters, Paper presented to the Indian Science Congress, Indore (Plenary papers).

Singh, H.H., P. Nag, V.K. Kumra, *et al* (eds) (1986), Geography and Environment: Issues and Challenges, *Concept Publishing Company,* New Delhi.

Singh, R.B. (ed) (1990), *Environmental Geography,* Contributions to Indian Geography Series, Heritage Publishers, New Delhi.

Sonnenfeld, J. (1978), Geography, perception and the behavioral environment, in Blunden John, P. Haggett, C. Hannett and P. Sarre (eds), *Fundamentals of Human Geography,* Harper and Row, London, New York, Hageration, San Francisco & Sydney, pp. 26-32.

[illegible] gess, Basil Blackwell, 1982.

[illegible], Kent (1990). [illegible] Business [illegible] issues and problems, [illegible] [illegible] pp. 14-20.

Hartshorne, Richard (1959). *Perspective on the Nature of Geography*, Rand McNally & Co., Chicago.

Harvey, David (1969). *Explanation in Geography*, Edward Arnold.

[illegible] (eds.) (1976), *Geographical* [illegible], IGU Commission on [illegible] Geography and New Zealand Geography Society, Christchurch.

Singh, [illegible] (1983) Perception of the environment: Issues and Challenges, *National Geographical Journal of India*, Vol. XXIX, pp. [illegible]

Singh, [illegible] (1991). Population and Natural Disasters, Paper presented to the Indian Science Congress, Indore (Pilot papers).

Singh, [illegible] (eds.) (19[illegible]). *Geography and Development: Issues and Challenges*, Concept Publishing Company, New Delhi.

Singh, R.K. (ed.) (1989). *Environmental Geography*, Contributions to Indian Geography Series, Heritage Publishers, New Delhi.

Sonnenfeld, J. (1972). Geography, perception and the behavioural environment, in [illegible] (eds.) *Fundamentals of Human Geography*, Harper and Row, London [illegible], pp. [illegible]

2

ENVIRONMENT AND DEVELOPMENT

SMITA SENGUPTA

In the modern age of science and technology the quality of our environment is deteriorating very rapidly. Industrialisation, urbanisation, exploitation and the consequent depletion of natural resources are leading to an exhaustion of irreplaceable capital assets. The deteriorating quality of the environment and the consequent imbalance and disruption in the ecological relationship has become a matter of an ever increasing concern throughout the world. While one can agree at a certain minimum standard of living for bulk of the humanity particularly in the third world countries which is below the poverty line, there does not seem to be any justification for limitless luxuries and fanciful goods. There is no definition of *enough* available in modern literature on economic growth. There is a growing insatiable desire of all kinds and multiplication of wants. The developed nations are being poorly imitated by developing nations. As a sequel there is rapid escalation of desire humanity into depths of moral depravity and un-quenched thirst for more. Since the supply of natural resources is finite, unlimited exploitation of nature's bounties has led to imbalances in ecosystem and environmental pollution and hazards of an unprecedented nature. The main driving forces are the exponential growth of people, production, power, places and pollutants. The impact of this growth on the water, air and land cycles of nature are cited.

Development and Environmental Approach

Man's urge for development is old. In the middle ages the European forests and woodlands were cleared in the name of development. The reason ascribed for the clearing was to make the world more habitable and comfortable for the human being. Development of technology helped the process of landscape modification, ultimately leading to massive urbanisation and industrialization.

After causing degradation of the natural environment over the explored land masses, the only left out frontier continent of Antarctica is now subjected to attach, in the name of research activities. Presently this continent has about 35 permanent bases, in addition to dozens of temporary summer camps staffed by 4,000 men and women.

Not only the landmass, the development has adversely influenced the atmosphere. The rise in the temperature through unprecedented carbon dioxide (CO_2) built up in the atmosphere is of prime concern. Carbon dioxide release into the atmosphere creates the greenhouse warming of the earth's atmosphere. The greenhouse effect prevents the radiation from the sun from escaping back into space. The rise of temperature due to greenhouse warming has caused rapid melting of ice caps in the polar region. This would raise the global sea levels, which in turn would flood many places and food producing areas of the present day. Further, the global warming will bring changes in the precipitation pattern with undesirable impact on agricultural and forestry systems. Various human, industrial and agricultural activities emit different basic elements, such as sulphur, nitrogen, and phosphorus into the atmosphere. Air conditioning, insulations, solvents, fire extinguishers, polyurethane foams and many other products use chlorofluoro carbon which causes the thinning of Ozone layer in the atmosphere. The depletion of the Ozone shield is increasing ultraviolet radiation and causing skin cancers.

The environment and economic systems have an intricate interconnections. The environment provides raw materials to the economic system and absorb the waste generated by the economic systems. Therefore, economic system will always be constrained by the production and the waste absorption capacities of the environment. Economic system have a more or less parasitic role in the biosphere. Thus, to control the environmental deterioration the end of economic or its sharp curtailment is required.

The modern agricultural system which uses chemical fertilisers, pesticides, insecticides etc. also have disastrous consequences in the rural ecosystem. The ever growing biotechnological revolution has shown potentials to improve the quality of human life, but it also causes serious damages to health environment and other socio-economic fields.

Development also pollute hydrosphere to a large extent. Effluents from the industrial units heavily contaminate the groundwater and rivers mainly in developing countries. On the other hand, industrial affluents affect the drinking water and ruins agricultural land. The presence of faecal material in the river water poses the most immediate hazard to human health since faecal material comes either from human or animal sources contain pathogenic micro-organisms. Diseases transmitted by faecal material are often caused by viruses, bacteria, protozoa or metozoa.

Another hydrological phenomena, i.e. flood, is also the result of the destruction of the environment in the name of the development. The deforestation in the upper catchment area is increasing erosion considerably. The main river channel and its tributaries are losing storage capacity due to sedimentation in their bed. Unplanned building of reservoirs and dams also trapping sediments in the river bed. All these activities cause flood in the lower part of the river basin.

The development and environment have a very strong relationship. The main's desire for development in the name of urbanization, industry, science and technology promotion is a great threat to the environment. Hence, it is essential to find out meaningful linkages between conservation, developmental programmes and technology. Ecologically, sustainable development requires common goals and understanding between people and their government and depends upon effective political support. Educational efforts are necessary to move the world into an ecologically sustainable mode of behaviour.

Global Interest in Environment

Global interest in environmental protection and conservation is less than two decades old. It is only after the Stockholm Conference on Environment in 1972, there has been a real outburst of interest in the problems of environment, ecological balance, eco-development etc. At the global level, the United Nations has established the United Nations Environment Programme (UNEP) which has given a catalytic role to urge appropriate action to halt and reverse environmental degradation in various parts of the world. The conservation Strategy launched world over is based on the theme of *development without destruction* and the concept of protecting eco-systems for providing the base for sustained development. During the past two decades the environmental study has shot into prominence all over the world. As a result, there has been a flood of researches on the environment. The main emphasis of the study of environmental scientists is related to the continuous deteriorating environment.

Another factor for ecological imbalance is the shrinkage of forest area. The unwanted population explosion and its pressure on land leads to encroachment of forest areas for farmland. Other environmental hazards ranging from acid rain to atmospheric disruption and destruction of the plants habitats. According to the projection of the population institute in Washington, the world population would reach 11 billion by 2050 A.D. and could result in crisis leaving the Third World wrecked by vast poverty and misery. The urbanization is the basic factor related to population growth that induces environmental pollution at alarming rates by way of industrial effluents causing water and air pollution. The problems of noise pollution and solid waste disposal more concerns urban areas.

The pollution is the major environmental crisis in the recent year. The term pollution implies the introduction of something harmful unpleasant or even toxic substance in our every day life. As a result of the re-awakening of interest in environmental matters, many countries have now consolidated and defied previous legislation and introduced new laws which endeavour to deal with the environment as a whole, for example, Pollution Control Act (1974) of the U.K.

Pollution, however, knows no barriers, and, although it normally only affects the country in which it is emitted but can be carried away by winds and water to other countries. With the greater development in international transport, mobile emitters too have brought in with them their own pollution problems. For these reasons a variety of international committees have been established in order to bring about further exchange of information and research with a view to drawing up of common standards for pollution control. The most important organisations in this fields are as follows:

European Economic Community has established an environment programme to improve the setting and quality of life and people's surroundings and living conditions. The aim of the programme is to promote studies into specific problems and the establishment of proposals for the improvement of the environment. Directives are binding on member states but the mode of implementation is left to the individual government. The programme is concerned with the establishment of international emission standard for some pollutants. A measure opposed by the U.K. as it felt that emission standards should be closely related to both what the environment can reasonably be expected to bear and to the use to which the environment is likely to be put.

Organisation for Economic Cooperation and Development brings together the major western industrialized countries providing a forum to discuss common problems and co-ordination of policies. The Environmental Committee, established in 1970, conducts investigations and proposes solutions to all common environmental problems, but at the same time it considers economic factors as well. It does not involve itself directly in research but promotes ideas which can then be taken up by specialised establishments.

Committee on the "challenges for modern society" established in 1969 by North Atlantic Treaty Organisation for the purpose of undertaking international studies of specific problems of the human environment. The objective is to bring the exchange of experience and technical knowledge and to put the findings into practice.

United Nations Economic Commission for Europe has been set up in 1971, involving E.C.E. Governments projects have been undertaken on various problems.

In addition to these, there are other bodies which have been established, consisting of non-governmental organisations with a primary aim of promoting the collaboration and exchange of information.

Environmental pollution is a very serious hazard in developing countries like India. It is necessary to provide legislative measures to ensure the protection of environment. The Indian Government passed the first national level Act to control water pollution in 1974. The Act took fourteen years to pass through the hurdles of parliamentary procedures. Passing of the Prevention and Control of Water Pollution Act was an indication of Government's awareness. The Act also created a Government Body for the enforcement of the legislation. The second legislation passed by Parliament was the Air Pollution (Prevention and Control) Act of 1980. The Department of Environment is now responsible for implementation of Water and Air Pollution Act. Department of Environment has to play a watchdog's role and to study and attract the attention of the government about the causes and consequence of environmental deterioration. It has not administrative responsibilities, like its counterparts in U.S.A., Japan and Sweden. So far the agency has only played advisory role.

Coming back to the discussion of law controlling environmental pollution in India, the Constitution (Forty-second Amendment) Act, 1977 is a historic document in the world constitutions which introduced detail provisions for the protection of environment. On the one hand it imposes in Article 48A a duty on the State "to protect and improve the environment and to safeguard the forests and wildlife of the country" on the other hand, Article 51A (g) imposes a duty on the citizen of India "to protect and improve the natural environment including forests, lakes, rivers and wildlife and to have compassion for living creature". There is no specific fundamental right in this regard but the court might include the right to live in a clear environment as a fundamental right under article 21 of the Constitution of India. In India, subjects directly or indirectly related to environmental pollution are distributed between Parliament and the State Legislatures separately and also jointly. But, environmental pollution as such does not find any place in the distribution of legislative powers.

The implementation of these measures are however not enough if not backed by executive authority. Legislation has to be made more comprehensive and precise. This is not possible through administrative measures alone but people's participation and co-operation is also necessary. It is true that no control measure or planning can get success without the support and the voluntary cooperation of the people concerned.

Policy Parameters and Holistic Approach

Certain basic policy issues need to be resolved at the national and international levels. What kind of development we want? What kind of facilities are worth the social, economic and political price that ought to be paid by us or by future generation for sustaining the present civilization and current life styles? A crucial question that needs to be asked weather the present civilization should be sustained by the production of synthetic products like pesticides, detergents, plastics and rubber necessitate the production of special kind of synthetic organic chemicals. This in turn call for a larger supply of chlorine and a large supply of mercury. It is revealing to note that mercury consumption has increased very much all over the world. The question therefore arises, should new technologies necessitate a rapid increase in the supply of poisonous substances? Are there not alternative paths of development which can progressively reduce the consumption of mercury, chlorine and heavy metals? Cannot we adopt appropriate technology which may prove reasonable in the use of resources? Are there not patterns of development which can reduce the incidence of poverty, provide basic minimum needs for the vulnerable groups with least damage to environment?

Development without destruction, calls for adaptation of a policy of golden mean: for avoiding the dogmatic outlook of a pure environmentalist and the rabbid commercial outlook of the down to the earth developmentalist. It is necessary to guard against the dangers of a rigid environmentalist's approach which can bring development to a grinding halt. It is equally important to bear in mind that some minimal levels of pollution are in escapable. Zero rate of pollution is neither a technologically feasible, nor economically sustainable. We should guard against the danger of such slogan. We should certainly continue to explore alternatives for adoption of cost effective appropriate technology for abatement of all types of pollution. Needless to add eternal vigilance is the price for maintaining a clean environment.

Environmental parameters are so many, so complex that till the beginning of the seventies, a holistic approach had not yet emerged. The tragedy of modern branches of knowledge is that they suffer from insularity or parochialism associated with compartmentalisation, departmentalisation, vivisection, dissection and dichotomy arising from narrow fields of specialisation and sub-specialisation. Laboratories and test tubes get isolated from the grim realities of social and economic situation.

Environmental concerns should worry not only departments of environment, but also the concern of all departments whose activities have a bearing on environment is some way or other. To sensitivise and resolve all major issues of environment, there is a paramount need for a grand orchestraction through super co-ordination.

There is only one world, in which every thing is related to every thing else. A truly holistic approach alone can save us from an impending disaster, bestow on us right vision, for meeting what now appears to be insurmountable challenges. The environmental parameters are being handled in a piecemeal, isolated and in effective manner by different agencies without adequate dialogue and interaction between them. There is no overseeing and no super coordination. As a result many environmental parameters got neglected or lost sight of whenever important development projects were formulated.

There is overall need for improving the capability of environmental impact assessment on land, water, air, space and for evolving a comprehensive balanced integrated energy policy; massive afforestation and greening programme, environmental education, ecological education, wild life education, mineral resources conservation exploitation policy; marine resources conservation exploitation policy; agricultural development, horticultural development; rural development policy; urban development policy; massive research and development programme in relation to each one of the environmental parameters.

Management within the Government including management of environment suffers from insulating parochialism. A comprehensive approach is an integrated, and holistic approach, unfolding inter dependence of variables. Holistic approach implies cohesion, synthesis, synergy overview and super coordination. Interdisciplinary approach built on a continual interaction and dialogue between different disciplines alone can produce spectacular results in a scenario of successive changes. Environmental planning and management call for team work among environmentalists, ecologists, physical planners, economists, financial analysts, geographers, sociologists, management experts, conservationists etc. There are no satisfactory organisational arrangements for coordinating eco-development activities in relation to ecology and environment.

It is urged that in the take off to the 21st century we should effectively utilise the powerful instrument of information technology for ensuring development without destruction.

Conclusion

The main problem is our inadequate knowledge about the extent of environmental hazards and their sources like, contamination of food and drinking water, air pollution, noise pollution, pollution from refuses and wastes. The second problem is the lack of knowledge about existing side effects, as a result, preventive steps are not undertaken. Protection of environment can only be achieved if sources of physical and chemical pollution of the environment are identified and taken into account.

REFERENCES

Allan, D. Mc. Knight and *et al* (1974), *Environmental Pollution Control*, George Allen and Unwin Ltd., London.

Basu, A.K. (1972), Some aspects of environmental pollution in and around Calcutta city, *Science and Culture*, Vol. 38, No.11.

Berry, Brian J.L. and Frank, E. Horton (1974), *Urban Environmental Management Planning for Pollution Control*, Prentice Hall, inc., Englewood Cliff, New Jersey.

Chapman, G.P. (1977), *Human and Environmental Systems*, Academic Press, London.

Dixit, J.M. (1972), *The Hazards of Environmental Pollution*, Proceedings of All India Seminar on Environmental Pollution, Bombay.

Kumra, V.K. (1981), *Kanpur City : A Study in Environmental Pollution*, Tara Book Agency, Varanasi.

Kayastha, S.L. (1982), Perspectives on environment and development, *N.G.J.I.*, Vol. 28 (1 & 2), March - June, 1982.

Walker, A. (1979), *Law of Industrial Pollution Control*, George Goodwin.

Webster C.A.R. (1981), *Environmental Health Law*, Sweet and Maxwell.

White, G.F. (1974), *Natural Hazards: Local, Global, National*, Oxford University Press, New York.

Zaidi, S.H. (1977), *Environmental Pollution and Human Health*, Proceedings of International Symposium on Industrial Toxicology, Lucknow, 1975.

3

DEVELOPMENT AND ENVIRONMENTAL CRISIS IN KUMAON HILLS

D.S. SINGH and A.K. SINGH

Environment is a matter of concern to every one, however its importance has been realised only recently, and efforts are now being made to understand ways and means for its preservation and improvement. It is being gradually realised that the developmental activities on this earth so far have been pressed at a rapid pace without due considerations to the maintenance of the fragile ecosystem. If a man cut his hand, nature immediately begins to restore the protective layer. Experienced medical care can help to speed up the process by inducing conditions favourable to nature. Similarly, when man damages the surface of the earth in a comparatively small way, nature steps in and creates a protective layer, first of grasses then shrubs and finally trees. The trouble is that man, in his haste to utilise his handworks, is unwilling to wait relatively long time so that these unaided natural processes conclude heeling operations. As a result, the environment has become one of the greatest challenges faced by the human race. Much harm has been done in the past but problems have been realised more-so-ever in time.

In the foregoing discussions, the authors have tried to put forward certain facts and figures regarding construction of roads in Kumaon hills (Uttar Pradesh), which is one of the main developmental activity in the region. Further, certain suggestions have been made to help in minimising the damages done to the environment. Fortunately, the discipline of geology is now widely recognised as an important aid to the civil engineering in this regard.

Kumaon Himalaya

Kumaon Himalaya is divided in to five well defined physiographic belts, each being a distinct geological unit. The submontane *Bhabhar* is a piedmont belt in the foothills. The outer Himalayan Siwalik range is built up of late tertiary sedimentaries, mainly sandstones, shales and dolomitic limestone rocks which exhibit a rugged and restive topography, characterised by steep hill slopes and deep valleys with crumbling walls scarred with landslides. Separated from the Siwalik by the *Main Boundary Fault* is the Lesser Himalaya. It has a comparatively mild and mature topography with gently slopes and deeply dissected valleys which suggest that the rivers and streams are still furiously at work. This part of Kumaon region shows the presence of varied group of rocks such as basic volcanics, quartzite, granite, phyleite, shale, slate and dolomites etc. The northern belt of great Himalaya is characterised by precipitous scarps and vertical walled gorgeous valleys and tumbling and foaming rivers, this belt of very youthful topography is still tectonically active and are made up of granites, schists, quartzites and dolomitic limestones at places which are the oldest rocks of Himalaya.

A glance at the geological map of Kumaon Himalaya would make it obvious that a very large part of the lesser Himalayan region is made up of sedimentary rock formations lying in tangled confusion. Critical study of these lithological succession in many sections of the region has led to the identification of eight lithostratigraphic groups, belonging to four tectonic units, one autochthonous and three trust sheets (Valdiya, 1980). The crystalline rocks form a great doubled syncline, thrustover the Siwaliks, these tectonic activities gave rise to folding, faulting, jointing and fracturing in the rocks thereby decreasing their physical strength. Hence after exposure to natural agents they started weathering and lead to mass movement. Apart from natural agents, men became an agent to destroy the environment on a larger scale.

Damages to the Environment

Natural catastrophes, such as the collision of giant meteorites with the earth, have resulted in wide spread devastation. Lightning strikes have shattered forests and fires have wiped out whole vegetation, and filled the air with smoke and ash. Floods have wrenched millions of tons of valuable top soil away and changed the face of the land. Catastrophic weather changes have witnessed extinction of several animal species in past. These changes were rapid and sudden but several other comparatively slow natural processes are still continuing and posing threat to the environment such as mass movement, though this term includes variety of deformations on the face which may be briefly discussed here. Soil creep is a very slow process which is encountered in most latitudes occurring on slopes as gently as 2-4°. The most common feature observed in the region is landslides which is a slumping process. A long slice of caprock or surface material separates from a scarp face and moves down slope on curved slip surface. The strike of the slump blocks is controlled by the trend of the cliff. The backward tilt of the rotating blocks, ground surface creates a depression at the headward end of the tilted block which becomes marshy or the site of a pond. In this way this process is furiously active in the region damaging there by the environment.

Road Construction in Kumaon Himalaya

In near absence of railways and water transport in the region, roads are the only means of communication. While construction of modern roads was taken up in mid 19th century, practically no roads were constructed in hill areas except for connecting a few hill resorts. Inaccessibility of hill areas kept its population isolated from urban influences resulting in social, cultural and political backwardness as well as deprivation of incentives for economic growth. These areas are lagging behind in development than the plains. Hence, to fill up the gap, government took up a drive for speedy construction of roads without properly assessing the ill effects on the environment and great damage was done to it in the process. In order to check this menace, the Forest Conservation Act, 1980 was imposed on the development agencies by the government and environmental aspects are being considered now, but unfortunately no sincere effort could be made in this direction. However, it is learned that for the development of the entire area for economic growth, economic marketing of the produce there is pressing need to connect all villages and production centres, even the small and isolated ones, with some means of communication.

The construction of roads in the hills causes visible and invisible harms to the nature, there by creating disbalance in the environment.

Visible Effects

The visible effects are: (a) leaving scars in the greenery due to cutting of hill sides and deforestation along the alignment, and (b) accumulation of debris as a result of cutting and its disposal on the hill sides. The increase of slit in the catchment of the rivers may ultimately reach the river bed and cause siltation. The details are described in the following paragraphs.

Deforestation and Road Cutting

Hill roads necessarily involve some cutting of hill sides at times which results in slope failures. It is estimated that every square kilometre of this fragile Himalayan region shows at least two big landslide scars, and one more is added every 6 km^2 annually. Direct consequences is the loss of the land of the order of 120 m^2 for every square kilometre area, with a moderate depth of sliding assumed to about 1,000 m^3 of debris every square kilometre area. This process is activated due to cutting and blasting which results in heavy destruction to vegetation. At times, alignment involves dense forests and a large number of trees are cut. As per the forest statistics of U.P. (1983) during the period 1951-80 the areas transferred by the forest department for non forest development purposes include 1,022 km^2 for the agriculture, 49 km^2 for road and 234 km^2 for industries. This shows that the actual forest land involved in the road construction is comparatively less, though the total number of trees cut during that period is not known but some idea can be made from Table 3.1 indicating number of trees cut recently during the road construction in different situations.

Removing the tree cover has adverse affects, such as: (a) the erosional processes get activated, (b) disturbance of wild life, (c) alteration in vegetation pattern and soil characteristics, and (d) modification in the hydrological regime. Further, deforestation leads to minor changes in the day and night temperatures which may help in retreating of the glaciers and it may have some effect on climatic changes in the Himalayan region. It has been estimated that Milam glacier has retreated 1,350 m during the period 1949-57 and Pindari glacier has retreated 2,840 m during the period 1945-1966 (Vohra, 1981).

Debris Generation and Increase of Silt

In the process of construction the hill sides necessarily have to be cut to provide width to the road and its stability. According to the recommendations, the road information width in hill areas should be as follows (Table 3.2).

TABLE 3.1: Trees Cut During Recently Road Constructions (some examples)

Name of the Road	Type of road	Length of the road	Total trees cut	Remarks
Dwarson-Kakrighat	Motor Road	5 km	5	Alignment passes almost from barren lands and cultivated fields.
Kosi-Daulaghat-Korichhina	Motor Road	12 km	472	Some patches of thc alignment passes from dense forests.
Bans-Awlaghat	Motor Road	5 km	397	Road passes from medium dense forest.
Gurna-Gogina	Light Vehicle Road	5 km	95	Road passes along rare forests.
Chandak-Bans Bastey	Bridle Road	4 km	Nil	Cutting of trees avoided by giving sharp curves and steep gradients as and when required.

Source: P.W.D. Records.

TABLE 3.2: Recommended Road Width in Hill Area

Classification of Road	Carriage way width (m)	Shoulder width (m)	Side drain and parapet width (m)	Total formation width (m)
National Highways and State Highways				
(a) Single lane	3.75	2X1.25	2X0.6	7.45
(b) Double lane	7	2X0.9	2X0.6	10.00
Major district Roads and other district Roads	3.75	2X0.5	2X0.6	5.95
Village Roads	3	2X0.5	2X0.5	5.20

Source: S.K. Malhotra (1985).

Hence after cutting these widths, large quantities of loose rock and soil materials (debris) are generated. A part of this material is utilised in the filling of the road, while a large amount of this material has to be disposed off on the side of the road which rolls down the slopes and often reach the stream bed. It has been estimated that in the present road construction on an average about 30 per cent of the debris produced is consumed during construction itself, and the rest is thrown on the sides of the hills. Some idea can be gathered about the extent of debris produced from Table 3.3.

TABLE 3.3: Average *in situ* Cutting

Name of the Road	District	Cutting width (m)	Average *in situ* cutting in (m^3)
Dwarson Kakrighat Motor	Almora	5.95	15,753
Kosi-Daulaghat-Korichhina Motor	Almora	5.20	6,640
Wadda-quitar	Pithoragarh	6.00	10,986
Gurna-Gogina	Pithoragarh	3.00	6,049
Chandak-Bans Bastey	Pithoragarh	2.00	3,413

Source: PWD Records.

It is obvious from the Table 3.3 that:

(a) the quantity of *in situ* cutting is proportional to the width of the road, and

(b) the maximum volume cut at places are often multiple times the maximum volume cut, and at least three times in a similar section of the road of the same width.

Since the volume of the loose (disposed off) material after excavation and removing from the site generally, expands to at least the one and half times its the original volume, the swelled volume of 70 per cent of disposed off material would almost be the same as the volume of the *in situ* excavation in the road cut. It is observed that the volumes for a 5-6 m wide road would vary between 6,000 to 15,000 m^3 per km length of the road with an average of about 10,000 m^3 per km or say 10 m^3 per metre length of the road. This figure indicates that large quantity of debris are disposed off on the slopes in the catchment area of the stream which under unusual conditions may bring unprecedented floods and erosion in the down stream area causing havoc and loss to life and property. Further, at several places where dams have been built along the courses of these rivers, the increased amount of silt is drastically reducing the capacity and life of the reservoirs which needs to be calculated. Similarly other linear constructions, such as irrigation canals, air strips, and river valley projects are also contributing their share in upsetting the environmental condition of the region.

Invisible Effects

The environment includes not only the land, vegetation, and forests, but also the living insects, mamals and fishes that share the earth with human beings. But, in fact, they are ignored at all stages from planning to execution of the projects. After the construction of roads, traffic is introduced which pollutes the nature in several ways like noise, toxic wastes, vibrations, and so on. Unfortunately, even today, these invisible effects are talked and discussed on papers only, but sincere care is not being paid during planning and execution.

Measures Suggested to Reduce the Menace

While these disturbances cannot be totally avoided but sincere efforts can help in minimising the effects. In this regard, a general awareness in the engineering profession needs to be created, so that our environment is protected while planning roads in this hilly region of Kumaon Himalaya. Following measures are suggested here:

(a) The road alignment should, as far as possible, not pass through densely forested area in order to avoid cutting of large number of trees.
(b) The number of trees cut should be replaced with new trees planted in the adjoining area while doing so, only the wild species inherent to the area needs to be considered so that their survival and ecological balance could be assured.
(c) Compulsarily cut and fill method of road construction should be adopted and the debris generated at the site needs to be arrested as far as possible to avoid its rolling down the slope. For this purpose it would be proper to stack the larger pieces of the rock on the sides of the cutting and simultaneously depending upon the need of retaining walls of low height be constructed at distances of 20-100 m down the level of the road to check the rolling down of the material. In the lower part where these slopes coverage in to a valley/depression

small check dams be constructed so as to stop any material that might roll down the slope from entering the stream below.

(d) As far as possible the alignment should not have hair pin curves, as construction of such curves can lead to landslides and drainage problems. Such loops also involves maximum number of trees cutting and large destruction to vegetative cover.

(e) Since cutting should be done in slope ranging from 1:4 to vertical, depending upon the nature of strata and its angle of repose. The fissures develop and boulders get loosened after blasting, hence affected strata should be removed so that fissures or pockets created do not help in allowing water to percolate and disturb larger mass.

(f) Hard rocks are invariably met in the cutting of the hill road which requires blasting but heavy blasting disturbs the strata deeply. Unfortunately, in hurry the tendency among working parties to use heavy charges of explosives to cut the road faster creates more and more problems during subsequent maintenance. Therefore, only low charge blasting should be allowed where it is most required otherwise manual cutting should be encouraged.

(g) In softer soil and landslide prone areas, breast walls should be constructed simultaneously with construction of formation so that support is provided and earth masses above cut is not disturbed. Further, fast growing shrubs and grasses should be planted above the breast wall immediately after its construction for providing coverage to the soil so that percolation of water is prevented.

(h) Alternative means of communication should be planned in this region which may include the provisions of ropeways, bridle roads, light vehicle roads and helicopter services, and roads should be constructed only where it is very essential.

Conclusion

In recent years the emphasis on environmental conservation in the region has been considerable but meaningful linkages between conservation, technology and development are to be established. The first and foremost need is therefore to plan a strategy for coordinating the information between different development agencies. A data bank to store all the geological, topographical and environmental information should be created. Above all, a general awareness in the engineering profession needs to be generated so that our environment is protected while planning road development in this region.

REFERENCES

Agrawal, P.D. (1984), Ecological consideration in design and construction of hill roads, *Journal of Engineering Geology,* Vol. XIII No. 1 and 2.

Laggest and Karrow (1962), *Handbook of Geology in Civil Engineering,* R.R. Donneley and Sons London.

Malhotra, S.K. (1985), *Proceedings of the Seminar on Construction of Roads in Hill Areas,* Nainital, July 7-9.

Singh, H.H. *et al.* (1985), *Geography and Environment: Issues and Challenges,* Concept Publishing Company, New Delhi.

Valdiya, K.S. (1980), *Geology of Lesser Kumaon Himalaya,* Wadia Institute of Himalayan Geology, Dehra Dun.

4

IMPACT OF DEVELOPMENT ON THE ECOLOGICAL BALANCE OF THE DAMODAR VALLEY REGION

SWAPNA BASU

The Damodar Valley Corporation (D.V.C.) is truly speaking the vanguard of modernisation in India. With its inception, evolved a new era of modern technology for a comprehensive development of an area. There is no denying the fact that the D.V.C. has brought in its wake spectacular beneficial changes in the socio-economic spectrum of the region. But this development is not unimpaired as evidenced from the gradual encroachment of ravine and gully lands in once cultivated plots of the upper valley, permanent waterlogging in fertile agricultural fields of the lower valley increased water and air pollution threatening the hygienic condition of the densely habited parts of the region.

This ecological imbalance is not an ideal condition for healthy socio-economic development. Science and technology should ensure a healthy environment rather than creating problems in it.

The object of the present study is to identify the problems and attempt to decipher the more vulnerable ones according to the nature of their virulence. Some suitable strategies would then be evolved to promote a harmonious ecological balance wherein the economic and social development of the region may thrive in a well balanced ecological environment.

Preamble

The Damodar Valley region is one of the most important economic regions of India, both from agricultural and industrial potentialities.

This region was selected as the arena for launching the nation's comprehensive development programme through the scientific management of the Damodar river under the aegis of the D.V.C. By and large this massive undertaking has not failed to achieve its objective of an overall development of the region. The Damodar Valley Region, now-a-days is throbbing with the fervour of development both from its industrial and agrarian points of view. The mining and industrial development from Barkakana-Bokaro-Dhanbad to Durgapur is spectacular, while the green verdure of the valley further east largely depends upon the irrigation of the D.V.C. canal system. The entire physico-economic scenario of this region is now pulsating with an environment of growth and development.

Issues

It is however an undeniable fact that every growth and development particularly if that be of a great magnitude entails some changes that in the long run generally disturb the ecological balance of the concerned region. This region is no exception to this general rule. This apparent picture of economic development is largely undermining the ecological balance bringing about gradual degradation in the physical and socio-economic environment. The accelerated development in resource exploitation for implementation of the multipurpose development programme has resulted in the adverse effects of environmental degradation and the region shows the symptoms of negative feed backs threatening the very existence of the prosperous valley. The present study deals with the shortcomings which have been slowly degenerating the geo-economy of the region. It is however a difficult task to concise the myriads of problems of the valley within this limited scope. A comprehensive picture of the posing problems have therefore been selected and highlighted in the next course of discussion.

Methodology

The work largely depends upon various information collected from authentic Government agencies like the Block Development Offices of the districts, the headquarters of the D.V.C. and also from other sources. Areal photograph and satellite imagery had been consulted to identify the worst affected gully and ravine lands. These information were further substantiated by field checks and observations in selected sample sites of thc Damodar Valley Region.

Discussion

As has already been stated the massive development programme in the Damodar Valley region has brought in its wake a number of problems, endangering the ecological balance of the region in many ways. The problems may be grouped as (a) physical, and (b) socio-economic and are being discussed here in a nutshell.

(a) Physical Problems

Deforestation and resultant degradation of land in the Upper Damodar Valley

The massive industrial establishment of the D.V.C. in most cases necessitated the clearance of virgin forests of the area which even a few decades back covered considerable portion of the upper and middle valley. This has resulted in adverse consequences in the following ways :

(i) Exposing lands to weathering and erosion hazards.
(ii) Rapid loss of productive soil cover.
(iii) Formation of gullies and ravines thereby rendering once fertile agricultural land unproductive.
(iv) Silting of the river beds as well as of the dams reducing their water holding capacity.

An estimate of the D.V.C. Soil Conservation Working Plan (1978) shows that 31.9 per cent of the total land area of the upper Damodar Valley is affected by varying intensity of erosion. Out of this 32 per cent of the total arable land is moderate to severely eroded under sheet and gully erosion and if immediate conservative measures are not adopted in near future they would go beyond any productive use. About 44 per cent of the total forest cover is under active sheet and gully erosion of severe intensity. The virulence of erosion and the resultant soil loss have been found to be varying in different parts of the Upper Damodar Valley, depending upon their geomorphic location. The rate of increase of gullying in some areas of the Upper

Damodar Valley has reached an alarming state.

Apart from reduction of soil productivity the more serious problem created by this erosion hazard is the generation of enormous silt load that flows down the channels and is eventually deposited in the reservoirs resulting in their rapid siltation. Siltation of reservoirs though is a universal phenomenon, their unabated rate is definitely detrimental to the socio-economic development of the concerned region. There is no denying of the fact that siltation in the reservoirs as a corollary of soil erosion is signalling danger in the forthcoming days of the Damodar valley region.

Subsidence hazards and despoilation of land from mining in the Middle Damodar Valley

The Damodar Valley region hoards the most important coal and mica reserves of the country and feeds the ever-rising demands from the developing industries of the nation. Prolonged mining for past four decades has resulted in a number of problems both in physical and socio-economic environment. Large stretches of land originally occupied by forests or agricultural land had been deprived of their natural stands by the gradual encroachment of mining and other ancillary industries.

Mining in this region is generally carried out by two methods, viz. underground and opencast. Both the processes have created related problems of which opencast mining is said to be more harmful for creating environmental hazards. The serious problem of land defacement results as an aftermath of open cast mining. The resultant problems of opencast mining may be enumerated as follows:

(i) The thriftless unscientific mining methods along the coal outcrops results in loss of valuable land as well as coal deposits. In some instances the quarries are left abandoned after extracting the upper few seams. Thus potential reserves underneath remain unworked with no future scope for extraction since the entire area is considered to be unsafe for any further mining operation. Thus, the area remains as barren lands with heaps of scraps and occasional scars on the ground devoid of any vegetation or human settlements.

(ii) The over burden dumps of the operational quarries generally comprise of loose, unconsolidated rock materials and are usually unstable in character. The height of the dumps averages between 15-20 m. The slopes are very steep varying between 70°-75°. After heavy showers these soil dumps cause landslides, resulting in opening up of gullies on top of the over burden dumps.

(iii) At many places the natural courses of the nalas were diverted to artificial channels to facilitate mining operation. These quarries invariably extend up to the banks of the nalas and their overburden or waste materials are often washed off to the nalas chocking their courses and thereby endangering the surrounding locality by problem of water logging. The Singaran Nala and the Noonia Nala are two such examples whose courses have been partially blocked by old overburden of the nearby quarries.

(iv) Old abandoned quarries (which cover about 20 km^2 in the area) often result in the subsidence of the topography, and look like natural depressions. While filled with water they look almost like natural lakes like Dhunbad, Diguli, Pathargora etc.

(v) Due to pumping out of water from the mine, cracks and joints develop in the underground structure resulting in the recession of ground water in some parts of the coal belt. This together with over use of water has resulted in the lowering of water table in Asansol, Hirapur, Sitarampur, Dhanua etc.

The problems with underground mining of the region is even more serious. The continuous exploitation of minerals has left quite a number of abandoned collieries (which account for about 30 km^2 in the middle Damodar Valley) without proper sand stowing which has been resulting in gradual subsidence of land. The very existence of Raniganj, and Kulti towns along with a number of surrounding villages is at stake and it is apprehended that a slight change in the underground water table may lead to catastrophe to the surrounding locality.

The mine fire is yet another problem which is causing atmospheric pollution apart from destroying huge amount of valuable coal deposits.

Formation of Water Logged Areas in the Lower Damodar Valley

Although the four dams of the project have ensured flood protection in the mineral rich industrial belt of the upper and middle valley, it on the other hand has exposed some section of the lower valley to annual flood devastations. The recent official data shows that the D.V.C. is responsible to cause flood about 31 per cent of its valley by releasing excess water in late monsoon from the reservoirs upstream, which the constricted flow of the Damodar and its tributaries cannot hold within their banks. The south central part of the valley is the most severely affected, since this stretch lies within the principal canal network zone which intensifies the flood hazards to a great extent by carrying water deep inland. In fact sections of Arambagh, Khanakul, Pursurah, Tarakeswar, Udayanarayanpur, Amta and Jagatballavpur police stations have become endemic zones of flood ravage. The amount of loss due to submergence of crops often runs into millions of rupees.

The damages during the post-dam periods are much higher than those occurring prior to this. Available reports also show that peak flow exceeding 11,000 cumec during the pre D.V.C. period has occurred at Rhondia only in 1933 and 1949, while the same have occurred eleven times between 1958 and 1978. It tends to prove that the intensity and frequency of floods have increased during the post-dam period and the lower valley is now practically exposed to flood at any flow over 18,000 cumec.

(b) Socio-Economic Problems

Hazards of Water Pollution

The Damodar river is being polluted to a perilous level by the industrial and domestic effluents drained to the harnessed flow of the river with a minimum discharge capacity. The Bokaro Steel Plant has in fact transformed the clear water of the river to a greenish turbid mass of waterbody by disposing off its waste materials into the river and thus polluting it. The Sindri Fertilizer discharges a huge effluents to the Damodar containing alkalies, ammonia, phenol, naphthalene etc. The B.O.D is as high as 25 mg per litre at the junction of the Damodar and Tamla nala. Industrial waste in the form of nitrogen, phosphate, acid mine drainage, washing from herbicides and pesticides, oil spills etc. together with domestic refuses are said to have polluted the Damodar water beyond the limit of healthy use upto the Durgapur Industrial Complex.

Water pollution occurs in two principal ways : (i) domestic

sewages and agricultural fertilizers enrich the nutrients content of water which promotes the growth of micro-organism that ultimately reduces the dissolved oxygen of water essential for thriving of fish and higher animals and (ii) poisoning of water by toxic chemical refuses from the industries which destroy living organisms and even eliminate all forms of life.

Beyond Burdwan the Damodar water suffers from another kind of pollution. Below the bifurcation point at Jamalpur the slow percolation of water results in the luxuriant growth of water hyacinth and other halophytic vegetation which gradually depletes the mineral nutrients and salts of the water thereby reducing their fertilizing effect. The large scale utilisation of water for irrigation seems to be a infractuous as they are already robbed of nutrient values.

This vast stretch of marshy tract provides ample scope for mosquito breeding ground while the polluted water gives rise to diseases like dysentery, typhoid, diarrhoea, hepatitis etc.

Hazards of Air Pollution

The major pollutants of the area originate from domestic and industrial smokes, smokes and chemical effusions from the locomotives and automobiles as also from garbage and other excreta. The huge urban industrial complex supporting three massive iron and steel factories at Bokaro, Burnpur and Durgapur, together with 50 other big and small factories, coal and mica mining, coal washeries, refractories, thermal power plant, the working population of over a lakh is facing the atmospheric pollution problem every moment. The remarkable development of smog over the long stretch of land from Barakar Thermal Power Plant to Asansol-Durgapur Industrial Complex is the aftermath of the emission of carbon dust and other injurious gases endangering the entire biotic life. The increased concentration of carbon-di-oxide (CO_2) in the upper atmosphere results in trapping more infrared radiation of the earth increasing gradually the surface temperature. Acid rainfall interferes with the pH levels of the soil.

The big factories like DSP, Thermal Power Plants, M.A., M.C., A.C.C., Vickers, Phillips Carbon, Asiatic Oxygen etc. are the principal culprits for causing air pollution in the locality. Durgapur Coke Oven Unit emits a high percentage of SO_2 and CO_2 together with grit and dust which amounts to about 75-100 tonnes on a daily basis. The

Durgapur Steel Plant emits a high amount of SO_2 and carbon gases from its blast furnace and eight open hearth furnaces. Another very dangerous pollutant is the Phillips Carbon which emits smoke with small particles of less than 3 micron in size. Besides, the burning of coal emits CO, CO_2 and SO_2 adding more to the pollution problem.

Normal atmospheric state is seriously disturbed through smoke, dust, carbon monoxide, sulphur di-oxide, aerosol etc. which are detrimental to public health and comfort. The undesirable gases and particles affects the lower level of the atmosphere up to 30 m. This air pollution results in incidence of diseases like cough, headache, nausea, asthma, lung cancer, skin cancer etc.

Conclusion

These are some of the multiple problems which are associated with the multifaceted development of the Damodar Valley. It is therefore, evident that the economic progress of the valley does not indicate an overall development. On the contrary such a development brings in its trail various environmental problems of different magnitude requiring immediate attention.

The problems of the Damodar Valley Region are many fold. While the upper valley suffers from serious problems of land despoilation, the middle valley suffers from land subsidence and the lower valley is being threatened with waterlogging.

The untreated chemical refuses are contaminating the farm output while the effusion of poisonous gases are viciating public health. Science and technology should ensure a healthy environment rather than creating problematic ones. Scientific management programmes are therefore to be evolved to restore the ecological balance in order to promote a development of true order.

Some of the suggestions to improve the present situation may be enumerated as follows :-

(a) Increasing forest cover in the upper catchment,
(b) To ensure minimum discharge to the Lower Damodar to carry on regular flushing of water,
(c) To increase water holding capacity of the reservoirs,
(d) Proper measure of mining and treatment of derelict land after mining operation has been abandoned to be enforced with strict abidement of laws.

(e) Enactment of pollution control in industries,

(f) Pollution front to be given special attention by monitoring the air, land and river pollution and their consequences on health and ecology with preventive measures, and

(g) The liquid wastes from domestic and industrial quarters to be discharged after proper treatment. River water utilised for irrigation should also be made pollution free.

These are some of the strategies which if implemented without further delay would promote a harmonious ecological balance wherein the economic development will go hand in hand to a well balanced environment.

REFERENCES

Basu, Swapna (1987), Soil Conservation in the Upper Damodar Valley, *Geographical Review of India*, Vol. 49, No. 4.

Bhattacharya, A. (1985), Impact of development on physical environment in B. Banerjee (ed), *Impact of Development on Environment*, Vol. 1. Geographical Society of India, pp. 1-5.

Damodar Valley Corporation (1981), *Implementation of Watershed Management in D.V.C. Watershed wise Achievement during 5th Plan & 1979-80.*

Mukherjee, Rina (1987), Surface mining and land degradation in Ranigunj coalfields, — *Geographical Review of India*, Vol. 49, No.2, pp. 69-77.

Sen, P.K. (1985), Environmental Change and Degradation in the Damodar Basin in B. Banerjee (ed), *Impact of Development on Environment*, Vol. 1, Geographical Society of India, pp. 6-17.

5

ENVIRONMENTAL DEGRADATION, ASSOCIATED FAMINES AND RELIEF MEASURES : A Case Study of Tribal Belt in Rajasthan

KISHAN S. RAJPUROHIT and DAKSHA C. BARAI

Aravalli mountain range known as the greenline and skyline of Rajasthan; divides the state into two regions, i.e. western and eastern Rajasthan, and is inhabited by Proto-Australoid tribals all along its length. It has also supported the state's rich and valuable dry deciduous forests, receiving adequate rainfall, much higher than other parts of the state. The tribal belt of southern Rajasthan comprising Banswara, Dungarpur and Udaipur districts extends to the southern fringes of Aravalli mountain range and is dominantly inhabited by Bhil, Meena and Garasia tribals. This tribal belt had almost virgin forest reserves about 40 years back and was receiving higher rainfall than other parts of the Aravalli mountain range. In recent years due to the extensive clearing of forests, multiple ecological problems such as disappearance of forest products and wildlife, heavy runoff, soil erosion, depletion of underground and spring water and desertification have crept into this region, which has led to severe droughts and famines. Famines of last five years, i.e. 1983-84 to 1987-88, were very severe and created acute scarcity of water, fuel, fodder and grains. This situation is forcing the state government to take new remedial measures in the form of various famine relief works to provide employment to the famine affected tribals. Most of the money given for the relief measures is diverted from the funds allocated for the planned developmental works under five year plan, causing disruption in the planned development of the state (Mathur and Rajpurohit, 1986).

In the present study an attempt has been made to assess the past and present environmental status of the tribal belt of the southern Rajasthan, and assess how degradation of forest belts in this part has led to the famine and associated problems. It is clear that famines and famine relief works are new to this tribal belt, keeping this in view, studies regarding nature and adequacy of famine relief works, availability of minimum essential facilities at the famine relief sites, working conditions of the tribal relief workers and amount of work allotted and executed by workers have been studied in detail on six famine relief work-sites during 1985-86. Eighty four famine relief works falling in four different categories, i.e. P.W.D. works, Irrigation works, RLEGP works and NREP works have been studied in detail for amount of work executed by tribal workers and payments made to them to find out their exploitation by above famine relief agencies.

Study Area

The study area, tribal belt of southern Rajasthan includes the districts of Banswara, Dungarpur and Udaipur, a part of the Aravalli mountain system. This hilly tract still represents almost half of the Aravalli's forest area, i.e. 2,779 km^2 out of 5,486 km^2. Due to the extensive damage caused to the green belt, precipitation is below normal for the past five years resulting in frequent famine and drought. It results in scarcity and unemployment among the tribals forcing state government to take adequate famine relief measures basically to provide employment to the tribal families, most of these lie below the poverty line, i.e. family having annual income below Rs. 3,500.

Environmental Situation of the Tribal Belt of Southern Rajasthan

The tribals of Proto-Australoid group were living in southern Rajasthan for ages and were able to manage by the utilization of forest resources for their daily needs, such as food, fodder, fibre, firewood, fertilizer, house construction, herbal medicines and the like. They also practised agriculture activities and had love and affection for the forest life. Most of their social, economic and cultural traditions were also related to the forests and surrounding natural environment. At the time of independence and even up to the promulgation of Jagir Abolition Act in 1954-55 in Rajasthan, most of the forested land as well as cultivable and waste lands of southern Rajasthan were under the direct control of former Maharajas of Banswara, Dungarpur and Udaipur princely states and their Jagirdars. Laws enforcing death penalty or heavy punishments for felling forests were enacted. After 1954-55 some of the acquired land from Jagirdars and princely states was allotted to the tribals for agricultural purposes and forested areas were declared state's asset, to be looked after by the state forest department. Due to the liberal policies of the state government, each tribal family was allowed to take a head load of firewood a day from surrounding state owned forests to meet their daily firewood requirements and earn some cash through sale. Some of the luxuriant forests of the region were leased to the contractors for exploitation of timber and other forest products to earn revenue for the state. This helped tribals to learn the value of revenue forest resources, and tribals in collaboration with the state forest department employees and contractors started illegal felling of trees to earn more money for better life causing great damage to the forest ecosystem.

At present, as per Rajasthan State Forest Department, only 9 per cent of the total geographical area of the Rajasthan is under forest, of which only one-third of the area is covered with thick forest, i.e. three per cent (Dhabriya, 1984). This is much below the national average of 33.30 per cent, the minimum required for maintaining an ecological balance. The Aravalli Hill Ecosystem which is considered as tribal belt was once rich in forest cover and is almost deforested at present. As per the results computed from the satellite imageries at a ten-year interval (Table 5.1) for the three tribal districts, i.e. Banswara, Dungarpur and Udaipur (southern Rajasthan) are now almost devoid of forest cover and out of 4,923 km^2 forest cover (1972-75) only 2,779 km^2 (1980-84) is left, indicating 43.55 per cent change during the period. The tribal belt of southern Rajasthan had been affected due to an extensive damage to the green belt. This damage has led to the multiple ecological problems such as disappearance of wild-life and forest products, heavy runoff, soil erosion, depletion of water resources, i.e. underground and spring water, loss of soil fertility and desertification. Due to the extensive damage to the ecosystem, droughts and famines have increased in frequency in the recent years in the tribal belt of southern Rajasthan. Droughts and famines of last five years (1983-84 to 1987-88) are so severe that they have created persistent problems of water, fuel, fodder and grain shortage. To deal with the famine situation in this part, state government has set-up famine relief works by diverting valuable funds allocated for development works. Famine relief works were not adequate forcing tribals to migrate neighbouring Gujarat state for alternate employment abandoning waste lands (Fig. 5.1).

TABLE 5.1: Decline in Forested Area in Dungarpur, Banswara and Udaipur Districts of Southern Rajasthan

District	Forest Area 1972-75 (km^2)	Forest Area 1980-84 (km^2)	Percentage Change during 1972-75 and 1980-84
Dungarpur	309	167	45.96
Banswara	348	197	44.40
Udaipur	4,266	2,415	43.39
Total	4,923	2,779	43.55

Famine Relief Programmes

Government of Rajasthan is providing relief to the famine affected tribals of southern Rajasthan particularly falling below the poverty line by generating employment under four major programmes, i.e. (i) Public Works Department works, (ii) Irrigation Department works, (iii) Rural Labour Employment Guarantee Programme (RLEGP) and (iv) National Rural Employment Programme (NREP).

Tribals and Famine Relief Works in Southern Rajasthan

Adequate number of labour force working on six relief works, i.e. (a) Kushalgarh-Bajana Road, P.W.D., Kushalgarh, Banswara, (b) Nanabhukhiya Land Management Work, D.P.A.P., Anandpuri, Banswara, (c) Anandpuri-Dokar Road, P.W.D., Anandpuri, Banswara, (d) Sunderpur Stone Work, P.W.D., Bichiwara, Banswara, (e) Sulpore-Boruk Road, P.W.D., Bichiwara and (f) Piplada Stone Work, P.W.D., Dungarpur, Dungarpur. The supervisory staff, engineers and district administrative authorities were interviewed through a set of questionnaire to obtain informations about nature and adequacy of relief measures provided, working conditions and facilities provided at relief work-sites and duration of work allotted and workers capacity to execute the allotted work during 1985-86.

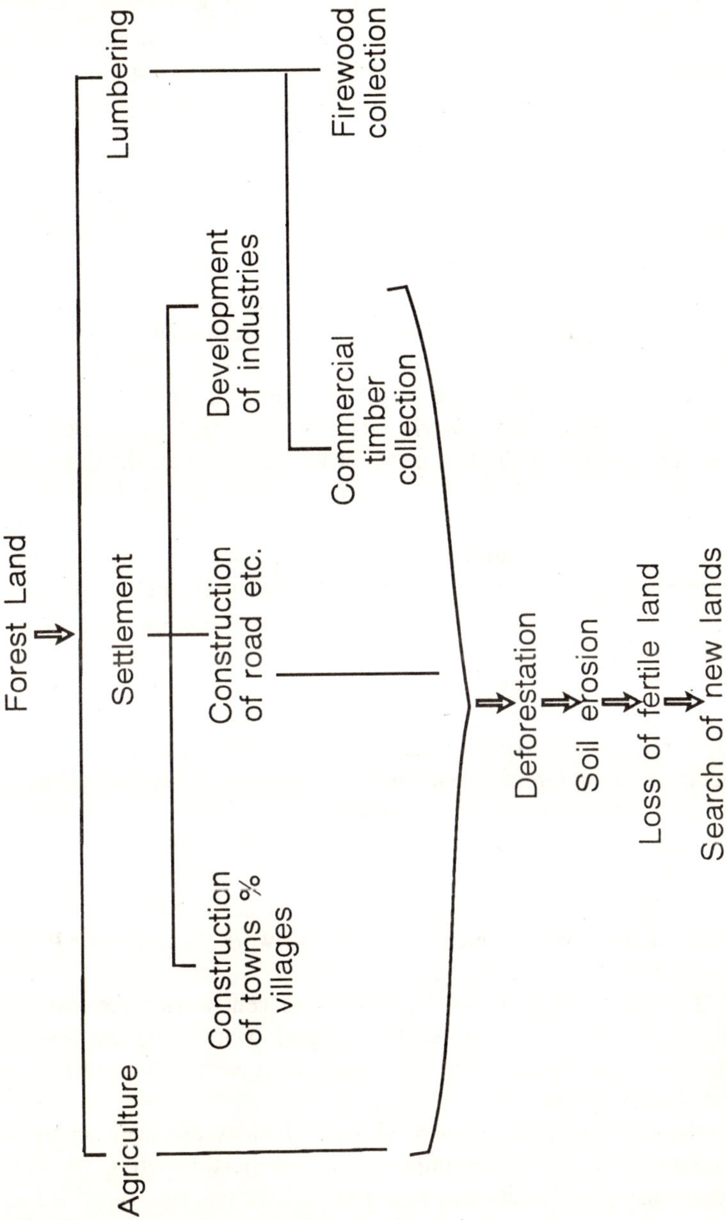

Fig. 5.1: Altered tribal Environment by Government Administration in Southern Rajasthan

Eighty-four cases of famine relief works of different categories, i.e. Public Works Department (13), Irrigation Works (66), Rural Labour Employment Guarantee Programme (3) and Rural Labour Employment Programme (2) executed during 1985-86 famine year were examined for the amount of work executed by workers and payments made to them.

Findings of the study reveal that most of the work was executed and being executed during 1985-86 were oriented to the irrigation and road construction works, respectively. Looking at the worst famine situation arising through the ruthless degradation of forest wealth in the past three decades, the state government has not taken adequate measures. Not much attention had been given to the afforestation of the deforested bare hills vulnerable to the soil erosion under the famine relief works. This should have been taken on priority basis under the famine relief works to restore the degraded environmental condition of the southern Rajasthan's tribal belt.

There was unanimity in opinion of administrative officers, engineers, supervisory staff and famine relief workers that measures taken by the government for the famine relief work are not adequate to provide relief to tribals of southern Rajasthan considering at the severity of the famine. It was felt that employment of one person per family was not adequate because most of the tribal families are having more than five members. It has been observed that Rs. 11 per day earned by a member of the family is not sufficient to feed all the members of the family, forcing tribals to migrate to the neighbouring Gujarat state for the search of employment.

Due to severity and persistence of the famine and the poor mental and physical state of famine relief labourers employed from families falling below poverty line, it is very difficult for them to execute the given work, which is equal to the work allotted during normal years.

Table 5.2 indicates that 9.25 to 25 per cent work force was engaged in other activities as unit leader, waterman, napatiwala choukidar and dakman and their work load is being shared by the rest of the workers.

Essential facilities such as weekly paid holiday, adequate supply of drinking water, working implements, medical facilities, shelter, weekly payment facility and ration shop have not been provided at any relief work side for the workers (Table 5.3). One paid holiday is provided once in a fortnight, working tools and implements were provided to the workers at only two sites and those were also old and ineffective. Salaries were being made two to three months after the execution of work. Due to all these problems faced by the famine relief workers they were able to execute only 75 to 80 per cent work at five out of six relief works (Tables 5.2 and 5.3).

Data of the eighty four samples of the famine relief works of different categories, i.e. Public Works Department, Irrigation Department, RLEGP works and NREP works conducted by the state government to provide relief employment to the tribals of southern Rajasthan have been computed and indicated in the Tables 5.4, 5.5 and 5.6. Table 5.4 shows that on an average 74.49, 72.95, 87.74 and 86.36 per cent of the work was executed and Rs. 8.19, 8.02 9.67 and 9.50 wages were paid on an average by P.W.D., Irrigation Department, RLEGP and NREP respectively. In majority of the cases only 60 and 80 per cent of the work was executed by workers (Table 5.5) and instead of receiving Rs. 11.00 per head/day wages, they have received only Rs. 7.00 to 9.00 (Table 5.6).

Conclusion

The following conclusions can be drawn from the findings of the present study:

(a) About three decades back, before the inaction of Jagirs and merging princely states of Banswara, Dungarpur and Udaipur into Rajasthan, forests of the tribal belt of southern Rajasthan was in good condition and tribals, i.e. Bhil, Meena and Garasia lived here with a properly managed utilization of the forest resources for food, fodder, firewood, fertilizer, house construction material, herbal medicines, etc.

(b) Post-independence faulty policies of the state government, leasing the forests to earn revenue for the state, allowing tribal families to cut a head load of wood have led to extensive damage to the forest wealth and nature based tribal economy of southern Rajasthan. It is also promoted the illegal felling of the trees.

(c) Ruthless and indiscriminate cutting of forests is evident from the facts that there is 43.55 per cent reduction in the forest wealth between years 1972-75 and 1980-84 which indicates the severety of the situation in the tribal belt of southern Rajasthan.

TABLE 5.2: Daily Work Given, Work Executed and Unproductive Labour Force per unit on Different Famine Relief Works in Southern Rajasthan

Name of the work	Duration	No. of workers/ unit	Daily work per head	Percentage of work executed per head	No. of unproductive labour force counted as productive unit						Total
					Unit Leader	Water Man	Napti Man	Choukidar	Dakman	Others	
1	2	3	4	5	6	7	8	9	10	11	12
Kushalgarh-Bajara Road, P.W.D., Kushalgarh, Banswara	16.11.85 to 30.11.85	20	52.94 Cu.ft	100	1	1	—	—	1	—	3(15%)
Nanabhukhiya Land Management work, DPAP Anandpuri, Banswara	-do-	50	100 Cu.ft	80	1	1	1	1	1	—	5(10%)
Anandpuri-Dokar Road P.W.D. Anandpuri, Banswara	-do-	20	52.94 Cu.ft	80	1			1		3	5(25%)

(Contd.)

TABLE 5.2: *Contd.*

1	2	3	4	5	6	7	8	9	10	11	12
Sunderpur Stone Work, P.W.D. Bichiwara, Dungarpur	-do-	41	0.65 Cu.ft	75	1	2		1		—	4(9.75%)
Sulpore-Boruk Road, P.W.D. Bichiwara, Dungarpur	-do-	61	3.5 Cu.ft	75	1	2	—	2	1	5	11(18.33%)
Piplada Stone Work, P.W.D. Dungarpur	-do-	20	12.36 Cu.ft	75	1	1	—	—	—	—	2(10%)

TABLE 5.3: Essential Facilities Available at Different Famine Relief Works in Southern Rajasthan during November 16-30, 1985

Name of the work	Weekly paid holiday	Drinking water supply	Implements for working	Medical facility	Weekly payment facility	Shelter huts	Ration shop
Kushalgarh-Bajana Road, PWD, Kushalgarh Banswara	Fortnightly paid holiday	At the cost of workers	Nil	Nil	Payments in two to three months	Nil	Nil
Nanabhukhiya Land Management Work, D.P.A.P., Anandpuri, Banswara	-do-	-do-	Nil	Nil	-do-	Nil	Nil
Anandpuri-Dokar Road, P.W.D. Anand-Puri, Banswara	-do-	-do-	Nil	Nil	-do-	Nil	Nil
Sunderpur Stone Works, P.W.D. Bichiwara, Dungarpur	-do-	-do-	100%	Nil	-do-	Nil	Nil
Sulpore-Boruk Road, P.W.D. Bichiwara, Dungarpur	-do-	-do-	50%	Nil	-do-	Nil	Nil
Piplada Stone Work, P.W.D. Dungarpur, Dungarpur	-do-	-do-	Nil	Nil	-do-	Nil	Nil

TABLE 5.4: Work Done and Payment Received on Different Types of Famine Relief Works in Southern Rajasthan

Type of famine relief work	Work sites taken	Daily work executed (in percentage)	Payment made (Rs./day)	Less wages received than minimum wage Rs. 11.00 per day
P.W.D. (Public Works Department)	13	74.49±14.28	8.19±1.57	2.81
Irrigation Department	66	72.95±11.49	8.02±1.25	2.98
RLEGP (Rural Labour Employment Guarantee Programme	3	87.74±10.27	9.67±1.15	1.33
NREP (National Rural Employment Programme)	2	86.36±0	9.50±0	1.50
Total	84	74.04±11.61	8.14±1.27	2.86

(d) Damage of the precious forest resources in the region have led to the multiple environmental problems such as disappearance of the wildlife and forest products, heavy run-off, soil erosion, depletion of river water, underground water and spring water, loss of soil fertility and desertification.

(e) Famines have become common feature of the region after 1983-84 and state government is forced to take relief measures for the famine affected tribals.

(f) Famine relief works are neither adequate nor directed to improve the degraded environment of the tribal belt.

(g) Scarcity, inadequate famine relief works and emergence of wastelands have forced these tribals to migrate to neighbouring Gujarat state for alternate employment.

(h) Daily work given to the tribals for execution is equal to that allotted in normal years. Work load of unproductive workers such as unit leader, waterman, napatiwala, choukidar, dakman and others on workers and lack of the essential facilities such as weekly paid holiday, supply of drinking water, working implements, medical facilities, shelter, weekly payment facility and ration shop have led to the less execution of the allotted daily work by tribals and deduction in their minimum daily wages of Rs. 11.00 fixed by the state government indicating an exploitation of tribals.

TABLE 5.5: Work Executed in Percentage on Different Types of Famine Relief Works in Southern Rajasthan (in %)

Frequency Classes of executed work	Types of Famine Relief Works							
	P.W.D		Irrigation Department		RLEGP		NREP	
(%)	Number	%	Number	%	Number	%	Number	%
40-50	—	—	1	1.51	—	—	—	—
50-60	1	7.69	8	13.64	—	—	—	—
60-70	5	38.46	15	22.73	—	—	—	—
70-80	4	30.77	19	28.79	—	—	—	—
80-90	—	—	19	28.79	2	66.67	2	100
Above 90	3	23.08	3	4.54	1	33.33	—	—
Total	13	100	66	100	3	100	2	100

TABLE 5.6: Payment Received on Different Types of Famine Relief Works in Southern Rajasthan

Frequency Classes of Payment Received	Types of Famine Relief Works							
	P.W.D		Irrigation Department		RLEGP		NREP	
(in Rs.)	Number	%	Number	%	Number	%	Number	%
4-5	—	—	1	1.52	—	—	—	—
5-6	1	7.69	—	—	—	—	—	—
6-7	1	7.69	16	24.24	—	—	—	—
7-8	6	46.15	16	24.24	—	—	—	—
8-9	2	15.38	17	25.76	2	66.67	—	—
9-10	—	—	15	22.72	—	—	2	100
10-11	3	23.08	1	1.52	1	33.33	—	—
Total	13	100	66	100	3	100	2	100

Suggestions

In order to avoid further damages to the natural environment of the tribal belt of southern Rajasthan, to improve the natural environment, and to provide adequate famine relief to the tribals, following suggestions are made here:

1. All the famine relief works and tribal employment works should be oriented to the following objectives:
   a. Preservation of existing forest resources and wildlife,
   b. Reforestation of degraded and barren hills,
   c. Social and farm forestry on community land and land owned by tribals,
   d. Proper managed use of both underground and surface water resources, wherever necessary, water conservation practices shall be followed, and
   e. Soil conservation works.
2. Additional employment should be generated under various schemes of rural employment programmes and famine relief works giving priority to the works given in above (1) in order to improve the quality of environment. This will also help/divert the attention of tribals from cutting the remaining scarce forests, and help in curtailing tribal migration to the neighbouring Gujarat state.
3. Better working conditions and facilities such as weekly paid holiday, adequate supply of water at the famine relief work sites, improved and new working implements, medical facilities, shelters at the working sites, weekly payment facility to buy the essential commodities and ration shop facility should be provided, and
4. Looking at the physical and mental status of famine affected tribals certain concessions should be given to them in the amount of daily work to be executed, so that they can earn daily minimum wage Rs. 11.00.

REFERENCES

Dhabriya, S.S. (1984), Major environmental problems of Rajasthan and their remedies, *Science & Technology Research Bulletin* (BISR), 1(4): 1-6.

Mathur, P. and Rajpurohit, K.S. (1986), *Rajasthan Ke Akalgrasta Kshetron Mein Rahat Kariyon Ki Sanchalan Prakriya Aum Niyuntum Vetan Bhugtan ki Vartman Sathiti: Ek Sarvekshan*, Social Policy Research Institute, Jaipur, pp. 54.

6

JHUM CULTIVATION AND ENVIRONMENTAL DEGRADATION : A Case Study of Meghalaya

R.K. RAI

Jhum or shifting cultivation represents a distinct stage in the human history of economic development of mankind. In the beginning human beings used fire for clearing the forests and its under growth. Farmers are no doubt decision-makers and what commodity out of the practicable range will be grown or produced from the field is the result of the past inherent experiences with which they are contended in a set of natural environment. Even then the various agricultural systems and traditional practices of farming in different parts of the world are very largely dominated and controlled by the natural environment (landforms, geomorphic processes, soils, climatic conditions, run-off of precipitation and vegetation cover). The agricultural practices like *Jhum* cultivation, terrace cultivation, and settled cultivation etc. conspicuously controlled by the geomorphology of a region.

Shifting cultivation is an age old method of agriculture, practised in all areas inhabited by tribals and primitive population. It is differing only in minor details and called by different names. This is a method of producing human requirements of food developed in the dawn of human civilisation with the shifting of their habitations as well as their cultivation plots. Fire was the only weapon known to them which they could employ to clear the vegetation for raising a crop of edible plants. Even today the basic characteristics still remain the same wherever shifting cultivation is practised. We can in short give the explanation of shifting cultivation as a method of raising crops on a fire cleaned field for a short period of one or two years which is then left to be clothed again by nature with vegetation. Thus, in the beginning of human history of a region the human communities living over there were in the state of ecological balance with their environment. The number of persons were quite small and the land was in plenty to support the population of that particular region. But now with the increase of population and being somewhat staked down to smaller areas a shifting cultivator has got much less choice to shift about.

In the present study an attempt has been made to analyse the existing agricultural practices and its impact on environmental conditions of the Meghalaya state of north-east region of India.

In hill areas of north-eastern region, specially in Meghalaya, Mizoram, Nagaland, Tripura and Arunachal Pradesh, the shifting cultivation and other economic activities are posing difficult environmental problems. Particularly in state like Meghalaya, where most land is community owned and people are free to cultivate where they want and as much they want. In the present type of study, it is the biggest problem to get the reliable data on land holdings and land under different uses. No survey of land classification, land holdings and land records have yet been undertaken. Thus it becomes very difficult to analyse the characteristics of landuse of the state.

Geographical Background

Meghalaya is one of the youngest states of India and came into existence as a separate state on 21st January, 1972. It was carved out of Assam from two districts, i.e. Garo Hills and Khasi and Jaintia Hills districts. From 1972 it attained its full statehood. Now it has five districts viz. East Garo Hills, West Garo Hills, East Khasi Hills, West Khasi Hills and Jaintia Hills districts. Meghalaya is a hilly state having about 80 per cent of its area in the hills. About 80 per cent of the population of the state are scheduled tribes namely Khasi, Jaintia and Garo.

The state is basically an agricultural state. It has a total geographical area of 22,429 km^2 with a population of 13,35,819 (as per 1981 Census). The total estimated forest area of the State is 8,514 km^2 of which only 722.36 km^2 are directly under the control of the State Forest Department. The remaining areas are managed by respective District Councils of Khasi Hills, Jaintia Hills and Garo Hills as per provisions of the Sixth Schedule of the Constitution. Except the reserved forest areas, and protected forests in and around Shillong (being managed by the Forest Department in arrangement with the District Councils), the rest of the forest areas are subjected to the primitive agricultural practice of shifting cultivation slash and burn method. However, there are few pockets of undisturbed natural forests till in existence which will account to about 1000 km^2 and is being protected by the tribals as *Sacred Groves* due to some religious belief. The state, however, contains areas of very beautiful scenery with evergreen forests and waterfalls as well as unique floral and faunal varieties.

The forests managed by the respective district councils have three kinds of forest: the old unclassed state forests which are directly under district council control, the forests owned by the clans or communities, and the private forests. For the second and third categories the district councils have very little control except for collection of royalty when they export the timber outside their own area for trade.

The Forest Department also collects royalty on all minor produce and share with the district councils on the basis of 60 per cent for the council and 40 per cent for the state management cost. But in case of other forest produce the district councils themselves collect the royalty and manage the forests.

Correct survey of forest areas under the district councils have not been done as yet. But an estimate was done with the help of the Forest Survey of India (formerly the Pre-Investment Survey of Forest Resources). Ground surveys of certain rich forest areas have been done by the Forest Department on behalf of the district councils through the Forest Resources Survey Division. Recent assessment made with the help of LANDSAT imagery indicates that the total percentage of forest cover has gone down very much.

Geology

The Meghalaya Plateau is mainly composed of (a) Archaean granites and gneisses, (b) Shillong series, (c) Lower Gondwana rocks, (d) Sylhet traps, and (e) Cretaceous, Tertiary sediments. The distribution of above rock formations shows that in central and northern parts of plateau the gneiss and granite rocks are present, while in south these rocks are concealed beneath cretaceous and tertiary deposits. In eastern and central part of the plateau the Shillong group of rocks are exposed. The lower Gondwana rocks have been found in Garo Hills, while some peat coal deposits have also been reported from Khasi and Jaintia Hills. The sylhet traps are exposed along the southern border of the Shillong Plateau. The Cretaceous Tertiary rocks occupying the southern part of the plateau are thick and extensive in nature. Around Cherrapunji and Mawsynram extensive limestone deposits have been noticed.

Geomorphic Features

The Meghalaya Plateau is an eastward extension of the massive block of the peninsular shield. The height of the plateau varies from 160 m-1960 m. The Shillong peak is the highest part of the region (1961 m). The central part of the plateau marks the major watershed zone of the region. The northern hilly terrain gradually slopes down towards the Brahmaputra valley while the southern part of the plateau stands in the form of steep scarp overlooking the Ganglasesh plains. On the whole the plateau has been dissected into the hills, deep valleys, scarps and gorges etc. It is one of the most important example of multicyclic landscape. The plain land available for agriculture is only about 15 per cent.

The ideal conditions for shifting cultivation prevailed in those areas which have extensive land resource but sparse population. Here the *jhum* cycles are long and nature has sufficient time for regeneration of fertility. The *jhum* fields are mostly along hillslopes. Hence, there is more surface run-off and so much water can not get stored in the soil. Thus *jhum* cultivation depends entirely on distribution of rainfall and moisture retention capacity of the soils.

As long as the pressure of population in the state was low, the cycle of shifting cultivation was more than 10 to 15 years. It did not have any serious impact on deforestation and degradation of environment.

The Meghalaya state in the north-east India has been exposed to outside contacts for a very long time. Thus the pressure on land for various non-agricultural uses is ever increasing with increase in population. Some of the areas especially in Garo Hills which were put under *jhum* are going out of cultivation because of extreme soil erosion and losing soil fertility. The net result is that the *jhum* cycle which was having 10 years or more interval has come down to 3, 4 or 5 years or even less in certain areas of the state. The experience and scientific studies have shown that when *jhumming* is practised on areas with average fertility and average rainfall conditions the shortest *jhum* cycle, the land can stand is about 7 years. Thus *jhumming* with 7 years cycle plus can remain productive with some damage to the soil and vegetation.

The *jhum* area in Meghalaya state can be broadly divided into two agro-ecosystems — the low elevation agro-ecosystem and high elevation agro-ecosystem. The low elevation agro-ecosystem is representative of the general shifting cultivation system. The *jhum* cycle in these areas have been reduced to about 4-5 years. The size of *jhum* plots varies from about one to three hectares. The clearing operation here begins during the winter months say between December and January. The larger trees are not felled but their lower branches may be looped. Smaller trees and bamboos are cleared and under growth is completely slashed. The dried debris is burnt in the months of March-April.

Cropping Pattern

The crop mix in the *jhum* in Meghalaya varies considerably according to the length of the cycle. There may be as many as 10 to 15 variety of crops in the long cycles. The maize and paddy are major constituents. As the cycle becomes shorter the number of cereals is reduced.

The *jhum* practice changes significantly in high elevations. The main reason is that the vegetation at those heights is not as luxurient as in the lower elevation. The pine trees are rather sparse. The vegetational undergrowth is abundant. The clearing process begins in December. The lower branches of pine trees are hacked and undergrowth is completely cleared. The twigs, leaves etc. of this clearing are arranged in parallel rows along the slopes. The dried plant material is covered by earth with narrow furrows and then it is set on fire. A wide mixture of crops are sown in these *jhum* fields. However, tubers like potato are now the most important crop. Each *jhum* field may be used for one year or two years at a time depending on the soil conditions. Thereafter it is abandoned for natural regeneration.

Growing of tuber crops like potato and ginger is largely for the market. Consequently in a bid to have quick gains and larger incomes the land resources are being intensively utilised unmindful of its long-term implications. There is substantial run-off in high rainfall areas like those in Cherrapunji and soils get depleted at a very fast rate. As the soils along the steep slopes of the hills are exhausted in this process, the farmers move to the lower slopes. This process continue till the entire ridge becomes barren land. The result is that the rich forest area in this region are now reduced to near desertification.

The hill slops along which potato cultivation takes place, soil absorb about 35 per cent of annual precipitation. The loose sediment yield washed out in the form of soil in the areas of potato cultivation is about 39.3 tonnes/ha. Thus, over exploitation of and resources for commercial crops has become indemic near urban centres.

Tables 6.1 and 6.2 show the area under reserved forests in Garo, Jaintia and Khasi hills of Meghalaya.

TABLE 6.1: Area Notified as Reserved Forest

Reserved Forest	Notified as Reserved Forest vide	Forest Area in ha.
Garo Hills		
1. Chimabangshi	No.28 dated 19.6.1883	2,331
2. Dhima	No.28 dated 19.6.1883	2,072
3. Dilma	No.28 dated 19.6.1883	259
4. Rajasimla	No.28 dated 19.6.1883	1,813
5. Ildek	No.28 dated 19.6.1883	259
6. Darugiri	No.28 dated 19.6.1881	1,036
7. Rongrenggiri	No.28 dated 19.6.1883	3,626
8. Dambu	No.18 dated 12.3.1885	1,813
9. Songsak	No.29 dated 1.10.1885	2,331
10. Emanggiri	No.44 dated 7.1.1881	829
11. Rewak	No.44 dated 7.11.1883	647
12. Siju	No.44 dated 7.11.1883	518
13. Baghmara	No.12 dated 24.2.1887	4,429
14. Angratoli	No.3 dated 26.1.1888	3,001
15. Dibru Hills	No.28 dated 10.6.1883	1,502
16. Tura Peak	No.For 10.75.32 dated 23.6.82	419
	Total Garo Hills	26,895
Jaintia Hills		
17. Saipung	No.26 dated 25.7.1876	15,035
18. Narpuh I & II	No.3978F dated 17.6.1909	16,110
	Total Jaintia	31,145
Khasi Hills		
19. Riathkhuran	No.806R dated 5.3.1892	391
20. Nongkhyllem	No.4692F dated 23.7.1909	12,591
21. Umsaw	No.GFR 234/46/3 dated 16.12.1946	44
	Total Khasi Hills	13,026
	Total (Meghalaya)	71,066

Source: Government of Meghalaya.

TABLE 6.2: Landuse annd Forested Area: 1983-84

A

Land Utilization	Meghalaya	India
Geographical Area (km^2)	22,489	32,80,480
Agriculture (km^2)	2,238	15,22,900
Forests (km^2)	8,514	7,45,890
Area affected by shifting Cultivation. (km^2)	2,600	N.A.
Others (km^2)	11,677	N.A.

B

Forest Area	Area in km^2	Percentage to total forest area	Percentage to total geographical area	Functional classification (only state owned)
Reserved Forests	710.66	8.35	3.17	Productive forest area 710.66 km^2
Protected Forests	11.70	0.14	0.05	Potentially exploitable productive forests 11.79 km^2
Community Forests and Private Forests	7791.83	91.51	34.74	–
Total	8514.83	100.00	37.96	722.36 km^2

C

Divisions	Geographical area km^2	Reserved Forests km^2	Protected Forests km^2	Total km^2	Percentage to geographical area	Remarks
Garo Hills	8167	268.95	–	268.95	3.3	Area covered under working plan 269 km^2 Area for which working plan under preparation. 531 km^2
Khasi Hills	1043	130.26	11.70	141.96	1.4	Area not covered
Jaintia Hills	3819	311.45	–	311.45	8.2	7714 km^2
Total	22,429	710.66	11.70	722.36	3.2	8514 km^2

The Forest Destruction

The principal factors of forest destruction have been population explosion, both human and livestock leading to enhanced requirement of fodder, timber and fuel wood. Even in hill areas like Meghalaya the preservation of forest cover had been given a very low priority in face of agriculturisation, urbanisation, mining and quarrying etc. The Meghalaya state is quite rich in coal and limestone deposits. Extensive coal mining is going on in Jaintia hills. Large areas in Cherrapunjee and Jaintia and Garo Hills districts have been cleared and laid barren as a result of open surface mining, destroying some of the irreplaceable Pine forests. Commercial exploitation for timber has been almost unrestricted. The individual rights have also given encouragement to deforestation in large scale. The management practice related to forests are widely divergent ranging from forest owned by individuals, village communities, district councils and forest department. Thus advantage is being taken by numerous vested interest people on this uncertain situation.

The construction of new hill roads mostly along river valleys and also in ecologically most fragile belt of the state have paved the way for environmental degradation. During rainy season there are frequent landslides along most of the roads. During field study it was noticed that between Mawsynram to Balat, there were about 32 spots where frequently landslides occur. In the same manner, on other roads landslides occur frequently blocking roads completely.

Under *jhum* control programmes the measures undertaken were mainly introduction of cash crops into the traditional *jhum* areas. The crops initially tried were coffee, rubber, cashewnut and cardamom etc. *jhum*ias were given all the planting stock required and the financial help for raising their plantation. The subsidy was 50 per cent on loans. But this programme has also not been of a great success.

Another programme for control of shifting cultivation comprises preparation of terraced fields for settle agriculture. This programme has been taken up by soil conservation department of state in many areas of the state. The valley bottom lands were reclaimed fully at Government expense or by subsiding the cost up to 50 per cent to 75 per cent. The gentle slopes up to 30° with good soil depth are decreased. The survey of terrace cultivation indicates that the soil structure of most of the hilly areas is quite fragile and loose. The result is that even when the terraces are prepared with care and their edges are protected by raised bunds, the terraces are badly affected by soil erosion during heavy rains and gradually they shrink in size. In about 5 to 6 years time even good terraces may not remain viable for cultivation. In some cases the terraces may destabilize entire hillslope leading to frequent landslides. The net result of all these factors is that terrace cultivation in the state is economically non-viable.

It has been noticed that in the state during the past three decades the pressure of population has tremendously increased. Due to this increase in population there are certain areas which have suffered heavily in relation to their ecological conditions. Specially the areas close to urban centres like Shillong, Tura, Jowai, Nogstoin, Nongpoh etc. due to construction of houses, roads, administrative buildings, hospital, educational institutions and industrial infrastructure, the ecological landscape has been very much disturbed. The areas which are covered under dense Pine forests are now totally devoid of pine forest, which has caused great imbalance in forest ecology. Specially the Shillong town being the capital of Meghalaya the urban population has increased manifold between 1971 to 1981. Along Gauhati-Shillong road concentration of new settlements have taken place due to economic reasons and thus it has very much effected the ecological landscape. The construction of new roads have further affected the ecology of the region. The environmental problems which the state is facing today at different scales reflects the stage of the environmental degradation. The environment and development should go hand in hand.

In the tribal areas the state can give financial support to the new technology adopted for the use of land which are marginal for settled agriculture, social forestry, tree culture, and animal husbandry. The shifting cultivators can attain a significant higher level of living within short time. This may also help in afforestation programme to bring sufficient land under forest cover, which will help to maintain ecological balance.

7

COMMERCIAL VERSUS TRADITIONAL AGRICULTURE : An Issue in Environmental Planning

ANJANA P. DESAI and SULBHA KHANNA

In rural environment man remains as a part of nature as other living organisms are. The crisis is felt when man tries to introduce inputs to the local system, which are not natural and do not fit in the system. The impact of such a human activity is perhaps very dramatic on environment at industrial locations. In the rural front, the environmental crisis is also wide and deeply rooted. Human interventions are related to deforestation leading to soil erosion, drought and flood; increasing use of ground water for domestic, agricultural and industrial purposes. In rural ecology, the crisis is far more dangerous on a long-term basis and it is not yet perceived under the disguise of commercial farming.

This study brings out the pros and cons of traditional and commercial farming and also the impact of commercial farming on rural ecosystem both from physical and human points of view.

Ecosystems in Traditional and Commercial Farming

In a traditional farming system, a variety of crops is raised by farmers on small holdings. The ecosystem is stable as it is based on diversity of crops which are suited to the local ecosystem based on organic fertilizer and traditional technology. In this ecosystem the level of biomass is high, having a wide range of species and habitat for zoological life. The different crops often are complementary e.g. legumens are providing nitrogen to grains. The landuse appears to be repetitive as all agricultural holdings have a similar combination of crops which are consumed locally. People live on subsistence level without any specialization, exchange or sale of crops.

In contrast, commercial agriculture is based on the concept of profit and is meant for sale. Specialization of crops is done on pattern of mono-cropping. In this ecosystem, the productivity per unit area is increased, but the variety in biomass is highly reduced due to lack of diversity of crops. Every effort is made to have the highest production based on modern technology with the help of high yielding variety (HYV) of seeds. A combination of machine and an introduction of non-biological inputs has certainly made man to progress and to have a better control over certain hazards and problems related to pests. This system of production is associated with major structural and functional changes of landscape as commodities need to be exported to distant markets. Hence connectivity and accessibility of a region are enhanced to a great extent along with a growth of a flourishing trading community. This particular style of farming has been thus equated with fast economic growth and development in such a way that it is difficult to point out several problems that are related to the system in a serious manner. Some of these problems are physical, and some are human in nature. It takes time to perceive the darker side of commercial agriculture as the problems are not so obvious as in the case of localised industrial pollution.

On the physical side, the trading-cum-farming community has ruined the maturity and the stability of natural ecosystem by changing them from a traditional multiple cropping pattern to a mono-cropping one. As the complexity of biomass is reduced in a mono-cropping system, its vulnerability is also increased. The HYV are susceptible to sudden mass destruction due to uniformity of crop and gene. The environmental degradation is created not just in terms of distortions of biomass, but damage is also done in terms of soil itself. The increasing use of non-biological inputs of chemical fertilizer and pesticides silently damages the biological ecosystem over a period of time in the disguise of economic development. The motivation of profit among commercial farmers is strong and they are not concerned about the destruction of the ecosystem on a long-term basis as they are eager to reap profit out of the soil in a short period of time. The change of market from a local to a distant one has made commercial farmers less aware and less sensitive to their own ecosystem. Moreover, the government also encourages the export of cash crops to earn foreign exchange. The change in occupation from farming to trading has also made farmer-traders less attached to the soil and environment. Hence they are not concerned about the maintenance or conservation of environment for future generations. This process of production with an emphasis on non-biological inputs has serious effects on the exploitation of man as well as on the physical environment.

The commercial agriculture is associated with exploitation of unorganised labour. Generally local labourers are aware of the rate of wages and other benefits of job. So commercial farmers hire labourers from distant areas who are unorganised and are unable to bargain for proper wages. As a result, an accumulation of wealth takes place in this system of production leading to exploitation of poor workers. It has enhanced the socio-economic gap between man and man in terms of property and life style. A new type of *colonialism* has evolved within the rural community. A territorial division of labour by supply and demand, has created the growth of a new capitalist trading community in agricultural sector.

Further, the commercial system of agriculture has enhanced the regional variation in productivity. The total production of food or other crops has increased no doubt, but the skewness in the distribution of food grain has widened. A small section of the community who are rich, has been benefited. They can afford to buy the extra amount of grain with their high purchasing power. But the per capita consumption of food grain has not changed for the common mass. In 1965, it was 480 grams per person per day in India, and the figure was the same even in 1984. The minimum requirement is 493 grams per person per day for a total daily calorie of 2,100 to 2,400. The high yielding variety of seeds is used mainly for wheat and not for pulses. There has been a reduction in consumption of pulses from 70 grams per person per day in 1965 to 39 grams in 1984. This has affected the health of farmers for not having a balanced diet. Further, in commercial farming, farmers tend to grow more non-food crops which can fetch more price in the market as those serve as raw materials for industries. This has reduced the total area meant for food crops. Moreover, the cost of inputs in commercial farming per unit area is higher than that of traditional farming. Hence the price of food grain has also increased. The net result is that the consumption of food grain for a common person per day has remained static during 1965-84. The increase in production does not guarantee equal distribution or upliftment of the poor.

Environmental pollution, in general, is very much related to commercial farming. The high yielding variety of seeds are vulnerable to pests, diseases and climatic irregularities. In other words, they have a low level of resistance. They need large quantities of fertilizer for growth and pesticides for pests and plant diseases. As chemical fertilizers generate heat in the soil, the HYV seeds need a constant supply of water through irrigation facilities. Commercial agriculture is often associated with large irrigation projects which cause environmental degradation in terms of deforestation and soil erosion. If there is a failure in monsoon rain, farmers have to draw more water from the underground resources. Some HYV seeds, especially the variety of Mexican wheat, need much more irrigational water which may lead to the salinity of soil and the over exploitation of ground water. Moreover, the use of pesticides on HYV crops leads to health hazards, as they are sprayed directly on vegetables, fruits and grain. Evidences prove that milk gets contaminated if cattle graze on grass having pesticides. Similarly, drinking water in wells and rivers may not be safe if pesticides are used heavily in the neighbouring fields. The drained water from fields will contaminate the drinking water.

Environmental Planning

In this discussion, it is realised that commercial farming has increased the total production but not the per capita consumption of food. Hence it has enhanced the economic gap between man and man, and between regions with and without commercial farming. Moreover, the motivation of making profit has ruined local ecosystems with the introduction of non-biological inputs to an unstable, vulnerable system having low biomass variety.

The issue is : whether is it possible to reverse the trend ? Certainly it is not so easy to go back to the stagnating traditional system of low production in spite of having many problems with the modern commercial agricultural system. It is a vital issue in environmental planning as commercial farming trends to enhance the gap between man and man, man and nature in the name of economic development. The objective of planning is to reduce the disparity of any kind. The decision is ours—the way we want to live today and tomorrow. Planning may not be of old style, but past experience certainly helps in understanding a system which has gone wrong. Perhaps a new technology is needed today to strike a balance between an increase in production and conservation of environmental stability. The development of an ecosystem should be such that the benefit of production is shared by the majority of the local people who are closely attached to their environment. The sense of belonging to environment will certainly be the base of its protection and maintenance. The issues of planning and implementation are very much dependent on the value judgement of the society concerned. The outlook of sharing the growth and development and the conservation of environment for the present as well as for the future are certainly critical issues to be thought of. It is to be decided whether we have the right to damage an ecosystem which will be the habitat for the next generation to come.

REFERENCES

Armstrong, Warwick and T.G. McGee (1985), *Theatres of Accumulation,* Methuen and Co. Ltd., New York.

Arnon, I (1981), *Modernization of Agriculture in Developing Countries— Resources, Potentials and Problems,* John Wiley and Sons Ltd., New York, Brisbane, Toronto.

Desai Anjana (1988) Energy and environment, published in the Proceedings of the Workshop on Energy Education, V.A.S. Community Science Centre, Ahmedabad (India).

D'Monte, Darryl (1985), Temple or tombs ? Industry versus environment : Three controversies, published by Centre for Science and Environment, New Delhi.

Rappaport, Roy A, (1971), The flow of energy in an agricultural society, *Scientific American,* September.

Scott, Allen J and Michael Storper (eds) (1986), *Production, Work, Territory,* Allen and Unwin Inc, Boston.

8

ANTHROPOGENIC IMPACT ON LANDSCAPE SYNTHESIS IN KASHMIR HIMALAYAS : A Study of Land Suitability and Capability Classification

R.B. SINGH and A.A. PIRAZIZY

Information regarding man-environment interactive systems in highlands is limited and fragmented. This aspect in return hinders the assessment of consequent environmental changes. Highland environments are generally understood to have uneven topography, inadequate soil cover, which is further impoverished by both autogenic and allogenic forces, ununiformity of precipitation, diurnal variation in temperature, and high altitude which renders the complex highland environments more fragile. Moreover, factors of low carrying capacity, high fragility of environment and vigorous downslope movement of material also determine the extent of interactive systems in highlands.

Resources utilization trend have been the main causal factor in attracting man to abide in mountains and adapting himself to severe environment of highlands. The technological achievements of man at various stages have determined the extent of his interaction with environment in general. Land has been the foremost concern of man. He has sought its utilization to sustain his livelihood in one form or the other. Besides, various processes have acted on land and resulted in subsequent landscapes. Hence, the evaluation and perception of landscape depicts itself in two specific functions, i.e., hazard avoidance which implies to minimise human disaster and loss of capital, and guidance on suitability for major kind of landuse. The latter aspect refers to segregation of land for its various appropriate uses. Landscape evaluation however is a very important factor in improving resources inventory, carrying capacity decision, and environmental impact assessment.

To deal with complex environmental problems in general and India in particular, there is a need for continuous monitoring assessment and long term prognostication at various levels, as Erickson (1979) said:

> "It must be acknowledged that a specific impact occurred as a consequence of one particular area is no guarantee that the same impact will occur or have the same significance as a consequence of another and similar activity in another and similar area".

Sustainable Land-Use: Measurement Through Carrying Capacity

The concept of sustainable landuse may be understood in terms of carrying capacity. The carrying capacity is used to describe the ability of the land to support a population/organisms on a sustainable basis. Moreover, there is no standard approach as how the carrying capacity should be calculated. The concept refers to any land unit or ecosystem that can tolerate a certain amount of pressure under normal conditions beyond which the sustainable utilization ceases to exist and degradation commences. Each land unit has its own unique carrying capacity in relation to anthropogenic activities such as production (agriculture, horticulture), construction (road, building), etc. The variation in carrying capacity is spatio-temporal phenomenon. For example, in India the high productivity of Indo-gangetic plains had been the main attraction to many settlers. Once fertile lands of the plains are now threatened by floods and droughts. Hence, it is imperative to consider such variations and adjust the population and their relative activities accordingly.

The above concept may be simplified as the formula (Vogt, 1948) follows :

$$C = B : E$$

Hence, C stands for the carrying capacity of any area of land i.e., its ability to provide basic needs such as food, fuel and fodder. Moreover, C becomes more complicated when considered in case of modernization. B means biotic potential, the ability of land to produce vegetation which is an important life support base. E stands for environment (including the part which is contrived and complicated by man), places on the biotic potential or productive ability.

India is physically a diversified land mass, and limitations to carrying capacity of its various environments is well depicted by following statement given by Michael and Straus (1955);

> "In this land of wealth and poverty are a variety of climates and soils, ranging from the perpetual snows of the highest [peaks] in the world to tropical alluvial plains ....
>
> Through this land, course mighty and wonderous rivers such as the Brahmaputra, Ganga, Indus and others. They are ... India's hope for the future".

Population pressure determines the carrying capacity of land under anthropogenic utilization, and is a dimension which must be taken into account while measuring carrying capacity.

Landscape Evaluation in Kashmir Himalayas

Vulnerability of morphology in landscape types facilities the identification of sensitive landscape features. Moreover, landscape features are outcome of interaction of geology, drainage, soils, and slope with natural processes and man's development and economic activities.

Despite the complex and diversified nature of highland environments yet they represent a unity which can be said that ecosystems of highlands prevail in severe conditions. Considering the above factor two tahsils of Gulmarg 33° 50′ N to 34° 05′ N and 74° 10′ E to 74° 40′ E, and Pahalgam 33° 50′ N to 34° 20′N and 75° 05′ E to 75° 40′ E of Kashmir Himalaya have been studied for empirical evidences of landscape evaluation, by considering geology, drainage systems, slope and soil (within the limited physical sphere).

Geological Structure

The tactonic processes responsible for creation of mountain systems are also connected to the physical configuration and genesis of the study area. The area consists of shales, limestones, sandstones, quartzites, mudstones, slates, agglomeratic and pyroclastic products, phyllites, schists and bafliaz volcanics (Shah, 1978), which are mostly fossiliferrous in case of Pahalgam tahsil and unfossiliferrous in case of Gulmarg tahsil (Wadia, 1934). All these formations are covered with glacial and lacustrine deposits of Pleistocene. Moraine materials are the major surface feature of both the tahsils.

The evidence of folding, faulting and tilling in the region is the result of simultaneous upheaval in Himalayas. These processes have increased the geomorphological complexities of the region and affected the formation of new land features.

Drainage Systems

Owing to varying nature of rock material and slope characteristics, drainage pattern of the region is diversed. Pahalgam tahsil has an antecedent drainage of dendric nature, which in upper reaches due to divergence and convergence of systems give drainage a radial pattern. Gulmarg tahsil in upper parts shows a dendric drainage pattern, while in lower reaches, streams have almost parallel configuration. The constant shift of channels in Gulmarg is caused by rapid deposition of sand and gravel.

There are 281 stream segments in Pahalgam tahsil with the total length of 679 km. The main river system of the tahsil is Lidder river, which till its deboucher into the Jhelum river (the main river system of the Kashmir Valley) has a mean fall of 31 m per km. Table 8.1 reveals that Gulmarg tahsil has 279 stream segments, out of which 203 with the total length of 411 km belong to Sukhnag river which has a fall of over 60 m per km, and 76 stream segments with the total length of 232 km. belong to Ningle River (Raza, et al., 1978).

TABLE 8.1: Drainage Characteristics of River Basins

Drainage Basin	Area(km^2)	Drainage Length(m)	No. of Streams
Lidder	1243	679	281
Sukhnag	932	411	203
Ningle	538	232	76

Slopes

The area under study is divided into five generalised slope regions

(Raza, et al, 1978). Comparing slope map with soil, natural vegetation and landuse maps, the areal differentiation in characteristics based on slope variation is identified (Table 8.2):

(a) Region of gentle slope (5°-10°) obeys the valley floor land-use types and except some land under settlements and roads, the area is devoted to cultivation of crops such as rice and vegetables.
(b) Region of gentle to undulating slope (10°-20°) is a transition between valley floor and the hills and has both landuse types. A good portion of land is under cultivation and equally size-able amount is under forest cover. At higher altitudes forest is the only landuse.
(c) Region of moderate slope (20°-30°) is totally hilly, with forest and grazing landuse types prevailing, and as the angle of slope increases the preference is given to grazing; and forest cover becomes sparse. This region is an attraction to trans-humants and their herds.
(d) Region of steep slope (30°-40°) due to very steep nature of region no specific landuse is practised. However, occasional grazing is found on slopes facing the rivers. The region otherwise is devoid of vegetation.
(e) Region of steep slope and escarpment (above 40°) is usually covered with perpetual snow. The region is the ultimate source of water in these tahsils.

Soils

The soils of the region vary in origin from alluvial to lacustrine and glacial. The main cause of variation being climatic factor. Hence, there is a little relation between the soils and parent rocks. It may be said that a long geomorphic history interrupted by alternative fluvial and glacial phases are responsible for their evolution. The soils of the region are divided into two types (Kaul, 1956), on the basis of physio-graphic divisions. One is the valley basin and the side valleys type (up to 1850 m) : this region embraces 50.87 per cent of Gulmarg and 12 per cent of Pahalgam tahsils, and is located in flood plain and paddy belt. The other type is the highlands (1850-3500 m), this type includes 87.08 and 49.13 per cent of Pahalgam and Gulmarg tahsils respectively. This is a region where high degree of soil erosion takes place. This region is mostly covered with immature mountain soil. Texturally, these highland soils are mostly silty, sandy and clay loam.

TABLE 8.2: Relationship Between Slope Categories and Physical Features

Slope Region	Slope (Degree)	Area (km^2)		% of area		Soils	Natural Vegetation	Main Landuse Type
		Pahalgam	Gulmarg	Pahalgam	Gulmarg			
Gentle	5-10	—	21.5	—	11.9	New alluvial	Crops	Cultivable
Gentle to undulating	10-20	48.49	79.25	11.92	39.68	Old alluvial	Crops	Cultivable
Moderate	20-30	114.05	66.37	28.17	34.50	Old alluvial to coarse Soil	Dense vegetation	Not cultivable
Steep	30-40	171.6	28.12	42.18	14.63	Rocky exposure	mixed	Not cultivable
Steep slopes & escarpment	Above 40	72.6	—	17.73	—	Rocky exposure snow cover	Open grass	Not cultivable

Capability and Suitability Classification

The main resource of the area under study is its high aesthetic and scenic landscapes which form a good resource base for tourism industry. Tourism related use of land appears to be the only profitable in the region. However, any overemphasis on one type of landuse, may cause a chain reaction with negative effects on the others. In order to avoid the degradation of environment subsequent development and practice of tourism in the ecologically sensitive areas, a capability and suitability classification is very important. In the region, generalized capability cases, based on site, properties of soils, proportion of land under different slopes categories, exposure to erosion hazards, and dominant landuse type have been identified.

With modification of Bonnet's method (Bonnet, 1955), the first step has been to classify the land suitability for cultivation and those unsuitable for cultivation (Fig. 8.1). Except a few tracts along valley floor, most of the land in the study area is not suitable for cultivation which can been further sub-divided into following three zones:

(a) *Forest zone* : This zone consists of steep slopes and is predominantly used for both grazing and forestry.
(b) *Exposed bedrock zone* : This zone consists of steeper slopes and poorer soils which almost have no capability to support vegetation.
(c) *Glacial zone* : This zone consists of the higher slopes of the mountains which are covered by snow.

After having classified the region according to its capability, the next step was to evaluate land suitability for its main economic use, i.e. tourism. For evaluation of highland areas of the Kashmir Himalaya, with specific reference to Gulmarg and Pahalgam tahsils, the modified form of the Dent and Young's (1981) classification was adopted, and four classes were selected to assess the suitability (Fig. 8.2). The classes are as follows:

(Sh), Highly Suitable, with no significant limitations to sustained application of tourism.
(Smo), Moderately Suitable, having limitations which in aggregate are moderately severe.
(Sma), Marginally Suitable, in aggregate severe limitations for sustained application of tourism.

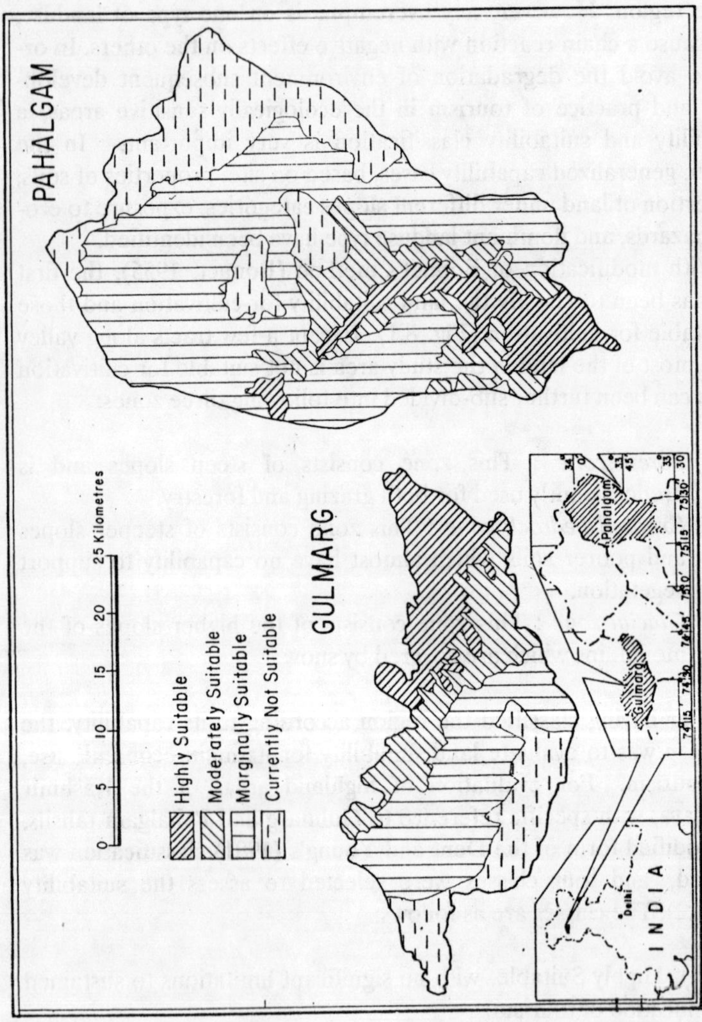

FIG. 8.1: Land Suitability for Tourism

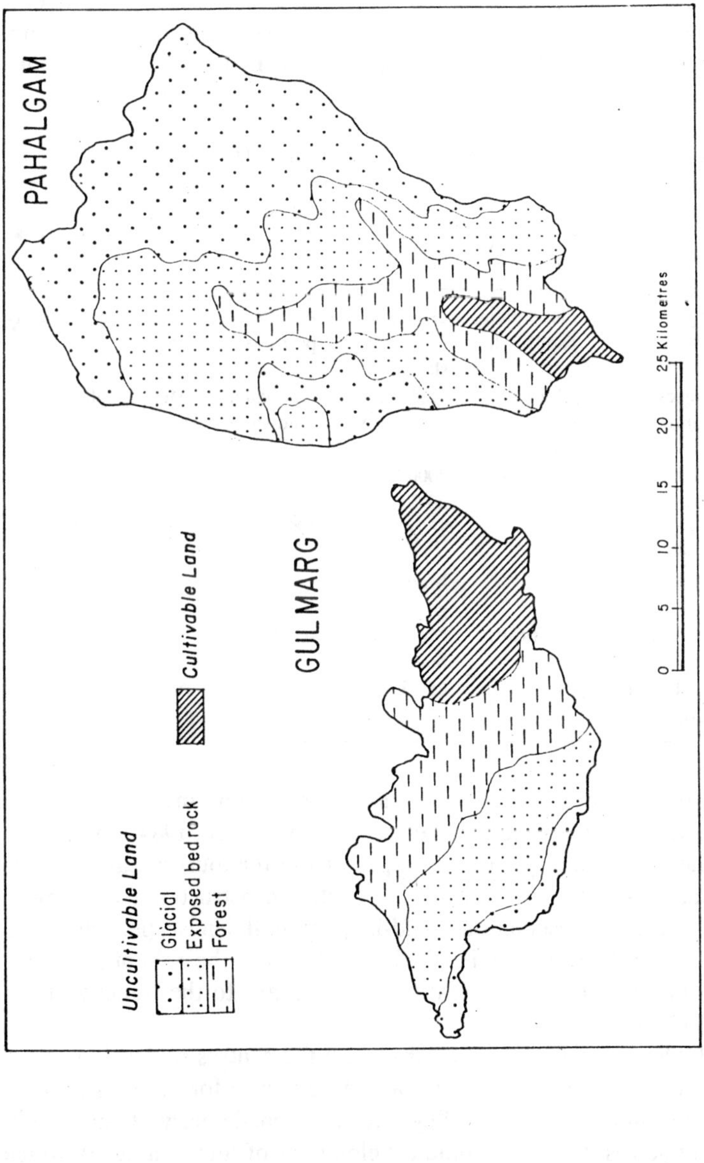

FIG. 8.2: Land Classification

(Scn), Currently not suitable, limitations are very severe and cannot be restored with existing knowledge.

These four classes were assessed on the basis of geomorphologically related items such as slope, soil, geology, drainage, altitude, etc. to ascertain the suitability of both the tahsils for tourism development (Table 8.3).

TABLE 8.3: Rating Criteria of Suitability for Tourism Development

Geomorphic Related Items	Suitability Groups			
	Sh	Smo	Sma	Scn
Topographic zone	Flood plain, paddy belt	Dry farm-land forested slopes	Alpine pastures	Rocky slopes glacial heights
Geological formations	Older alluvium, Karewa	Shale, lime-stone, clay	Quartzite, sand-stone	Metamorphosed rocks
Altitude (m)	1500-2000	2000-2500	2500-3500	> 3500
Slope (%)	5-20	20-30	30-40	> 40
Area under each slope type (km^2)	149.25	180.42	199.72	72.6
Natural drainage	Poor	Moderately good	good	Very good
Soil erosion	Practically negligible	Moderate to heavy	severe	Bed-rock exposed
(%) of top soil depletion	25	25-75	≃50	Nil

Human interaction with environment on any landscape is depicted in the type of landuse activities (Khoshoo, 1984). In the case of the area under study, four types of use for tourism can be identified, i.e. settlement (hotel, housing, etc.), construction (roads, highways, etc.), preservation (wild-life, natural landscape, etc.) and production (agriculture, horticulture, etc.) with the help of the above factors, land uses were classified with regard to the fragility of the area (Table 8.4).

Gulmarg and Pahalgam tahsils are the centres of developing urban regions and have become a service centre for growing tourism industry in the Kashmir valley. However, inadequacy of land use in the region is due to dynamic development of tourist industry which has forced concentration of its facilities to easily accessible areas. The rapid growth of the industry with little or no regard to suitability and capability of the tahsils to support the present activities, has imposed negative environmental effects. Moreover, this aspect has influenced the improper functional division of land. For instance, areas which by mere nature should be preserved for recreation have been spared for developmental activities, subsequently resulting in encroachment of natural landscape by pressurizing the fringe areas which have a limited carrying capacity.

TABLE 8.4: Classification of Land-use Type with Regards to Fragility

Use Type	Suitability Groups			
	Sh	Smo	Sma	Scn
Settlement	Good	Not recomm-endable	Dangerous	Impossible
Construction	Good	Not suitable	Dangerous	Impossible
Preservation	Poor	Moderate	Good	Very good
Production	Good	Moderate	Not feasible	Impossible

Conclusion

The degradation of Himalaya through anthropogenic impact, however, is a supposed assumption of various land use pattern, which in case of the present paper is contemporary tourism and related mismanaged developmental activities, such as road and building construction, to promote further growth of the industry. The ecosystems of highlands are very fragile and have evolved in the absence of man, hence, they are vulnerable to the presence of man. There is a high degree of incompatibility between anthropogenic use and the maintenance of the integrity of these ecosystems. The original ecosystems have already been degraded by such activities in the region, yet this degradation is a accelerated and spreading.

To mitigate the forthrising negative aspects, the community appearance ought to be controlled in the interest of preserving landscapes and their characters, new developmental policies having functionally harmonious relationship with the environment must been encouraged, and a continuous monitoring assessment and long term prognostication at various levels must be enacted.

REFERENCES

Bonnet, H.H., (1955), *Element of Soil Conservation,* New York.

Erickson, P.A., (1979), *Environmental Impact Assessment — Principles and Applications,* Academic Press, New York.

Dent, D. and Young, A., (1981), *Soil Survey and Land Evaluation,* George Allen and Unwin, London.

Kaul, M.L., (1956), "Study of Soils at Different Altitudes in Kashmir State", *Unpublished Assoc. Thesis,* SSAC, Indian Agricultural Research Institute, New Delhi.

Khoshoo, T.N., (1984), *Environmental Concerns and Strategies,* Natraj Publishers, Dehradun.

Raza, M., et al, (1978), *The Valley of Kashmir — A Geographical Interpretation,* Vol. I, Vikas Publication House, New Delhi.

Shah, S.K., (1978), "Facies Pattern of Kashmir within the Tectonic Framework of Himalayas", in Saklani, P.S., *Tectonic Geology of the Himalayas,* Delhi.

Singh, H., (1958) (edited), "Land Erosion in Kashmir Valley", *Indian Forester,* Vol. LXXXIV, No. 10.

Straus, M.W., (1955), *Why Not Survive,* Simon and Schuster, New York.

Vogt, W., (1948), *Road To Survival,* William Sloane Associates, New York.

Wadia, D.N., (1934), *Geology of India,* Tata McGraw-Hill Publishing Co., New York.

9

DEFORESTATION IN CENTRAL HIMALAYA : Issues and Challenges

S.K. SINGH, R.K. PANDE and SANTOSH VERMA

The recent programmes of construction of road networks, laying power lines and construction of hydroelectric units in the Central Himalaya and Kalagarh Dam (Pauri-Garhwal) has submerged 500 ha of forested area. It is expected that about 200 ha of forest land will be submerged by the proposed Tehri Dam (Fig. 9.1), in the central Himalaya which will result in considerable deforestation, especially from the middle and lower ranges. The forest cover left today is only 18-19 per cent in Himachal Pradesh, 28.7 per cent in Garhwal-Kumaun region and even 23 per cent in Nepal. The forest canopy over the Himalaya has dwindled to mere 35 per cent against the optimum of 60 per cent. All these are the cumulative consequences of acts of man for the ghetto development. A perusal of working plan of the East Forest Division, Kumaun Circle reveals that prior to the 1906, there was no commercial exploitation of forests but when the staff of the Forest Department was increased after the introduction of forest rules in 1903, the expenditure had also gone up and the Government of Her Imperial Majesty, considered it desirable that revenue should balance the expenditure and commercial felling started. Increase in population has also meant increasing pressure on forest resources (Table 9.1) because an increasing population has to be provided with food and energy to sustain itself and for this the population tends to :

- cut forests for fuel and timber,
- cut forests for bringing larger area into cultivation,
- extend crop cultivation into the higher altitudes, and
- promote overgrazing on existing pastures and forest vegetation.

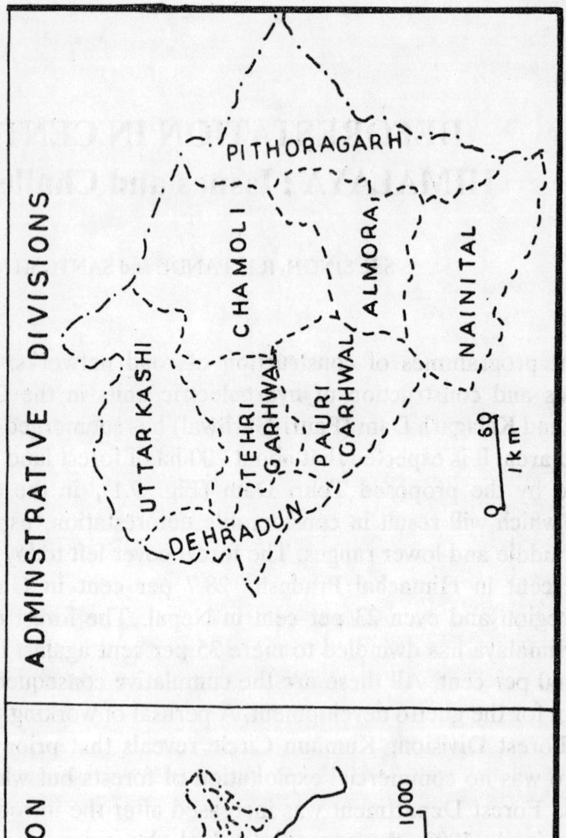

FIG. 9.1: Uttarakhand region of U.P.

The need for exploitation of forest wealth and enriching the government exchequer through the Forest Department causes rapid and brutal felling of trees to convert vast areas to waste.

TABLE 9.1: Forest Cover in Kumaun : A Comparative Statistics

	ALM	NTL	PITH	Kumaun
Geographical Area (km^2)	5,385	6,794	8,856	21,035
Forest Cover (km^2)^A	3,944.32	4,026.40	3,302.89	11,273.61
Forest Cover (km^2)^B	1,780	3,026	2,359	7,165
% short from NFP	26.94	15.46	33.36	25.94
Per Capita Forest Area^A	0.521	0.354	0.675	0.473
Per Capita Forest area^B	0.235	0.266	0.482	0.301

Source: (A) U.P. Forest Statistics, 1983
(B) Landsat Imageries

TABLE 9.2: U.P. Himalaya : Areas Available for Afforestation and Grassland Development

	Area in km^2
Unworkable	1,993.57
Plantable areas	667.58
Sandy, reverberls, boundary etc.	308.23
Grassland	1,299.33
Degraded	300.16

Source: U.P. Forest Statistics, 1983.

Distributions of Forests

Over the years, broad-leaf forests have been replaced by more hardy and commercially important species, like pine. The plantation of pine in place of oak has also created an environmental problem. Oak leaves give rise to a nutrient rich water absorbing humus. An oak forest holds into water for a long time and releases it slowly. This gives rise to springs around which hill villages have been established. Traditionally, the village settlements were constructed in the middle portion of the hill slope. The upper reaches were left as forest and pastures which provide springs for drinking water and irrigation and the lower slopes were terraced and cultivated. The excessively high rate of vegetation cover and the forest cover destruction generates a natural vicious circle — the sequence of less vegetation and high soil erosion and low water retentivity. The mountain soils become unproductive and so the higher productivity claimed through increasing area under cultivation becomes self-defeating. Natural factors enhance the destructive effects, seasonal rainfall can be observed by intact soils and slowly released, instead, we have floods at one point of time and droughts at another point of time, 550 m^3 of debris, produced by landslides and rockfalls from construction of one km long road annual, generates about 24 X 10^6 cubic metres of sediment to slide down the slope, killing vegetation, smothering springs and choking streams.

People's Consciousness and Possible Participation

For the hill people, deforestation means more and mere misery. According to Kawosa if the present rate of deforestation, i.e., use of forests for timber, fuelwood. . . continues, the forest will reach almost a dead end by 2020 AD (Kawosa, 1987). It is necessary to control felling of trees and also following a scientific afforestation policy. Undoubtedly it is a man-made problem. This was brought to focus by the 'Chipko Andolan' of 1973. 'Chipko' literally means 'embrace'. The 'Chipko Andolan' was inspired by Gandhian ideals in which an agitator 'hugs' the tree which a contractor and his labourers are trying to cut. The success of the 'andolan' was based on the realisation of the fact by the villagers that their lives are dependent on their forests. One factor which the enthusiastic villagers overlooked was that along with the contractor their own increasing numbers and needs were contributing to the problem.

Commercial felling can be regulated incidentally, the pine supplied to Star Paper Mills, Saharanpur, from Kumaun Himalaya was 15789.20 metric tonne in 1977-78 and 20706.43 metric tonne in 1981-82, an increase of 983.446 metric tonne per year. Similarly in 1979-80 1372 m^3 of pine was supplied from Central Himalaya for railways sleepers, which increased to 12,499 m^3 in 1980-81. The yield of resin has decreased by more than 300 per cent, but the greed of state government has increased with the rise in resin prices in the market. The needs of increasing human and animal population have, however, to be placed on a different footing because their very existence is dependent on the forest. For the hill people it is net so much a question of *rights* or *necessity* (the villagers were poor because the forests they once lived off have been turn down in the name of development) (Times of India). In the absence of shops selling timber, both for building houses as well as for use as fuel (in the eight hill districts of Central Himalaya the per capita out turn of timber and fuel is 0.08 m^3 and 0.28 m^3 respectively), if necessary permission is not granted to the villagers, they will be forced either to bribe a local official or to illicitly fell a tree. Similarly, their cattles need forage. If the community land in the villages is not sufficient, cattles will drift into the reserved forest area. The number of forest guards is also very limited to check such illicit activity over the vast forest areas in such a difficult terrain. The forest cover is, therefore, gradually receding around the villages. There is, however, one redeeming feature. The 'Chipko Andolan' has made the villagers, as well as the government, conscious of the needs of preserving the forest wealth. There is also, in these areas, a need to strengthen the institution of the *forest panchayat*. It is interesting to note that at present only 4.63 per cent of the U.P. Himalaya is under *panchayat* forest. A group of villagers can resolve to form a *forest panchayat* which is managed by *surpanchy*. All the members of *panchayat* get rights in the *panchayat forest* i.e., rights of fodder and timber. It has been observed, in some cases, that villagers prefer to fell a tree in a reserved forest rather than in their own forest. Some of the *panchayats forest* are, therefore, in a better condition than reserved forests. The reason is very simple. Authorities are very far away and if caught in illicit felling the villager is fined by a court which is very nominal. Illicit activity in the *panchayat forest* is very difficult because the villagers are all around and every villager is a forest guard, punishment is also immediate and dispensed by the *sarpanch*. Over the years *forest panchayats* have also been encouraging a limited afforestation. These *forest panchayats* can now be used as the medium through which institutional credit can be obtained for the afforestation programme provided it is linked to economic activity which can provide the necessary returns to repay the loan.

Possible Approaches

The villagers make a demand on the forest wealth because of the needs of survival. If an alternate economic activity can be provided, and it can be demonstrated that these economic activities are not only more remunerative but are also dependent on the preservation of forest wealth, the villagers themselves can be moved to preserve their surrounding forests. If for example, animal husbandry, cattle rearing, milk production, meat, wool production and tussar cultivation (in the oak forests) can be introduced the rate of return would be sufficiently high from these activities to provide an important subsidiary occupation.

In the Himalaya, a very limited area falls under the watershed definition of the government. However, civil, soyam forests, panchayat forests, pasture land and roads and river sides contain (Table 9.2) an extensive deforested belt. In addition to it, due to the migration of hill people for better opportunity, a considerable proportion of cultivated land is lying as fallow land. It will be better if all these areas can be utilized for plantation of fodder and fuel species (Tables 9.3 and 9.4). It has been observed that in Himalayan district of Uttar Pradesh about 90 per cent of the land available for plantation, has no vegetation of any type. Afforestation from plantation on these lands takes almost 10 years to develop. Therefore, it will be better to develop these areas as pasture lands of good quality grasses before plantation for complete afforestation. Replacement of existing grasses by improved grasses (high protein and quick growth) like *kikuyu, cock's foot, clouer, pangola* etc. will have even higher yields. This technique will have three dimensional ecological benefits:

a. preparation of environment for afforestation,
b. soil conservation, and
c. fodder production.

The protection of saplings is as important as the plantation. Every year plantation in Himalaya is done in lakhs, but due to the lack of proper management the rate of survival is limited only upto 20 to 30 per cent. It has also been noted that if proper protection is provided for at least five years the survival rate can increase upto 80 to 90 per cent. However, the Forest Statistics of Uttar Pradesh (1983) produces some interesting facts about the area planted under planned scheme:

a. No plantation of fast growing species was done in the hills of Uttar Pradesh till 1981-82,
b. Only 11.40 per cent of degraded forests were rehabilitated till 1981-82,

c. No conversion of low value forest to more valuable forests has been done,

d. Only 22.85 per cent of land has been planted under soil conservation scheme.

TABLE 9.3: Afforestation Fodder Development Plan of Almora District for the year 1987-88

Unit	No. of Plantation (in lakhs)	Grass Plantation (quintal)
Forest Panchayats		
Baramandal	1	50
Ranikhet (I)	1	50
Ranikhet (II)	1	50
Bageshwar	1	50
Soil Conservation		
Almora	1	25
Ranikhet	1	25
Bageshwar	1	25
Total	7	275

Source : U.P. Forest Statistics, 1983.

TABLE 9.4: Suitable Species of the Grasses for Various Altitudinal Zones

Sl.No	Scientific name	Common name	Altitude (in metre)	Protein (%)
1.	*Pennisetum clandistynum*	Kikuya	800 – 1,800	12-14
2.	*Digitaria dermbense*	Pangola	800 – 1,800	7-10
3.	*Cynodom plectostachum*	Giantstar	800 – 1,800	10-12
4.	*Setaria sphiacilata*	Setaria	800 – 1,800	7-10
5.	*Brachiaria mutica*	Para	800 – 1,800	6-8
6.	*Panicum maximum*	Guinea	800 – 1,800	8-10
7.	*Chrysopogon montanous*	Goria	800 – 2,500	5-7
8.	*Heteropogon contortus*	Speargrass	800 – 2,000	4-6
9.	*Apluda mutica*	Bhanjura	800 – 2,000	5-7
10.	*Rennisitum oriental*	Bhimaulsa	800 – 2,000	5-7
11.	*Arundinella setosa*	Arundinella	800 – 2,000	4-6
12.	*Dactylis glomerata*	Cocksfoot	2,000 – 3,000	8-10
13.	*Pestuca arundinacea*	Tall Fescue	2,000 – 3,000	8-10
14.	*Lolium perennie*	Rye grass	2,000 – 3,000	7-10
15.	*Triplium repens*	White clover	2,000 – 3,000	15-16

Conclusion

It is now a recognized and established fact that socio-economic development of the mountainous districts requires a regional approach with an integrated planning, contrary to the previous system of development approach without recognising the specific agro-socio-ecological requirements. The fertile river valleys are capable of being converted into granaries, whereas the mountain range lands and slopes could be developed into animal husbandry based on forest and pasture lands. Therefore, a proper understanding of the problem and scientific-ecological measures will, no doubt, strengthen the forest-ecosystem management and also the village communities which have to play an even more important role in the lives of the hill people. The restoration of the ecological balance will lessen the miseries of the hill people as well as of the people in plains. Afforestation of barren slopes has to be adopted as the basic strategy for development of hill areas.

REFERENCES

Government of Uttar Pradesh (1983), *Forest Statistics*.
Kawosa (1987), in *Times of India*, 28th November, New Delhi.

LANDSLIDE MECHANISM IN CENTRAL HIMALAYA (INDIA): Causes and Controls

RAVINDRA K. PANDE and ANIL K. SINGH

Our earth system consists of a totality of matter, supplied by diastrophism with an additional trivial net increase of mass from meteoritic infall, and energy, derived from solar radiation, gravitation, rotational inertia and internal heat. Slopes convert gravitational energy into kinetic energy and simple diastrophic structures into complex denudational features such as landslide, rockfall, slumping, creeping etc.

THE MECHANISM

Work and Energy

The basic framework of landslide reflects the interrelationships between work produced by exogenetic forces and the potential energy. When a body moves due to the action of a force, the force is said to do some work. The work done by a constant force is equal to the product of the force and the displacement of its point of application in the direction of the force. If due to the action of force F, its point of application O moves to O', then work done = $F.OO' \cos\theta$ where θ is the angle between the force and the displacement (Fig. 10.1). This can be written as either F X projection of the displacement OO' along F or displacement OO' X component of F along OO$'$.

If the force is variable and the displacement is along a curved slope, let F be the component of the force along the tangent at any point and let ds be an element of displacement at that point which is evidently along the tangent to the curved surface of slope, then the work done for this elementary displacement is Fds and the total work done is $\int F\, ds$ when the integration is to be taken along the whole displacement.

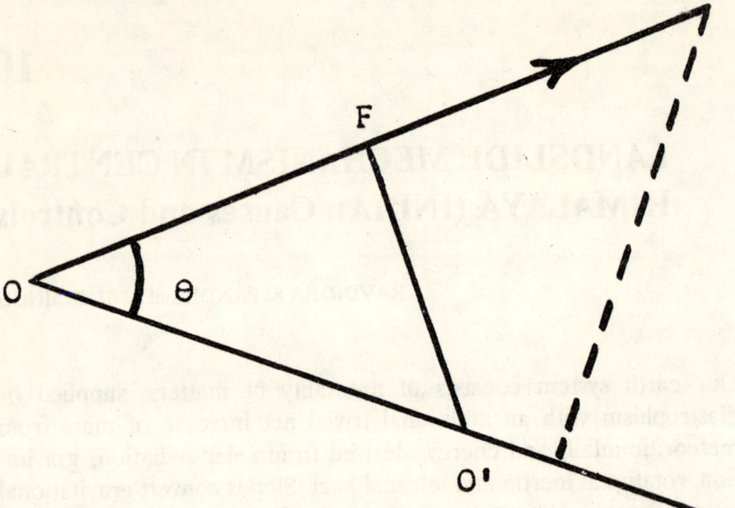

FIG. 10.1: Displacement of point of application of a force (F)

The potential energy of a body, such as rock or debris, due to a given system of forces, is the work the system can do on the body as it passes from its present equilibrium to some standard equilibrium. Thus, the potential energy of a loose rock particle or a debris mass of weight w at a height h above the ground is wh.

Considering the attraction of the earth, supposed to be a sphere of radius α, to be μ/x^2 at a distance from its centre so that $g = \mu/\alpha^2$ we have potential energy at a height

$$h = \int_{\alpha+h}^{\alpha}\left(-\frac{\mu}{x^2}\right)dx = \frac{\mu h}{\alpha(\alpha+h)} = \frac{\alpha g h}{\alpha+h}$$

The Forces

In the universe as a whole, gravity is the dominant form of energy, greatly exceeding heat, light and nuclear energy (Bloom, 1978). The gravitational force (g) generates shear stress (F_1 and F_2) on the loose debris (having the mass m) resting on the slope surface i.e. tan θ (Fig. 10.2).

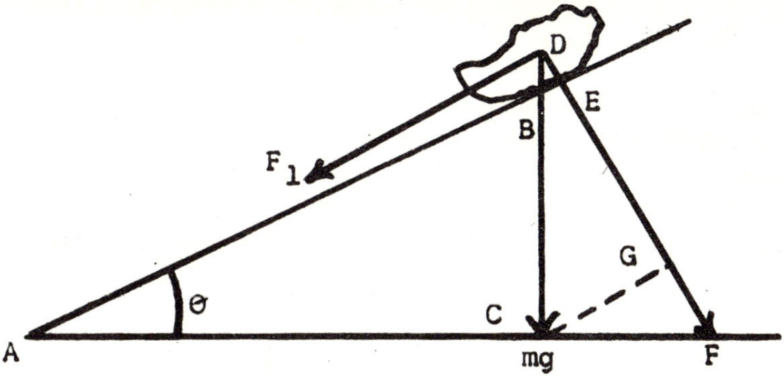

In Δ AEF

suppose $\angle AEF = 90^\circ$
$\angle EAF = \theta$
$\angle AFE = 90^\circ - \theta$

similarly in Δ DCF

$\angle DCF = 90^\circ$
$\angle CFD = 90^\circ - \theta$
$\angle CDF = \theta$

$\sin \widehat{CDF} = \sin \theta = F_1/mg$
$F_1 = mg \sin \theta$
similarly $F_2 = mg \cos \theta$

FIG. 10.2: Slope Surface

However, the frictional force (*Sr*) will also operate against stress components (F_1 and F_2). This will be called shear strength which will resist the motion (Fig. 10.3).

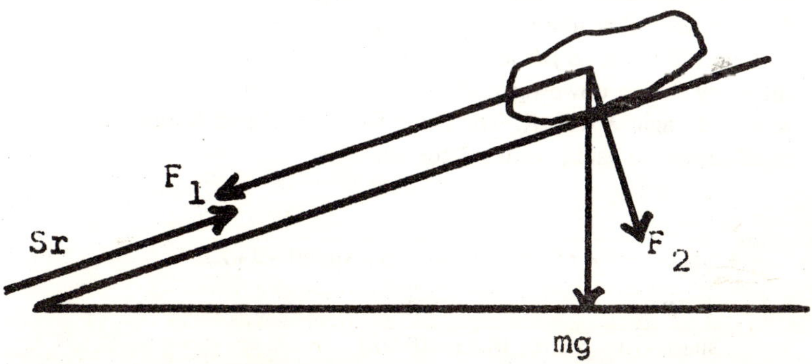

FIG. 10.3: Stress component

F_1 = stress component parallel to surface (mg sin θ)
F_2 = stress component perpendicular to surface (mg cos θ)
SR = shear strength (F_2 tan θ)

The body of the debris or rock (A) is in the contact of slope surface (B) at a point O (Fig. 10.4). The normal ON at O to the common tangent plane to the surfaces is a generally called the common normal. Let R be the mutual reaction between the two bodies and let it makes an angle θ with ON. Then the force of reaction will be resolved into two components: $R \cos \theta$ and $R \sin \theta$ along a line in the tangent plane at O. The second component is called the force of friction.

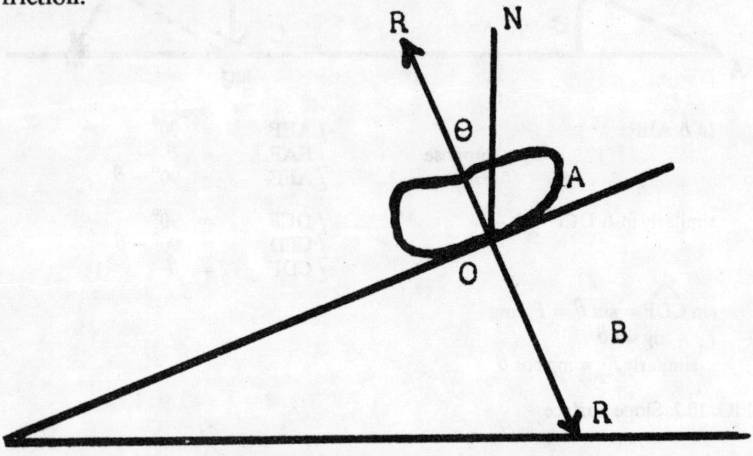

FIG. 10.4: Common Normal (N) and the Force of Reaction (R)

Let P be a force which will prevent the mass resting on the slope surface from sliding down. It makes an angle θ with slope surface. If R is the normal reaction, then in the position of limiting equilibrium, friction will equal to μR and will act up the slope surface to prevent the body sliding downwards (Fig. 10.5). The resultant reaction S will make an angle λ with R. The three forces S, P and W maintain the equilibrium; so using Lami's theorem:

$$\frac{P}{\sin(180 - \alpha + \lambda)} = \frac{S}{\sin(90 - \alpha + \theta)} = \frac{W}{\sin(90 - \theta + \lambda)}$$

Or $$\frac{P}{\sin(\alpha - \lambda)} = \frac{S}{\cos(\alpha - \theta)} = \frac{W}{\cos(\theta - \lambda)}$$

$$P = W\frac{\sin(\alpha - \lambda)}{\cos(\theta - \lambda)}$$ which defines the value of P for any given θ

The least value of *P* corresponds to the greatest value of cos $(\theta-\lambda)$ which is 1. The least value of *P* is thus *W* sin $(\alpha-\lambda)$ The direction of *P*, when *P* has its least value, is given by $\theta = \lambda$. In that case the resultant reaction *S* will be at right angles to *P*.

If $\alpha = \lambda$, it is seen that $P = O$; no external force is required to maintain equilibrium. So if a body be on the point of sliding down an inclined plane under its own weight, the inclination of the surface will be equal to the angle of friction.

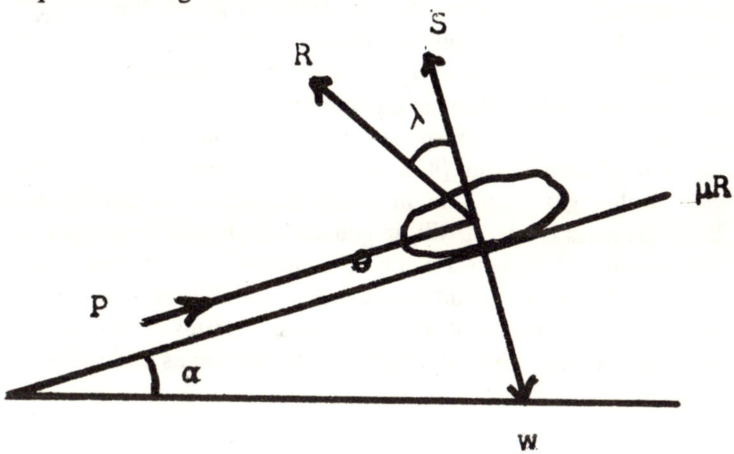

FIG. 10.5: The Equilibrium - $S = P = w$.

Whenever, the shear stress (stress components) will exceed the average shear strength along the critical angle or angle of repose the mass resting on the inclined surface will become unstable;

df = shear strength *x* shear stress^{-1}. The product value will range:
if df is greater to 1 = stability
if df is equal to 1 = critical stage (Eq.)
if df is less than 1 = instability

The Equilibrium

A mass is said to be in stable equilibrium (df=1) when, if it be slightly displaced from its position of equilibrium, the forces acting on the mass tend to make it return towards its position of equilibrium; it is in unstable equilibrium when, if it be slightly displaced, the forces tend to move it still further from its position of equilibrium; it is in neutral equilibrium, if the forces acting on it in its displaced position are in equilibrium.

A mass is, in general, in a stable position of equilibrium when the centre of gravity is in the lowest position it can take up. For if the mass is slightly displaced from its position, the centre of gravity as it can not descenal any further will tend to rise and consequently the mass will be brought back to its original position. Similarly the mass is in unstable equilibrium when its centre of gravity has the highest position. If the mass is slightly displaced from this position, the centre of gravity will descend and the mass will be consequently pulled further away from its position of equilibrium.

However, in a geomorphic system there may be more than one position of equilibrium (it is due to the open system pattern of geomorphology), the one in which the centre of gravity of the landslide mass is lowest will, in general, be the stable position, and, that in which it is highest will be the unstable one. Hence, to determine the positions of equilibrium and the stability, express the height of the centre of gravity of the sliding mass in terms of the variable θ, the angle through which the body is inclined from its position of equilibrium above a fixed point.

Thus $z = z(\theta)$

The mass will be in equilibrium if z is either a maximum or a minimum.

Therefore,

$$\frac{dz}{d\theta} = 0$$

The values of θ obtained from the above equation will give the position of equilibrium. Hence, $d^2x/d\theta^2$ must be positive for stable position and negative for unstable position.

Some Dominant Causes of Landslides in Central Himalaya

Valdiya (1971) explained some of the geologic causes responsible for landslide activities which arc given as:

(a) Highly thrusted, faulted, folded and crushed Himalayan rocks constitute the weak slopes and are highly sensitive to even a minor shock.

(b) Chemical degradation of rocks is a common phenomenon in the Himalaya. For example pyriteris shales, a common rock type in the lesser Himalaya, suffer oxidation of pyrite under climatic variation. This process formulates limonite and sulphuric acid. Whenever this sulphuric acid comes in contact of limestone converts calcite into gypsum-like lubricant material.

(c) The repeated occurrences of deep seated earthquake resulting from the internal adjustments and readjustments of rock sequences are also responsible for breaking the rocks and causing occurrence of landslides.

It has also been noted that in the Central Himalaya most of the major landslides tend to occur during the monsoon. The scientific cause behind this mystery seems to be the increase in pore pressure due to rise of water table in monsoon period (Fig. 10.6).

$$s = c + (p-u) \tan \phi$$

Where : s = shear strength
c = cohesion
ϕ = angle of internal friction
p = normal stress on the slip plane,
u = pore water pressure on the slip plane.

The construction activities (roads or building construction) have greatly enhanced the frequency of landslides. Ill-conceived, mismanaged, unrestrained and unplanned construction activities have given rise to a large scale cutting of the hill surface. The cutting of slope can disturb the angle of repose and landslide can occur. In cases where small amount of rock mass is involved, if the load on the lead is added due to ill-conceived dumping of cut soil or by removing the material from the toe for construction of roads or buildings slope instability increases. The construction activities also produce colossal volumes of debris and downslope disposal of the excavated material causing damage and destruction of the vegetation and the springs on the slopes. In USA the cutting of road has increased incidence of landslide by a factor varying from 25 to 345 in the Rocky Mountain (Swanston and Swanson, 1975).

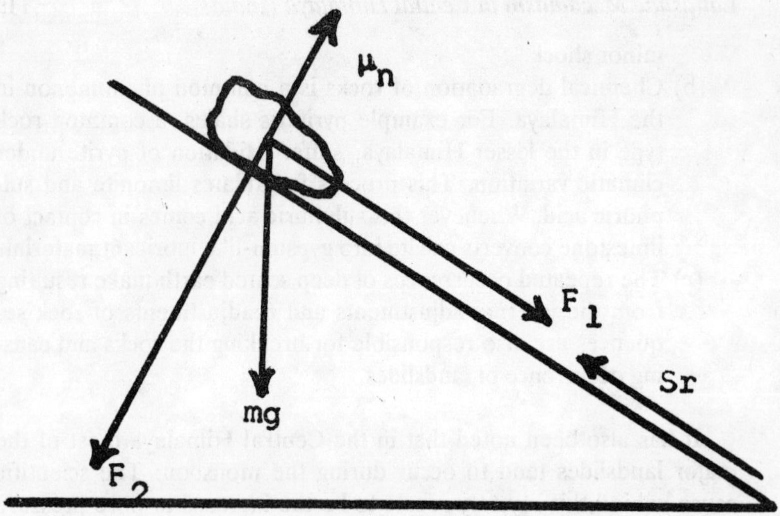

FIG. 10.6: Component of seepage pressure (μ_n) normal to the surface, counteracting F_2 ($F'_2 = F_2 - \mu$) and reduces Sr. The body will move because $F_1 > Sr$.

The large scale reservoirs often trigger local earthquake which generally occur at shallow depths under the reservoirs where the strain in the rock would increase by extra load of the water placed on the earth's crust. This may exceed the value for local faulting and displacement. Another possibility is that reservoir water, percolating through fractures below the newly filled reservoir, would increase the pore pressure in the rocks by a few tons of bars which would be sufficient to reduce the frictional resistance along the lines of weaknesses, thus allowing for sudden slip to take place.

The pressure of population has driven the people to the slopes of rugged mountain terrain, thus bringing under plough the steeper lands which were densely forested tracts until recently. Most of these villages and their farm land are located on loose unconsolidated fragmentary materials (talus accumulation or debris of old landslides). The inhabitation of these geomorphic features has caused extensive furrowing and gully erosion along with local water logging due to inappropriate agriculture practices and drainage system. Deforestation and overgrazing have also reduced the compactness of the soil and it is no longer capable of absorbing rainwater. Therefore, much of the precipitation runs off, eroding the hill slopes at furious pace.

Apart from these above mentioned factors, steepness of the hill side, development of steeper slopes as in excavations where there is abrupt change of slopes, presence of load of loose material resting at critical angles, squeezing out or lateral spreading of the soft rocks underlying the larger ones or within the succession of rocks, removal of soluble material such as earborates, gypsum, salts etc. are some other causes which seem to be responsible either for the increase in shear stress or for the decrease in shear strength.

Some Controlling Measures

It is a well known fact that landslides occur naturally and it is rather impossible to stop them. Landslide process is an important component of earth's dynamic behaviour, which always functions within the framework of two famous laws, i.e., law of least work and maximum possible entropy and law of isostasy. It is therefore, instead of wasting our energy in fighting against the natural phenomenon, we should concentrate on how we can minimize the severity, frequency and magnitude of these hazards.

Seminar on *Strategies for Environmental Regeneration in the Himalaya* (1986) at Nainital recommended the constitution of a professionally competent experts to prepare a consolidated hazard zone map of central Himalaya.

Valdiya (1984) also recommended some preventive and remedial measures, i.e., long term strategy and short term strategy. Mathur (1975), Prasad and Mathur (1976) and Kumar *et al* (1977) suggested some conservation measures, viz., afforestation, construction of check dams and toe protection walls etc.

In the national seminar on *Construction of Roads in Hill Areas* (1985) special attentions were paid towards the environmental considerations in selection of alignment of roads in hilly regions, road drainage practices and landslide problems. The authors are in agreement with the ideas and strategies formulated by the earth scientists.

However, it has been experienced that slope instability problem can not be handled by a generalized conservative approach. As the causes and effects of landslide processes vary from region to region and from time to time, the problem should be dealt with specific approaches suitable for a particular region and operationable in a particular time. For example in the landslide zones of increased ground level and pore water pressure, it is confirmed that water is reducing the shear strength. In such cases, it has been experienced that electro-osmotic stabilization is quite helpful (Kumar, 1981). The installation of electrodes in the suitable locations in respect to the potential slide and application of a voltage potential difference across them will divert the soil water flow from anode towards cathode and will drain the soil. Similarly, in cases where small amount of rock mass is involved, the load on the head can be released by removing the rocks or by adding the material to the toe to increase the stability.

It has been experienced that though landslides are natural hazards, however, anthropogenic encroachment, viz., random falling of trees on hill slopes, increasing agriculture pressure, overgrazing, ill-conceived construction activities accelerate the mechanism. It is therefore, certain measures should be implemented, such as, the subject of location of a construction site (road, building, dam, reservoirs, tunnels, barrages etc) is multidisciplinary and must be discussed in-depth before it is planned and executed.

On the steep slope surfaces, specially in geodynamically fragile and ecologically vulnerable areas, forest cutting and overgrazing should be absolutely banned.

The aforestation policy also needs a drastic rethinking. Because, plantation needs a scientific approach. The forest department and planners must be selective about the species for specific areas according to the need and nature of plantation surface.

Apart from these above mentioned conservative measures, specific techniques can be used for the specific problem, such as, anchoring of rock, construction of protective nets, drainage control, plugging or grouting of cracks, fissures, etc. with hydrated lime-slurry, cement, sodium silicate solution or bitumen (Valdiya, 1984), releasing the build up hydrostatic pressure from the landslide zone, remodeling the slopes, construction of retaining walls and buttress below the slip surface, installation of prestressed anchor pads at the toe of slope failure, etc.

It has also been felt that the public policy on landslide hazard management should encompass decision making on the issues: when, where and at what cost the various mix of preventive and ameliorative of palliative measures be applied to mitigate the hazards of mass-movement. This aspect involves regulatory and legislative steps. Since damage consequent to occurrence of landslides affect one and all in one way or another, there should be a mandatory regulation governing the procedure of hazard management and for providing guidelines in the matter requiring fixing responsibilities for causing or escalating landslides, promising deterrents to activities detrimental to environmental balance and awarding costs of damage by the law courts.

REFERENCES

Bloom, A.L. (1978), *Geomorphology: A Systematic Analysis of late Cenozoic Landforms,* Prentice-Hall, Inc., New Jersey.

Kumar, R., R.C. Kanwar and S.P. Kapila (1977), On the landslide problem and proposed realignment of the national highway between Udhampur and Chenani, Jammu and Kashmir, India, *Bull. Indian Geol. Assoc,* 10(2).

Kumar, A. (1981), Geological hazards and related problems of Himalayan villages: Their direction and control, In J.S. Singh *et al* (eds), *Science and Rural Development in Mountains,* Gyanodaya Prakashan, Nainital.

Mathur, H.N. (1975), Landslides in Himalayan ranges — their control, *Indian Farming.*

Prasad, C. and H.N. Mathur (1976), Study of landslips along the main boundary fault near Rajpur, Dehradun. *Himal. Geol.,* VI.

Swanston, D.N. and F.O. Swanson (1978), Timber harvesting, mass-erosion and steep-land forest geomorphology in the Pacific coast, In D.R. Coates (ed), *Geomorphology and Engineering,* Halstead, Stroudsburg.

Valdiya, K.S. (1984), Landslides in U.P. Hills: Geological Aspects, *DGM Report* — GM 366/84, Lucknow.

11

IMPACT OF DEFORESTATION ON RURAL LANDUSE : A Case Study of Budhawale

D.G. PHONDEKAR and V.S. PHADKE

Man, in spite of his technological feats, has to depend on plants for his daily food packet as they are the only autotrophs on the earth. It is true that man's dependence on natural vegetation for food has considerably reduced after domestication of plants, yet his other needs continue to come from the natural vegetation or forests. Moreover, forests influence human life in a number of other ways.

Forests play an important role in the national economy. Nair (1967) speaks of imponderable and ponderable values; the former refer to concrete or direct gains from the forests while the latter are invisible benefits. If exploitation of forest products is important for short-term economic gains, the conservative role of forests is significant form the viewpoint of the very stability of human species. Therefore, man has to work for survival of the forests in order to make his own standing possible.

Although people from ancient India did not fully realize the protective function of forests, there was a high regard for trees partly, due to close dependence of our ancestors on forest products and in part, due to teachings of Hinduism reflected in rituals such as *banyan festival* (Dwivedi et al, 1984). Hence destruction of the forest habitat was never such a serious problem before the colonial settlement which, in its wake, has brought in growth of cities, increase in demand for forest products and the resulting depletion. This forced the British rulers to start commercial forestry that has changed *rights* of the people, especially tribals, first to *privileges* and then to *concessions* (Chaudhury and Chaudhury, 1982). Even in post-independence period, the colonial tendencies continued and today, there appears to be a very serious threat to forests and along with them, to the rest of the environment. This has caused severe degradation on mountain slopes which are highly vulnerable (Shah, 1982) creating a number of unprecedented problems, such as flash floods and landslides (Thomas, 1978). All of us are aware of the problem associated with deforestation of the Himalayan slopes and the popular dissent in the form of *Chipko movement* (Mishra, 1979). Today, however, the problem is not restricted to the Himalayas. Almost every village in the country is suffering from the absence of forests to satisfy daily needs like fuel, fodder and construction timber. Other types of costs such as, gullying on slopes and resulting deposition of eroded material in agricultural fields are some of the common problems. Rural dwellers have been forced to change the pattern of landuse due to lowering of watertable. The studies related to impact of deforestation exist at macro and meso levels (White, 1978; Parson, 1981; Meher-Homji, 1983 and 1988; Thomas, 1978; Allen and Barnes, 1985; Chattopadhyay, 1985; Dogra, 1987). However, there is a general lack of micro level studies. Hence, some of the physical problems, such as changes in the landuse pattern, menace of wild animals, due to the disturbance of their habitat, which can effectively be studied only at the micro level, have not been sufficiently highlighted.

During the recent visit to parts of Sindhudurg district some of these problems were brought to the notice of the present authors by local residents and hence it was thought desirable to investigate into them with the help of a micro level study. The primary focus of the present study is, therefore, on the impact of deforestation on rural landuse. Village of Budhawale in Malwan taluka of Sindhudurg district, in Maharashtra has been selected for the purpose.

Study Area

The village of Budhawale occupies an area of 8.90 km^2 and had population of 745 (1981). One of the major right bank tributaries of Gad river passes through the area (Fig. 11.1). While Malwan, the taluka place is located at a distance of 35 km to its southwest, Kankavli in the northeast, lying at a distance of 22 km, is the nearest urban centre.

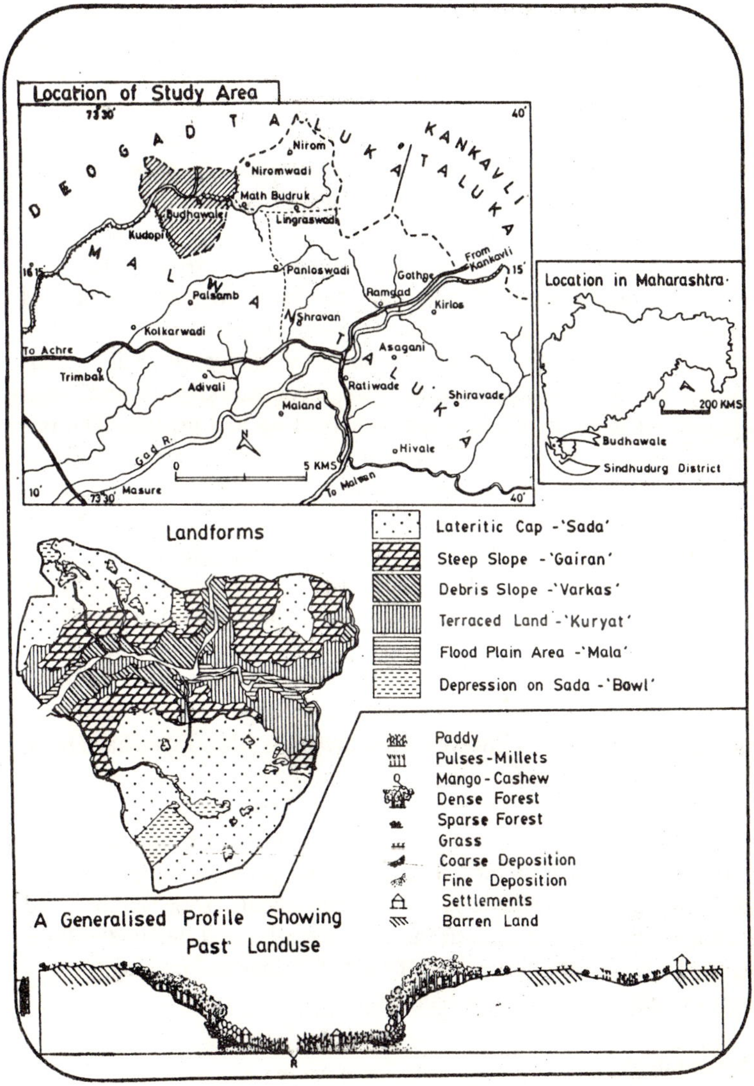

FIG. 11.1: Location of Village of Budhawale

Physiographically, the village possesses the characteristic lateritic landscape formed of a low level plateau or *Sada* which is dissected by two streams flowing in it (Fig. 11.1 inset). At the valley bottom, are the flood plain areas or *mala* lands flanked by terraced lands called *kuryat*. The edge of *sada* invariably depicts a scarp followed by a steep slope and later, by debris slope at the foot hills. The area comes under wet monsoon type of climate characterised by seasonal rains. Soils of the area are derived from laterite. They are typically reddish in colour and acidic in nature with a pH value ranging between 4.5 to 6.5. Except for the very low lying areas, they are not very fertile.

Warm, moist climate of the area is conducive to growth of natural vegetation which mostly occupies relatively steep slopes but is not restricted to them. As one moves against the slopes, the height, density and other attributes such as girth, progressively fall and, *sada* tops are associated with very sparse scrubland and grasses. The important phanerophytes in this area are Teak (*tectona grandis*), *Kinjal* (*terminalia alata*), ain (*terminalia crenulata*), jack fruit (*artocarpus integra*), *nana* (*largerstraemia parvifolia*), *sisam* (*dalbergia latifolla*) on lower slopes and *Khair* (*acacia catechu*) on higher slopes.

The landuse of the village has been greatly influenced by its physiography and underlying geology. While cultivation is mainly restricted to valley bottom, lateritic tops are mostly devoid of cultivation except for the depressions where some finer material accumulates; these are locally known as *bowls*. Slopes are generally associated with forests. However, in some parts, the slopes have been cleared to make room for the plantation of mango and cashew.

People of the area are mostly farmers and have generally small land holdings (Table 11.1). Farming is of subsistence type and there is no source of cash income; this limitation prompts people to earn hard cash by any available means. When agents from Bombay and Kolhapur reached the area in the late seventies and persuaded local people to lease out their private forests for charcoal making, people were too eager to accept the bait.

Methodology

Since the thrust of the present study is on understanding the impact of deforestation in landuse pattern of the village, it was imperative to compare two scenarios, namely, one before the process of deforestation was set in and the other, after completion of one cycle; the years

1975 and 1985 were, therefore, selected for the purpose. Incidentally, climatically, these two were normal years.

TABLE 11.1: Size of Land-holdings of the Households

Land holding (Acres)	Percentage of Households
Less than 2.50	15.59
2.50 — 4.99	14.67
5.00 — 7.49	22.01
7.50 — 9.99	10.09
10.00 —12.49	9.17
12.50 —14.99	0.90
15.00 —19.99	4.68
20.00 —29.99	3.66
Above 30.00	4.58
No response	14.75
All sizes	100.00

Source : Fieldwork

The data for landuse in 1975 were collected from 7/12 and 8A records of the local revenue officer (Talathi). Checks were also applied by collecting retrospective information from the farmers. The data for 1985 were not readily available and were, therefore, collected by field to field survey and enquiries with landholders. In addition to the quantitative data on landuse, information regarding the problems faced by the local inhabitants were also collected.

Although it was intended to contact all the revenue account holders, but about 45 per cent of them could not be contacted even after repeated visits. The partial enumeration was, however, compensated by information collected through neighbouring households.

Analytical Findings

The process of deforestation has induced a number of changes in the general landuse (Fig. 11.2, a and b). In the first place, the density of forests has declined both in densely as well as sparsely forested areas. Virgin forests have almost disappeared and correspondingly their place has been taken by secondary growth (Table 11.2). The field to field survey revealed that the composition of the forest community has also undergone a change. After depletion of virgin cover the

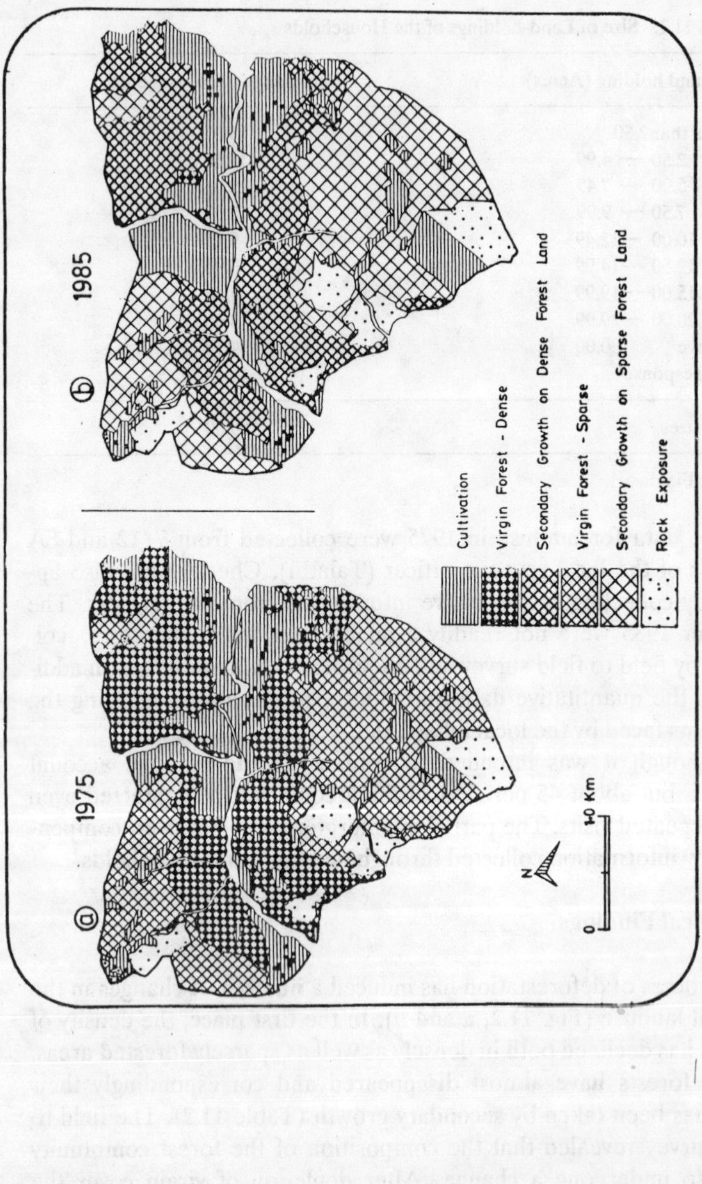

FIG. 11.2: Village of Budhawale: General Landuse

secondary growth seems to be dominated by thorny species like khair (*acacia catechu*), *karwand* (*carissa carandas*), *gehela* (*randia dumentorum*), etc. Due to exploitative depletion the older trees have not developed sprouts for the second time. Even the other trees have been badly damaged and thus secondary growth is also restricted. The lowering of density has led to an increase in damage caused by activities related to collection of forest products, deteriorating the poor vegetation cover further. The forest depletion has also resulted into marginal extension of the cultivated land in continuation with the already existing arable land (Table 11.2).

TABLE 11.2: General Landuse — 1975 and 1985 : A Comparison

Landuse	Percentage to Geographical Area		Change (%)
	1975	1985	
Cultivation	18.34	20.86	+ 13.74
Dense Virgin Forest	37.82	00.18	−99.53
Secondary Growth on Dense Forest Land	00.00	36.50	—
Sparse Virgin Forest	35.50	00.00	− 100.00
Secondary Growth on Sparse Forest Land	00.00	34.16	—
Rocky Exposure	7.40	7.22	− 2.43
Settlements	1.00	1.08	+ 8.00

The agricultural landuse depicts a considerable modification in its pattern (Fig. 11.3, a and b). The landuse map of 1975 shows six major categories of agricultural landuse of which three are suggestive of double or triple cropping where, even the second or third crop is water intensive, viz. paddy or vegetables. *Mala* lands have paddy as the main crop and paddy or vegetables as the second important crop; pulses are a transitional crop between the two. *Varkas* land is mainly under long duration pulses or millets or plantation.

Pattern of agricultural landuse for 1985 shows certain striking changes in relation to earlier pattern. Both the categories of land characterised by triple cropping, viz. paddy-pulses-paddy and paddy-pulses-vegetables have disappeared from the scene. Even the area under double cropping, growing paddy and vegetables, is highly restricted to pockets of *mala* land close to river banks taking

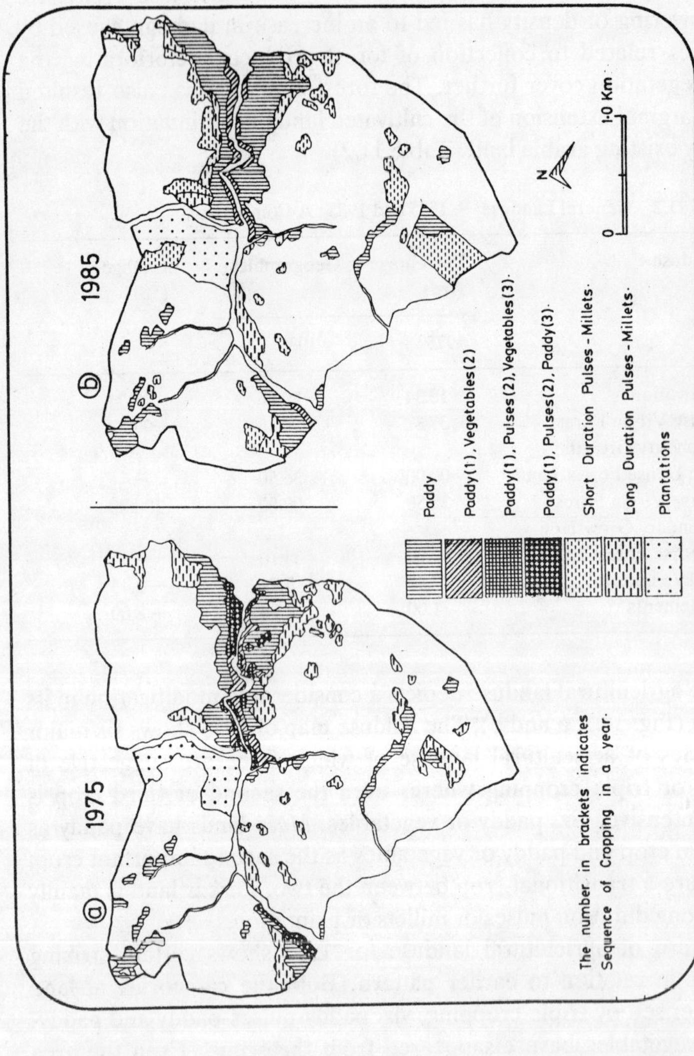

FIG. 11.3: Village of Budhawale: Agricultural Landuse

advantage of local pools of water. An enquiry with the local residents revealed that position with regard to river water has considerably changed. While in the past it is used to be abundant even during late summer; in recent years it is found to be inadequate for water intensive crop like paddy even by the beginning of summer. It was also pointed out that in the western part of the village the river was never fordable till March. However, it can be easily crossed these days in November itself. This suggests a fall in storage of underground water.

The other important feature is the expansion of plantation cover. This is due to non-availability of labour which has got diverted to tree felling activities starting immediately after the monsoon; plantations are less labour intensive.

An introduction of short duration pulses or millets, in place of long duration pulses, is yet another important change, especially on the upslope side of the *kuryat* area. It is reported to be due to increasing menace of wild animals which forced the farmers to cultivate short duration crops in *varkas* areas; such a practice enables them to harvest the crop along with the short duration paddy in these areas. In some areas, especially, in north-eastern section, some patches of paddy lands in upper part of *kuryat* have given way to pulses or crops like groundnut; this is mainly a result of deposition of material washed down the slopes as a result of deforestation. Cultivation of paddy has been made difficult on account of lower moisture retentivity of this material.

The foregoing analysis suggests that the process of deforestation has brought about some important changes in landuse in the village of Budhawale. Although area under cultivation has increased, cropping intensity has declined. Moreover, there is an increasing tendency towards raising of less water and labour intensive crops like short duration pulses, millets, and plantation. Although areas released from double and triple cropping have gone to single crop of paddy, the paddy lands have made room for pulses and millets. These changes are more vividly brought out by Table 11.3. It is interesting to note that there is not only a complete disappearance of triple cropping system but area under double cropping has also been reduced by over 25 per cent. Moreover, this has not substantially improved the position of paddy as a single crop as is clear from negligible increase in area under the crop; it has gone to pulses through a route already explained.

TABLE 11.3: Agricultural Landuse-1975 and 1985 : A Comparison

Crop/s	Percentage of Area to Total Cultivated Area		Change (%)
	1975	1985	
Paddy	40.14	41.40	+ 3.13
Paddy (1) Vegetables (2)	5.75	4.26	− 25.92
Paddy (1) Pulses (2), Vegetables (3)	6.04	0.00	− 100.00
Paddy (1) Pulses (2), Paddy (3)	5.61	0.00	− 100.00
Long duration Pulses-millets	28.20	10.66	− 62.20
Short duration Pulses-millets	0.00	28.55	—
Plantations	14.24	15.33	+ 7.67

Conclusion

The study aimed at understanding of the changes in the landuse brought about by deforestation in a rural setting. The analysis brings to light some significant features.

Deforestation has influenced landuse in a number of ways. Apart from the changes in density and composition of the forest cover, there is an extension of agricultural activities on lower slopes and even in *bowls*. But the lowering of water-table has resulted into a change over from more water intensive to less water intensive crops; paddy is nowhere grown as a second crop. Transitional crop of pulses have disappeared. Shortage of agricultural labour and menace of pests are also the factors contributing to such a change. But this is only a symptom of a deep seated problem that the villagers are likely to face in near future. The real issue to be clinched lies elsewhere and required some serious thinking.

The rising trend towards urbanization through concentration of population in large cities like Bombay has created unprecedented demands for certain utilities like fuel, and is affecting even distant areas because of the exhaustion of such materials from nearby sources. Deforestation in Konkan is an example of the process of meeting charcoal demands of the metropolitan city of Bombay which has spread its tentacles over 400 km. The city of Bombay has been benefited in the process through the availability of the necessary materials. The middlemen have got substantial benefit by way of profits through trade. Local residents seem to have been benefited through cash income, and employment in off-season. However, these benefits will last only over a short period of 3 to 4 years. From long term perspective, the process involves tremendous socio-economic costs. Apart from the exhaustion of forest potentialities, the rural residents are constrained even in their own traditional activities. They are forced to change the seasonal rhythm in agriculture and combination of crops; these problems would be on increase. Continuation of deforestation in this way may uproot the rural residents by snatching away the very source of their existence. And the tide of migration, which is a characteristic of this area, would swell in near future.

The city, unable to find its requirements from the area, would search for them elsewhere and manage to get them at any cost. The middlemen would move to other virgin areas and thus their activity would continue to flourish. In the rural areas alone, the benefits would shrink and costs would get exacerbated. This shows that the relationship between the city and its hinterland is not one of equality. While rural economies have limits of space, economies of urban areas like Bombay or those of middlemen have no restrictions on their operational space. Rural residents are thus losers in the process.

REFERENCES

Allen, J.C. and Barnes, D.F. (1985), The causes of deforestation in developing countries, *Annals of Association of American Geographers, 75* (2), 163-84.

Chattopadhyay, S. (1985), Deforestation in Parts of Western Ghat Region (Kerala), India, *Journal of Environmental Management, 20,* 219-30.

Chaudhari, B. and Chaudhari, S. (1982), On some tribal problems, in B. Chaudhari (ed.), *Tribal Development in India : Problems and Prospects*, Inter-India Publications, Delhi, 83-95.

Dogra, B. (1987), To whom belong the forests?, *Development, 1,* 40-43

Dwivedi, O.P., Tiwari, B.N. and Tripathi, R.N. (1984), Hindu concept of ecology and environmental crisis, *Indian Journal of Public Administration, 30* (1), 33-67.

Meher-Homji, V.M. (1983), Development-vegetation-environment link, *Geographical Review of India, 45* (3), 66-75.

Meher-Homji, V.M. (1988), Deforestation, ecological destabilization and droughts, *Social Action, 38* (2), 115-28.

Mishra, A. (1979), The forest cover, *Seminar, 257,* 26-28.

Nair, K.K. (1967), Rapporteur's report on seminar on forestry development in India, *Indian Journal of Agricultural Economics, 22(4),* 138-40.

Parson, J.J. (1981), Human influence on the pine and laurel forests of Canary Islands, *Geographical Review, 71* (3), 253-71.

Shah, S.L. (1982), Ecological degradation and future of agriculture in the Himalaya, *Indian Journal of Agricultural Economics, 37* (1), 1-22.

Thomas, T.V. (1978), Forest preservation in the Western Highlands of Guatemala, *Geographical Review, 68* (3), 417-34.

White, S. (1978), Logging in Peru, *Geographical Review, 68* (4), 394-416.

12

PERCEPTION OF THE RURAL ENVIRONMENT: A Case Study

J. SINGH and V.K. KUMRA

Although most of the developing countries are experiencing rapid urbanisation, it is perceived that overwhelming population will remain rural with dominant rural environment. In India 76.23 per cent population lives in about 0.57 million villages (1981) and therefore, rural environment has a very important place and role to play in the overall environment of the country. By and large, major share of the government's exchequer and resources are being cornered out for the development of urban areas, the rural areas are kept in backyard. But unless and until the fate of rural masses are improved, the development cycle of the country cannot move in a forward gear. Many of our developmental policies are framed by the urban elites so these are inherently urban biased. Consequently the gulf in the quality of life of urban and rural areas is widening day-by-day. Despite hue and cry for the upliftment of countryside in successive plan periods, nothing substantive progress could have been achieved so far. Apart from other reasons, the appalling environmental quality of villages are equally responsible for the large scale migration from the rural areas to cities and towns of the country. Therefore, it is very appropriate to investigate the varied environmental problems with which this large stock of human beings are confronting. It has been also attempted to assess among the ruralites about the awareness of those environmental problems. The perception of the rural people, if taken into consideration, will certainly help in improving the environmental quality of villages.

A Brief Account of Environmental Perception Study in India

Several studies (Karan, 1977, 1980; Kumra and Singh, 1979; Kumra,

1982; Kayastha and Ram Bilas, 1983; Nag, 1980; Desai, 1980, 1981, 1982, 1985; Singh *et al.*, 1988 etc.) have been made to know the level of awareness towards various types of pollution in urban centres but very little and insufficient work (Kayastha and Yadav, 1977, 1980; Tewari, 1979; Ambhire, 1979; Swaminathan, 1980; Swaminathan and Jothimani, 1987, 1988 etc.) has been undertaken so far to investigate the perception of the ruralites towards multifarious environmental problems. Nag (1985) has rightly concluded that urban centres, similar to the western experiences, have been the focus of attraction for perception research in India.

Keeping these facts in mind, the present study is attempted to know (a) the environmental problems faced by the rural masses, (b) the way they perceived the problems, and (c) their awareness towards these specific problems and solutions. For this purpose, a case study of three villages, lying in one of the most industrially backward district of Jaunpur (Uttar Pradesh) has been taken.

Study Area

The present investigation is confined to three villages namely Umarwar, Patail and Shivrampur Khurd which are located 5 to 6 km west of the state highway (No. 85) linking Varanasi to Azamgarh (Fig. 12.1). The villages under study lie at a distance of 30 km east of Jaunpur town, the district headquarters, and about at the same distance from Varanasi, the divisional headquarters. The sub-divisional and development block headquarters (Kerakat) lie at a distance of 5 to 6 km from the studied villages. The total population of these villages is 3,378 (1981 Census), with 534 households. The largest chunk of the working population (82%) derive their livelihood from agricultural pursuits. Altogether 22.23 per cent people are found as literates against the district average of 26.3 per cent.

Data Base and Methodology

The present investigation is purely based on the primary data which has been collected with the help of questionnaire. One hundred and fifty household heads (28% of the total households) in the sample villages were interviewed. The questionnaires on community attitude was designed in accordance with two criterion, first, it had to be short enough to be administered through personal approach; second, it had to elicit responses not only on attitudes towards pollution but also a behaviour in participating in pollution control programmes (Caris, 1978). The design of the questionnaire was prepared in such a way that each interview took 10-12 minutes to answer it.

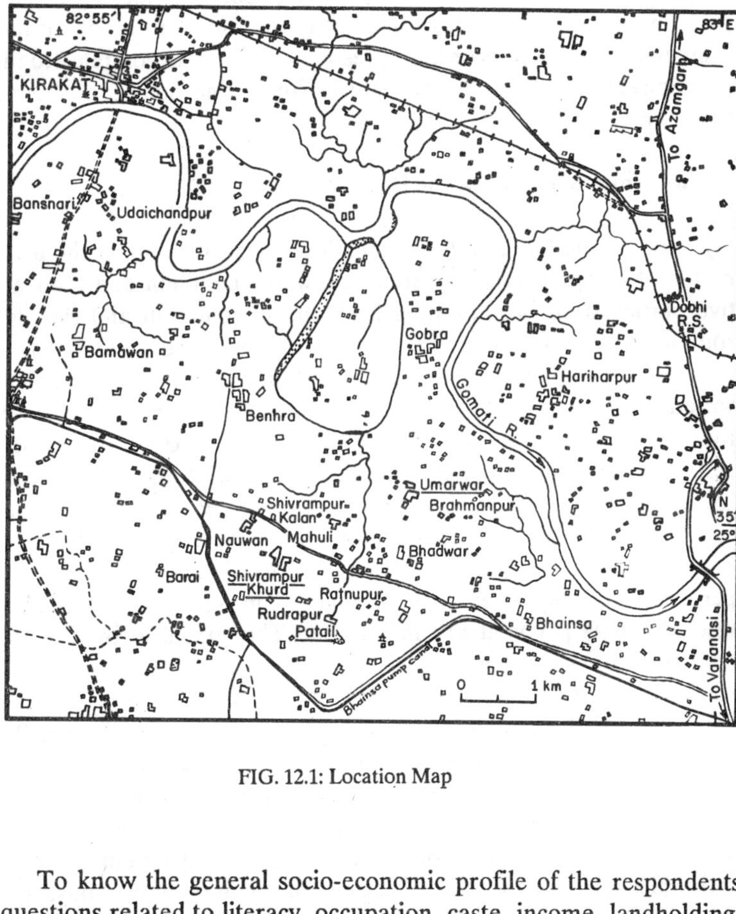

FIG. 12.1: Location Map

To know the general socio-economic profile of the respondents questions related to literacy, occupation, caste, income, landholding, type of houses, type of kitchen, lavatory facilities, etc. were asked. Investigations were also made on the nature of fuel used for domestic consumption and the problems arising thereof. Further, enquiries were made about the state of affairs related with essential amenities which includes type of water supply, sanitation, solid and liquid waste disposal system etc. To find out the level of awareness among inhabitants questions like 'Do you feel that rural environment is affected by any kind of pollution ?, if yes, then which functional activities have prompted to it ?', 'Do you give more importance to the problem of pollution over unemployment, food and housing ?'. Any problem based study concludes with the formulation of certain remedial measures. Accordingly, responses were also compiled to the question 'How will you participate in controlling pollution of rural environment?'.

Socio-economic Profile of the Respondents

Socio-economic levels of the respondents very much affect the magnitude of the perception/awareness towards pollution and their active participation in alleviating it. Housing condition and the surrounding environment have also an impact on the respondents' perception of the environment. In the study area respondents belong to varied castes in which Harijans are found in good number (33.01%), followed by Kshatriya (28.30%) and Brahmins (16.03%). Among the respondents about 51 per cent are literates. Out of the total literates 25.92 per cent have education up to primary level whereas 16.66 per cent respondents possess graduation and above degrees. The main occupation of the respondents is agriculture (63.20%). Besides, 16.03 per cent are daily wage earners engaged in agricultural activities, whereas 16.98 per cent are in services. The annual income of the respondents varies from less than Rs. 6,000 to more than Rs. 12,000. About 42.60 per cent respondents have an annual family income below Rs. 6,000 while about 18 per cent come in the category of Rs. 6,000-12,000. Nearly 39.4 per cent respondents have an annual income of more than Rs. 12,000. Income structure of these villagers reveals that a considerable number of families belong to poor class.

Further, dwellings are considered as the immediate environment with which a person interacts. The general condition of the house and the facilities available within it such as separate kitchen with smoke outlet, adequate ventilation and lighting, shelve etc. affects the home environment considerably. Most of the dwellings of the interviewers were found *kachcha* (56.2%) and without adequate provision of ventilation. About 14.2 per cent of the houses are mixed, while 29.52 per cent own *pucca* houses. Invariably one and two-room dwellings reflects the poor socio-economic status of the dwellers. Majority of the Harijans have such dwellings. Nearly 37 per cent of the houses are with one to two rooms. It shows that most of the domestic works are carried on in the same living space which is unhygenic in all respect.

Among all the households surveyed, 33.93 per cent have attached kitchen and without any provision of smoke outlet while nearly 62 per cent of the houses possess separate kitchen, of which 48.58 per cent are devoid of any provision of smoke outlet. Only 13.33 per cent of the houses have some sort of arrangements for exit of smoke generated in the kitchens (Table 12.1). The present survey evinced that higher castes such as Kshatriya, Brahmins enjoy the facility of separate kitchen whereas Harijans having mostly one to two rooms dwellings use the same room for cooking, dining, storage etc.

TABLE 12.1: Type of Kitchen According to Caste (in percentage)

Caste	Attached kitchen		Separate kitchen		Outside the house	Percentage of total respondents
	a	b	a	b		
	As per cent of respective castes					
Scheduled caste	Nil	72.09	2.32	2.93	4.64	35.83
Kshatriya	Nil	7.69	26.92	61.53	3.86	21.66
Brahmin	Nil	7.40	25.92	62.96	3.72	22.50
Kayastha	Nil	Nil	100.00	—	Nil	0.83
Ahir	Nil	33.33	Nil	66.67	—	2.50
Koeri	Nil	14.28	—	85.72	Nil	11.68
Others	16.66	33.36	Nil	49.98	—	5.00
Total	0.83	33.93	13.33	48.58	3.33	100.00

Note: a = Kitchen with Ventilation
b = Kitchen without Ventilation
Source: Compiled from field interviews, 1989

Residents Perception and Sensitivity to Various Environmental Problems

Perception and sensitivity both varies from person to person and from place to place. Accordingly, people's opinion has been obtained from varying socio-economic strata. Urbanites are more conscious about their neighbouring environment but the people of the countryside are not so sensitive towards the quality of their environment where they live and work, perhaps owing to their poor socio-economic base. Therefore, an attempt has been made to assess the level of perception and sensitivity to varying problems with which they interact some times knowingly and some times unknowingly.

Fuel Consumption for Domestic Purposes

The nature of fuel used for domestic consumption depends upon the availability, purchasing power of the consumer, utility value of the fuel, and the knowledge of the cost and associated benefits (Swaminathan and Jothimani, 1988). The villages under study lies in Jaunpur district where only 0.5 per cent area (1981-82) is covered with forest. The situation is more critical when one accounts for per capita forest as 0.0001 ha. So it is quite natural that people will use cowdung and crop wastes as fuel to meet their cooking energy needs. Further, majority of the population is below the poverty line, henceforth, their purchasing power is also low. They cannot afford fire wood and kerosene oil from market. From the study it has been observed that out of the total respondents, 64.21 per cent use cowdung as the main source of cooking energy whereas 30.52 per cent use cowdung and crop wastes for the same (Table 12.2). Only 4.22 per cent of the total respondents use firewood, kerosene oil etc. as fuels for domestic purposes. It is interesting to note that in spite of all efforts and encouragements (subsidies) from the government, only seven households of the study area have bio-gas plant which are being used for cooking and lighting. Out of these, six units of bio-gas are owned by Rajputs. During the course of the survey it was also found that out of total households surveyed 20 families have no cattles at all. Many others, particularly Harijans, have one or two cattle and very nominal land which in no way could meet the daily fuel requirements of their kitchen. It is common to see that the females and small children of these families devote on an average 2 to 3 hours in collecting either the crop wastes or cowdung lying here and there. It further reveals that there is scarcity of energy resources in these villages.

Cowdung and crop wastes emit more smoke and particulate matters on the one hand and generate less energy on the other. The poor design of *chullahas* and lack of ventilation for diffusion of air pollutants add the problem in a bigger way. It causes health problems of varying magnitude particularly among the women. In order to find out the effect of smoke on health they were asked 'how does kitchen's smoke affect you?' The enquiry revealed that working females of thc 16.10 per cent households complained the problem of breathing, and 11.01 per cent faces the problem of eye irritation. About 27.11 per cent users feel the problem of headache, whereas 32.20 per cent have responded to the combined effect on their health (Table 12.3). Rest of the families have stated no effect at all. It may be either due to ignorance or use of bio-gas and *chullah* outside their houses. It gets more space for diffusion of air pollutants.

TABLE 12.2: Fuel Consumption Pattern

Sl. No.	Source	Percentage of Respondents
1.	Cowdung and crop wastes	30.52
2.	Cowdung	64.21
3.	Bio-gas	1.05
4.	Others (Kerosene oil, firewood etc.)	4.22

Source: Compiled from field interviews, 1989.

TABLE 12.3: Responses to the Question 'How does Kitchen for Smoke Affect's You'?

Sl. No.	Effect of kitchen smoke	Percentage of respondents
1.	Problem in breathing	16.10
2.	Eye irritation	11.01
3.	Headache	27.11
4.	Combined affect	32.20
5.	No effect	13.58
	Total	100.00

Source: Compiled from field interviews, 1989.

A study of 36 houses of Anand district of Gujarat highlighted that, on an average, a women inhales during cooking 7,000 $\mu g/m^3$ of total suspended particles (TSP) against 150 $\mu g/m^3$, the prescribed standard of WHO. The survey further reveals that these working females also inhale on an average $3850 X 10^{-9}$ gm/m^3 (Nainogram) of benzopyrene (BAP) which is as injurious as a person smoke 20 cigarettes (1 cigarette emits BAP equivalent to 20 nainogram). It is worldwide known that cigarette smoking causes or aggravates lung cancer (Misra, 1988). This is the situation in which our females work for hours together in the kitchen. In the rainy and winter seasons the problem becomes more severe owing to the frequent use of wet fuels. Besides, inhalation of carbon monoxide (CO) causes reduction in haemoglobin contents in blood of such females which causes anaemia and reduces the blood circulation. It ultimately lead to cardiac problems. The women often complain about the problem of irritation in ear, nose and throat which is caused due to formaldehyde, the second most injurious pollutant found in smoke.

Villagers awareness is not only confined to pollution from smoke generated by *chullah,* but they have also noticed the menace of the problem by observing the damages to trees by the smoke of brick kiln. A brick kiln operating in the nearby village (Brahmanpur) had completely destroyed a cluster of mango, *mahua* and *kathal* trees. When asked from the villagers of the study area about the cause of dying of those trees, they replied that a brick kiln which operated there for about three seasons (1982-85) was the main source. On an average, 18 tons of coal was consumed for making. 0.10 million bricks in this brick kiln. The height of the chimney was nearly 14 m. The total production capacity of this plant was 1.5 to 2.0 million bricks per year. Not only this, even residents of village Umarwar have also blamed to neighbouring brick kiln operating in the midst of the same village since 1986 for drying up of some of the trees. So they are well aware about the pollution problems arising out from the smoke, generated in the kitchen as well as by the brick kilns.

Drinking Water Supply and Sanitation

Despite declaration of 1981-90 as International Drinking Water and Sanitation decade by the United Nations, a little progress has been achieved in India towards providing safe rural water supply and sanitation facilities. In 1981, 31 per cent of India's population (U.P. 7.2%) had only access to safe water supply (UNICEF, 1984). This statement is also true in the case of surveyed villages. The main source of water supply in the studied villages is dug wells, followed by handpumps and piped water supply (Table 12.4). It is notable that most of the privately owned handpumps and water connections are availed by economically sound people of different castes. It is supplemented by the fact that out of 61 handpumps, 43 are possessed by

Rajputs, Brahmins and Kayastha. Among all the households, 75.33 per cent use well water for domestic purposes whereas 20 per cent fulfill their requirements from handpumps. Only a handful persons of Shivrampur Khurd village avail the facilities of piped water supply from the nearby overhead tank (capacity 275 kilolitres) situated at Benhra village. This particular village comes under the Benhra village cluster drinking water supply scheme. This scheme of rural water supply is operative since 1983-84 and provides piped water supply to adjoining eight villages namely Benhra, Udaichandpur, Shivrampur Khurd, Shivrampur Kalan, Nuon, Barai, Bansbari and Bammavan (Fig. 12.1). It demonstrates that the majority of the inhabitants do not get safe drinking water because of contamination through bathing and washing clothes very near to the terrace of the wells and completely non-existence of covered roof over the wells. During the course of survey it was observed that people are aware and cautious towards getting safe water. It is evinced from the fact that they periodically used bleaching powder and potassium permanganate for treatment of well water. During the summer they also try to clean the surface of the well. Despite their all voluntary efforts, many other constraints come in the way. National Environmental Engineering Research Institute, (NEERI), Nagpur and State Health Institute, Lucknow, after intensive study of wells in different parts of the country came to a conclusion that there is not a single well in the country which could provide bacteriologically safe water as per the standards recommended by the Environmental Hygiene Committee. As a result, water related diseases like diarrhoea, dysentery, typhoid, cholera and jaundice continue to make infants and children as victims (UNICEF, 1984, p. 81).

The situation in respect of sanitation is more appalling. It is exemplified by the fact that the proportion of rural population with access to some form of sanitary waste disposal has not crossed 5 per cent (Padmanabhan, 1990). It is further stated that the poor performance is owing to the absence of wants and lack of awareness in the rural community about the benefits of hygiene and sanitary waste disposal. The survey of the three villages revealed that all the residents used to go to the nearby fields for defecation. The method of solid waste disposal is also improper and unscientific. Of all the respondents, about 59 per cent dispose their waste materials in some sort of depression within a distance of 10 m from their dwellings. Only 10.5 per cent respondents have stated that they dispose off wastes at a distance of 50 m and above (Table 12.5). It may be inferred from the foregoing information that improper disposal of waste materials at a short distance is unhygenic and invariably invites various diseases. The improper disposal of animal excreta also contribute in making the environment unhealthy. Wherever villagers find any sort of depression in their surroundings they use it as a site for waste disposal and remains uncovered throughout the year. Further, the enquiry reveals that about 74 per cent families partly use animal excreta as manure and dump it in the depressions close to their dwellings for composting. The existing form of preparing organic manure is neither scientific nor hygienic. It is also common to see the open *kachcha* surface drains full of stagnant mud and dirty water. At many places it is found spreading over the lanes of the settlements in the villages. *Pucca* drains are practically non-existant in the studied villages. It further adds to the magnitude of the problem. It is strange to note that not a single settlement of Uttar Pradesh had access to sanitation upto 1981 (UNICEF, 1984,). Ella Bhatt, Member of the Planning Commission has rightly remarked in a colloquium on Role of Women in Water Resource Management that decades come and go but people remain where they are.

TABLE 12.4: Sources of Drinking Water Supply

Sl. No.	Village	Source					
		Wells		Hand-pumps		Piped-water	
		Kachcha	Pucca	govern-ment	Private	Public tap	Private
1.	Umarwar	16	71	2	43	—	—
2.	Patail	09	03	5	9	—	—
3.	Shivrampur Khurd	10	8	2	9	3	7
	Total	35	82	9	61	3	7

Source: Based on field surveys, 1989

The situation reflected in the foregoing paragraphs demonstrates the crisis proportion in respect of both sanitation and safewater. At the same time people are aware about their ill-effects but the poverty and carelessness in personal as well community hygiene come in their way.

TABLE 12.5: Distance of Waste Disposal Site from Dwellings

Sl. No.	Distance (in metres)	Percentage of total respondents
1.	Below 10	59.27
2.	10-30	24.06
3.	30-50	6.15
4.	50-100	3.76
5.	100 and above	6.76
	Total	100.00

Residents Perception to Varied Environmental Problems

In the present study our main concern is with the appearance and feel of the rural environment. With the help of questionnaires an investigation was made. How far the people are aware of various environmental problems? Their perception, responses and priorities may be of great use in making decisions for successful implementation of the environmental policies. With this objective, their responses were recorded and analysed herewith in sequel.

Responses to the question: 'Do you feel that rural environment is being affected from any kind of pollution?, if yes, then by what sources?' It is interesting to note that even illiterates have also perceived the problem. From the responses it is found that 43.04 per cent respondents feel improper waste disposal as the main source of pollution in rural areas, whereas 21.30 per cent blamed to the usage of the pesticides and chemical fertilizers for raising the agricultural productivity (Table 12.6). In their views, pesticides and fertilizers while entering to food chain and water through soil prove toxic and inflicts the villagers continuously. They further opined that these have lessened the resistant power to fight with various diseases. Nearly 16.5 per cent of the respondents rated population pressure as the main culprit of pollution. Further, 19 per cent of the residents perceived the problem of pollution through smoke generated by the burning of fuels and contamination of drinking water sources. It is surprising to note that among the illiterates 44.44 per cent respondents have given priority to waste disposal as the main source of pollution followed by use of pesticides and fertilizers (24.24%) and increasing population pressure (17.18%). It is a good realization by them that population is the root cause of many environmental problems. It is evident from the table that with the increase in educational levels, the respondents have given more emphasis of pollution caused by using pesticides and fertilizers, burning of fuels and population pressure rather than improper waste disposal. It establishes that greater the level of education, more is the vividity in the response pattern.

Table 12.6 shows that the sources like contaminated drinking water, air pollutants through burning of fuel are not well perceived by the rural population. Therefore, awareness should be inculcated about these two invisible sources of pollution which cause more harmful effects than the rest stated earlier.

TABLE 12.6: Sources through which the Villages Perceive the Pattern of Pollution (in Percentage)

Literacy	Improper waste disposal	Contaminated drinking water	Burning of fuel	Consumption of pesticides & fertilizer	Population pressure	Total
Illiterate	44.44	6.06	8.08	24.24	17.18	43.04
Primary and below	44.11	8.82	14.70	20.59	11.78	14.78
Up to high school	51.02	12.24	6.12	14.28	16.34	21.32
Up to Intermediate	29.16	20.83	8.33	25.00	16.68	10.43
Graduate and above	33.33	16.66	8.33	20.83	20.85	10.43
Total	43.04	10.43	8.69	21.30	16.54	100.00

Source: Compiled from field interviews, 1989

In order to assess the relative importance of pollution over other significant problems of the day, they were exposed to the questions : Is pollution an important problem now-a-days ? If so, is it more important than unemployment, housing and food problems? The results were tabulated on the basis of different income groups ranging from less than Rs. 6,000 to above Rs. 12,000 per annum (Table 12.7). In response to this, a large number of interviewees in different income groups perceived pollution as a emerging problems of the day but not more important than unemployment, housing and food as these are the basic needs of the human survival. Table 12.7 indicates that only a mere 11.70 per cent respondents have given more importance to the issue of pollution than the above referred basic problems. And most of them (9.22%) belong to the income group of more than Rs. 6,000 per annum. It is interesting, to note that 31.66 per cent respondents from low income group (below Rs. 6000) rated housing as the most important problem of the day. It is closely followed by the unemployment and food. About 6.4 per cent of the total respondents could not have judged the priority. Out of this 66.7 per cent belongs to low income group, more particularly the daily wage earners. Table 12.7 amply demonstrates that ruralites are aware about the effects of pollution but they are more serious and interested in resolving the basic problems of human life.

TABLE 12.7: Priority to Pollution Over Unemployment, Housing and Food According to Income Group (in percentage)

Income per annum(in Rs.)	Pollution	Unemployment	Housing	Food	No response	Total
Below 6000	5.83	26.66	31.66	25.85	10.00	42.55
6000-12000	21.56	23.53	23.53	23.53	7.84	18.08
12000 and above	13.51	28.83	27.93	27.93	1.80	39.37
Total	11.70	26.95	28.72	26.24	6.39	100.00

Source: Compiled from field interviews, 1989

In order to over come any problem of the society, public participation is most desirable. Accordingly, the people's attitude towards making better environment is to be known. Of all the respondents, 33.73 per cent have opted community participation as the the best means of making better and healthy environment whereas 16.86 per cent have favoured voluntary action (Table 12.8). Nearly 23.5 per cent of the interviewees have a desire to make complaint to the government authorities about their pressing environmental problems. Further, 16.72 per cent of the interviewees were of the opinion that combined efforts is the most befitting approach in making better livable environment. However, 9.6 per cent of the interviewees have given no response as they have little concern for solving problems. Maximum number of respondents (94%) who have given no response are illiterates. However, responses to the attitude towards pollution control drive compiled on the basis of literacy groups does not portray any meaningful correlation.

TABLE 12.8: People's Response Towards Making Better Environment (in percentage)

Literacy	Through community partici-pation-	Volunt-ary action	Complain to the govern-ment	Combin-ed effort	No res-ponse	Total
Illiterate	27.86	21.31	14.75	11.49	24.59	36.74
Primary and below	33.33	20.00	30.00	13.34	3.33	18.07
Up to high school	38.88	16.66	19.46	25.00	Nil	21.68
Up to Intermediate	42.85	Nil	42.87	14.28	Nil	8.43
Graduate and above	36.00	12.00	32.00	20.00	Nil	15.08
Total	33.73	16.86	23.49	16.26	9.66	100.00

Source: Compiled from field interviews, 1989.

Conclusion

From the foregoing discussion it has been inferred that villagers are aware about their nearest environment where they breath, grow up and take shelter. But owing to their poor socio-economic base the problem remains unattended. Within a decade we shall enter in the 21st century but still one can easily notice the prevalence of age old long practices of constructing houses without adequate provision of ventilation and smoke outlet, use of traditional sources of fuel, old design of *chulha,* open fields as lavatories, open *kachcha* drains, improper waste disposal system etc. With the increase in the population, whenever the villagers find space, built the houses and allow the waste water to flow in puddles. Our women's are suffering from suffocating environment of the kitchen but again the economic constraints come in the way of adopting alternate sources of energy despite growing awareness. It is interesting to note that people are aware of their environmental problems and conscious of making better environment but they still feel that the basic problems like unemployment, housing, and food should be resolved first. Based on the findings of the present study, following remedial measures are placed for making a cleaner and greener environment in the countryside:

(a) Encouragements should be given to install bio-gas or gobar-gas plants who so ever possess sufficient feedstocks, e.g. cattle dung, domestic and agricultural wastes etc. It has double advantages: (i) gas will be used for cooking and lighting, and (ii) residue will be used as organic manure for crops.

(b) There are many families in the villages who do not have cattle at all, they should be provided with alternative source of energy like solar energy with solar cookers on cooperative basis. It will reduce the requirements of fire wood, cowdung etc.

(c) In the ongoing social forestry schemes, emphasis should be given to the plantation of those trees which generate less smoke e.g. *Babool, Kaitha* etc. Villagers should be encouraged to plant large trees wherever land is available. In long run trees yield more than cultivation.

(d) There should be made a proper provision for smoke outlet. Kitchen should be airy. It is the immediate and least expensive measure of abating the problem arising from smoke pollution which cause harmful effects on women's health.

(e) The campaign to construct smokeless *chulhas* with the help of common mud and ceramics should be made more popularised. It will improve the working in the kitchens.

(f) Public sanitary latrines on the pattern of *Sulabha Sauchalayas* should be provided by the village panchayat so that sanitation condition may be improved to some extent.

(g) A vigorous campaign should be launched to promote awareness towards sanitation and personal hygiene.

(h) For making any programme a success, it is necessary to rise the socio-economic conditions of the people. This will also facilitate in increasing awareness and adoption of innovations as well.

(i) Potable water supply should be made available in all the villages. It will improve the health status of the ruralites.

(j) Due care should be given to the efficient and judicious uses of fertilizers and pesticides.

(k) Efforts should be made from the governmental agencies and voluntary organisations so that some demonstration could be given to rural folk in making and maintaining their environment cleaner and greener.

REFERENCES

Ambhire, D.B. (1979), The drought prone areas of Maharashtra : A study of human response to the environment, in Reddy, N.B.K. (ed.). *Drought Prone Areas of India.* Rayalseema Geographical Society, Tirupati, pp. 167-73.

Caris, Susan L. (1978), *Community Attitudes Towards Pollution,* Research Paper No. 188. Department of Geography, The University of Chicago, Chicago.

Desai, A., (1980), The environmental perception of an Urban landscape : The case of Ahmedabad. *Ekistics,* Vol. 47 (2 & 3), pp. 279-85.

Desai, A. (1981), Differential perception of residents to environmental quality of an urban area : The case of Ahmedabad. *Geographical Review of India,* Vol. 43(2).

Desai, A. (1982), Environmental quality in the core city of Ahmedabad : A study in resident's perception. *The National Geographical Journal of India,* Vol. 28 (1&2), pp. 1-14.

Desai, A. (1985), *Environmental perception : The human factor in urban planning.* Ashish Publishing House, New Delhi.

Karan, P.P. (1977), Perception of environmental pollution in Chotanagpur industrial area, India. *National Geographer,* Vol. 12 (1), pp. 17-24.

Karan, P.P. (1980), Public awareness of environmental problems in Calcutta Metropolitan Area, *The National Geographical Journal of India,* Vol. 26 (1&2), pp. 29-34.

Kayastha, S.L. and Ram Bilas (1983), Public awareness towards river pollution along the ghats of river Ganga at Varanasi : A case study of environmental assessment and perception, Paper presented in the National Conference on *River Pollution and Human Health,* Varanasi.

Kayastha, S.L. and R.P. Yadava (1977), Human perception and adjustment to environmental hazards : A case study of three villages of the Ganga flood plain (Chandauli Tahsil), U.P. in R.C. Eidt *et al.* (eds). *Man, Culture and Settlement,* Kalyani Publishers, New Delhi, pp. 404-14.

Kayastha, S.L. and R.P. Yadava (1980), Flood hazard in Lower Ghaghara Plain (U.P.): A study in perception and impact on socio-economic development, *The National Geographical Journal of India,* Vol. 26 (1&2), pp. 19-28.

Kumra, V.K. and R. Singh (1979), Perception of air pollution : A case study of Kanpur city, *Indian Journal of Air Pollution Control,* Vol. 2(1), pp. 50-4.

Kumra, V.K. (1982), *Kanpur City : A Study in Environmental Pollution,* Tara Book Agency, Varanasi.

Misra, Anupam (1988), *Our Environment* (in Hindi), Gandhi Peace Foundation, New Delhi, pp. 108-9.

Nag, P. (1980), Perception of technological hazard in an urban environment: The case of power crisis in Calcutta (India), *The National Geographical Journal of India,* Vol. 26 (3&4), pp. 174-84.

Nag, P. (1985), Environmental perception in Singh, H.H. *et al.* (eds), *Geography and Environment : Issues and Challenges,* Concept Publishing Company, New Delhi, pp. 31-50.

Padmnabhan, B.S. (1990), When will be the Mirage End?, *The Hindustan Times.* September 9.

Singh, Onkar *et al.* (1988), *India's Urban Environment : Pollution, Perception and Management,* Tara Book Agency, Varanasi.

Swaminathan, E. (1980), Transformation of rural habitat through diffusion of innovation in Coimbatore Region in Singh, R.L. and R.B.P. Singh (eds). *Rural Habitat Transformation in World Frontiers,* National Geographical Society of India, pp. 233-9.

Swaminathan, E. and Jothimani, P. (1987), Awareness and adoption of healthcare measures : A case study of Sadayan Gounden Pudur, a rural village in M. Srivastava, and A.K. Singh (eds), *Rural Development in India,* Deep & Deep Publications, New Delhi, pp. 215-23.

Swaminathan, E. and Jothimani, P. (1988), Energy utilization pattern, awareness regarding energy sources and planning alternate energy resources in a rural village, in R.Y. Singh, and A.K. Singh (eds), *Planning in Integrated Rural Environment,* Deep & Deep Publications, New Delhi, pp. 344-352.

Tewari, A.K. (1979), Perception of drought hazard among the people of drought prone areas in arid lands of India — A case study, in N.B.K. Reddy, (ed.), *Drought Prone Areas of India,* Rayalseema Geographical Society, Tirupati, pp. 51-9.

UNICEF (1984), An Analysis of the children in India, U.N. Children Fund Regional Office for South Central Asia, New Delhi, pp. 81-85.

13

MIGRATION AND SOCIO-ECOLOGICAL ADAPTATION : A Study on the Nepalese of Sikkim

CHAMPA MITRA and MAMATA DESAI

The present population of Sikkim is primarily composed of three different ethnic groups. They are the *Lepchas*, the *Bhotias* and the Nepalese. These different groups have their own distinctive socio-cultural ethos, though they have been sharing the same institution, the state. Among these groups, Nepalese constitute a major portion.

The Nepalese of Sikkim, form a complete cultural unit, though they have some sub-stocks among them. They were migrated to Sikkim long after the Bhotia immigration and were mainly encouraged by the British.

Though they are migrated people to Sikkim, but because of their long co-existence with local *Lepcha-Bhotia* population, they have acquired a life style, a socio-cultural ethos, which as a whole is Sikkimese.

In this present study an attempt has been made to examine the history of migration of Nepalese in Sikkim. The study deals with the socio-cultural context of the Nepali immigrants and explains the socio-ecological adaptation in their new home.

The Nepalese of Sikkim are a complete cultural unit, and have considerable differences in physique, nature and customs. They are very hardy and industrious and responsible for most of the economic and regional development of Sikkim.

Who are the Nepalese?

Primarily Nepalese are divided into three major sub-ethnic stocks, the *Thakuris* or *Gorkhas* — mainly came from western and central Nepal, the *Newars* came originally from Kathmandu valley and eastern Nepal while the *Kiratis* came from the north-eastern and eastern Nepal. The *Gorkhas* or *Thakuris* are Hindus, and are proud of rough and tough martial traditions. They speak Gorkhali language.

The *newars* follows Buddhism and Hinduism and speak their own dialect along with Gorkhali. They are known for learning, art, craft, architect and literature. They also have special liking for trade and commercial transactions and as such mainly live in the urban areas. In Sikkim they occupy prominent positions in the states' administration and policy making circles, and hence enjoy a special social status.

The *Kirati* sub-ethnic group is said to be the autochthonous inhabitants of eastern Nepal. There are some tribal stocks, who may be counted among the *Kiratis*, e.g. Limbu, Rai, Magar, Gurung, Sherpa, Tamang etc. They speak their own dialect and like to maintain a more primival life style. The *Kiratis* may broadly be divided into three groups, viz., animists (Magar, Gurung and Limbu), Hindus (Rai) and Buddhists (Tsongs, Tamang and Sherpa). *Kiratis* are excellent soldiers and are used to live in harsh climatic conditions.

Limbu

Limbu is a branch of *Kirati* tribe. They live in this part of the eastern Himalaya for a long time. Intercaste marriages especially with Lepchas are very common among the Limbus. According to some anthropologists, they came from Varanasi, and as such are also known as Kashi gotras. They call themselves as *Yakthamba*. They have their own language, which is known as *Limbu Kura*. The language has been recognised by the state government as one of the state languages of Sikkim. They mainly live in western and north-western parts of the state.

Magar

The *Magar* is another animist sub-stock of Nepalese in Sikkim, whose society, customs and religious ceremonies closely conform to those of Hindus. They have their own language, known as *Magar Kura*, which belongs to Tibeto-Burman group. The *Magars* are sub-divided into seven more classes, who enjoy equal social status. They are Ale, Burathaki, Gharti, Pun, Rana, Roka and Thapa. Thapa constitutes maximum *Magar* population. Interclass marriage is allowed among the classes.

Gurung

The *Gurungs* are of Mongoloid in origin. Though they are animists, in religion they are Hindus, and also agriculturists. They have their own language the *Gurung Kura*. *Gurungs* are sub-divided into two groups: Char jat and Sara jat. Socially, the former is much superior. But marriage between the two groups is very common. The social differences among the sub-groups are now fast disappearing in Sikkim.

Rai

Açcording to religion the *Rais* are Hindus. Their marriage customs do not differ much from the *Kiratis*. The *Rais* have their own dialect. They are good writers and artists and agriculturists. In Sikkim, they have formed an association with *Kiratis*, the aim of which is reviving the *Kirati* culture and ethnic entity.

Tamang

The *Tamangs* are lamaists as well as Buddhists. Their language is similar to those of the *Gurungs*. They are divided into two sub-groups viz., The Bara *Tamang* and *Atharajat*. The Bara *Tamang* is socially more superior.

On the basis of socio-ritual activities, the Nepalese are mainly divided into two broad groups, the *Tagdharis* and the *Matwalis*. The former wear sacred thread, and maintain a level of purity while the latter are used to consume liquor. As such only *Gorkhas* can be counted among the former groups, and the latter is the joint stock of the *Kirati, Newari* and the *Thakuri* castes.

Though all these sub-stocks have their own dialects, on the whole they speak Nepali language and it is being treated as lingua-franca of the state.

History of Migration

Among the three main communities (viz. *Lepcha, Bhotia* and Nepalese), Nepalese constitute the major part of the Sikkimese population today. The British colonialization was the result of the expansion of the Gorkha Power in Sikkim during the latter part of the eighteenth century, when they received British favours. The Nepalese, especially from eastern and central Nepal, began to move towards Sikkim, as a result of over exploitation of cultivable land in Nepal and also because of the increasing pressure of population. The socio-cultural and socio-religious differences between the Nepalese and the Tibetans were sharp. Because Nepalese were Hindus and polygamous, while the *Lepchas* and the *Bhotias* were Buddhists and polyandrous.

Up to eighteenth century, no such census figure for the population of Sikkim is available. According to the estimation of Dr. Campbell, during 1840, there were 3,000 *Lepchas* and 2,000 *Bhotias* in Sikkim. Though *Bhotias* established blood-brotherhood with the *Lepchas,* yet their attitude towards *Lepchas* were not always docile. The *Bhotias* always occupied highest positions in the government, while *Lepchas* were being offered the lowest. They destroyed *Lepcha* literature, containing valuable mythological stories. The highandedness of the *Bhotias*, did not leave any immediate effect, but in due course, it often offended *Lepcha* sentiments which created two different factions — a *Lepcha* and a *Bhotia*, as a result quarrels occurred very frequently among them.

The British wanted to exploit the sentiment of the aggrieved *Lepchas* against *Bhotias*, who had pro-Tibetan interests. Due to the repeated *Gorkha* invasions, *Lepchas* were also aggrieved with the Nepalese. On the basis of the Treaty of 1861, British acquired *de fecto* suzerainty over Sikkim. But they found *Lepchas* and *Bhotias* were not numerically enough to serve British objectives of opening up of Tibet. British also found that the country was sparsely populated, so to populate the lower hills, enthusiastic, hardworking Nepalese were encouraged by the British to settle down in Sikkim. Besides, British also found local *Lepcha-Bhotia* population quite unsatisfactory for menial and manual labour. But for the development of this hilly region, it was necessary to import hard working people from the neighbouring countries. Maharaja Tsugphud Namgyal was often advised by his ministers and *Kazis*, not to allow to settle a single person from Nepal and Bhutan in Sikkim. The same advice was given to the next Maharaja Sidkeong Namgyal and his sister Dalma by the ministers during 1871. But the fact is that during his reign in 1867, a lease was accorded to a Nepali (Newar) for some lands. As a result, rioting was very frequent between the Nepalese and the local *Lepcha-Bhotias*. Some Sikkimese elites, especially *Lepchas* also had some interest in settling the Nepalese immigrants, as they wanted to acquire some benefits, like easy availability of agricultural labourers etc. The *Bhotias* dislike the Nepalese very much due to socio-ritual and socio-cultural assimilations. Being Hindus, Nepalese used to sacrifice animals during religious ceremonies, which is an anti-Buddhist practice. On this very particular issue, there were frequent clashes between them and at the order of the Maharaja, the Nepalese were repeatedly expelled from the southern and western parts of the country, particularly from the Namchi region. But they were always resettled by some local pro-immigration personalities. They also encouraged Nepalese to settle at Rhenok. At this point, natives living below Teesta, agitated and requested Maharaja for not allowing Nepali immigrants to be settled in the country, and also not allowing them to sacrifice animals during religious ceremonies. The agitation was led by Tatsang Lama. They also tried to get an order from the Lieutenant Governor at Darjeeling, prohibiting new settlers in Sikkim. Consequently, Deputy Commissioner of Darjeeling and Maharaja Thutob Namgyal were called to meet the Lieutenant Governor at Kalimpong and a decision was taken at the meeting that Nepalese might be allowed to settle to populate the waste lands, but they should never be allowed to become headman. The decision brought a temporary relief among the local natives.

But Nepali immigration was carried out in full swing from 1889, when White became the first political officer, and when also the Government of Bengal accepted the principle to allow the Nepali settlers in the wastelands of Sikkim, without hampering the interest of the natives. It was decided at a council meeting during 1889 to allow Nepalese settlement on the wasteland of Sikkim, under certain conditions e.g. not to allow Nepalese settlement above Teesta-Rangit meeting point. It was also decided that they might not be settled above the valley of the Bockcha chu and Dikchu, and especially *Lepchas* were to be encouraged to settle above Dikchu. As a result of this arrangement, about four hundred Nepalese with their families came to settle in Sikkim during the year 1889-90.

Thus, during 1897, Nepalese established, three-fourths of Sikkim's population. Due to heavy influx of the Nepalese, the political officer, White tried to open up the tract along the Teesta for the Nepali settlement. At this point Maharaja pointed out to the

Lieutenant Governor about the dwindling nature of the *Lepcha-Bhotia* population, and requested him not to allow more Nepalese to settle in Sikkim. But the political officer White did not give any importance to this apprehension. He opined that the *Lepcha* reserve in the Lachung valley was sparsely populated. And as such immigration continued right up to 1920. After that, though certain restrictions were imposed, they came in freely.

It appears from the census of Sikkim 1901, that though the population of Sikkim increased considerably, increase of *Lepcha-Bhotia* population was very negligible.

The census figures for the past eighty years given below explains demographic trends in Sikkim up to 1971 (Table 13.1).

TABLE 13.1: Census Figures from 1891 to 1971 in Sikkim

1891	Total			30,458
		Lepchas	5,762	
		Bhotias	4,894	
		Limbus	3,356	
		Nepalese	15,458	
1911	Total			80,000
		Nepalese	50,000	
		Lepchas, Bhotias & others	30,000	
1931	Total			1,09,808
		Lepchas	13,060	
		Bhotias	11,955	
		Nepalese	84,693	
1951	Total			1,37,725
		Lepchas	13,625	
		Bhotias	15,625	
		Nepalese	97,863	
1961	Total			1,62,189
		Lepchas & Bhotias	49,894	
		Nepalese	1,08,165	
1971	Total			2,03,000
		Lepchas, Bhotias & others	50,750	
		Nepalese	1,52,250	

Table 13.2 shows that Nepalese population has increased con-

tinuously from 1891 to 1971 in Sikkim. Further, the East district has the maximum Nepali population in the state.

TABLE 13.2: Districtwise Concentration of Nepali Population in Sikkim (1971 Census)

Name of the District	No. of Nepalese
East district	53,204
West district	38,451
North district	1,925
South district	40,655

Socio-ecological Adaptation

As mentioned earlier, among the three main ethnic groups, Nepalese dominate the population of Sikkim today. With more than two-thirds of the total population of the state, the Nepalese out number all other communities. They are the latest immigrants to the state, and are mainly Hindus in religion. Hence they have retained a distinct cultural identity. Apart from the British policy of immigration, which has already been mentioned, British also attempted to counter balance the Tibetan inclinations of the place, by flooding the state with Nepali settlers. The *Sikkim Gazetteer* can be quoted here (1894):

> *"Most of all will our position be strengthened by the change which is insensibly but steadily taking place in the composition of the population of Sikkim".*

The Nepalese settlers as most of them are polygamous, multiplied at such a remarkable speed that within one century they reduced the original inhabitants to sheer minority. They are phenomenally fertile people and till few years ago, it was not unusual to find four to five wives of a Nepali. To have more wives was a greater asset for a Nepali because of the availability of more working hands in the fields.

Most of the Nepalese of Sikkim are Hindus, and some are Buddhists, while a small fraction has been converted into christianity. The legacy of Hindu Buddhist traditions tampered with tribal beliefs gives the Sikkimese Nepali culture a vigour and colour which is no less impressive and exotic than that of the *Lepchas* and *Bhotias*.

The lives of the Hindu Nepali population of Sikkim follow the patterns which govern the Hindus living in other parts of India. However, a parallel religion indigenous to the Nepalese co-exists with their Hindu beliefs. They practice a form of Shamanism, in which *Jhankris* have an important role in the Nepali Society. The *Jhankri* is by profession a tailor, grocer, a carpenter or any common man. But, who through a special calling has the power to exercise evil spirits and witches, which according to belief, dwell in the forests and streams and cause illness or bring misfortune. The *Jhankri* uses ritualistic dance and goes into a state of trance during these ceremonies. The belief in this shamastic cult is quite wide spread among the Nepalese, and this tradition is, at the same time, enmeshed with the orthodox Hindu doctrines, giving their culture a unique flavour. These *Jhankris* are nothing but a substitute feature of *Lepcha*, *Bonthing*. The Sikkimese traditions thus have influenced the religious and cultural lives of Nepalese.

But all the important sacred centres of Nepalese in Sikkim have close links with Indian mythology. They have elaborate concepts of untouchability, caste hierarchy and priest-hood. The Nepali immigrants, especially *Brahmins* maintained their solidarity in Sikkim also.

There is no Hindu sacred centre in Sikkim except Thakurbari in Gangtok, which is a very small unit, and is looked after by an ill-educated priest. The Hinduism, which is followed by about two-thirds of the Sikkimese, does not present an organised and unified image. In this set up, the orthodox Hindus may feel disillusioned in the face of the lamaist ritual display.

Despite varied socio-cultural identities among the ethnics in Sikkim, there is no apparent social differences between them. Because, over the years, there has been cultural cross pollination among them, which has helped to create a distinctive Sikkimese ethos. One strong evidence of this acculturation is absorption of Mount Kanchenjunga into Hindu ceremonies. This mountain has been worshipped by the *Lepchas* from very early times and now it is equally worshipped by the *Bhotias* and Nepalese.

The Hindu and Buddhist rituals, the Nepalese and the *Lepcha-Bhotia* culture interwovened with colourful social customs and beliefs. Each community has their own folk songs and dances which are mainly connected either by agricultural cycle or any particular social occasion like birth, marriage, death etc. Though each community performs these in a distinctive manner, yet the participation in other's celebration is an accepted rule. Most of the Buddhist festivals are celebrated in the monasteries while Hindu festivals are mainly celebrated independently at home.

From the economic point of view the Nepalese of Sikkim may empirically be divided into three broad categories:

(a) The marginal farmers,
(b) The ex-soldiers, and
(c) The skilled and semi-skilled professionals.

(a) The marginal farmers are mainly agriculturists, and came in Sikkim mostly in search of new agricultural opportunities. The only capital, a marginal farmer carried with him from Nepal are his hardy hand and a *Khukri*. Though they came in Sikkim mainly for new agricultural land, yet very often they perform combine roles of dairymen, agricultural labourers, porters, carpenters and even as errand boys. When they first came in Sikkim they mostly located themselves on the unhabitated barren forest fringes.

But due to their perseverance and industriousness they have been able to make a living and transform the local economy because of their ability for any type of labour. Even the small children in the family are not liabilities, as they add to the work force of the family from their very infancy. Hence, it may be mentioned that the more number of wives of a Nepali increases his economic productivities, because more wives means more children. Previously, before the Nepali immigration, the *Lepchas* used to practice *Jhum* cultivation. But, the Nepali for the first time introduced intensive terrace farming and improved agricultural productivity. Previously, to the *Lepchas* and *Bhotias,* agriculture was subsistence type. Nepali first introduced permanent commercial agriculture, and as such cultivation of cardamom, ginger etc. started.

(b) Since the end of the World War II, the retired Nepali soldiers were encouraged to settle in some parts of Manipur, Assam, Nagaland and Sikkim. These new settlers have all the good qualities of a good peasant, along with disciplined nature of their previous service. As such, they easily acquired an image of legal citizens, which was again added by their handsome cash income in the form of the pension. In this way, they have been able to occupy prominent positions in the state administration and policy making circles, as well as they occupy a significant positions in the state's economy.

(c) Most of the persons in this group are not permanently settled. They move from one urban centre to another only because of the nature and characteristics of their professions, which are sometimes caste oriented, such as goldsmiths, carpenters, tailors, traders, etc. With the expansion of the different developmental activities, like construction works, electrification programmes, industrialisation, development of communication system etc. demand for skilled and semi-skilled man power has been on increase. For all the above mentioned requirements, the Nepalese are the best suited. In these types of hard works, in which Nepalese are engaged, there is very little competition from the local indigenous communities.

If we examine the Settlement Report of 1958, it reveals that the highest percentage of total cultivated land is owned by the Nepalese (66%) who are migrants to Sikkim. *Bhotias* and especially *Lepchas* are the sons of the soil, owning only 20 per cent and 14 per cent respectively (Table 13.3).

TABLE 13.3: Ethnicwise Landholdings and Excess Land likely to be Acquired

(in acres)

Ethnic	District	No. of holdings	Excess land likely to be acquired
A. *Bhotia*	East	90	903
	West	20	207
	North	23	333
	South	54	1271
	Total	187	2795
B. *Lepcha*	East	12	45
	West	10	49
	North	21	124
	South	17	425
	Total		
		60	645
	Total		
C. *Nepalese*	East	114	704
	West	27	146
	North	3	20
	South	57	480
	Total	201	1352

Source: Land Revenue Department, Government of Sikkim.

Again, migrated people are not only dependent on agriculture. They occupy a prominent position in non-agricultural enterprises, which are becoming much higher in number than that of agriculture. It is interesting to note that 80 per cent of the total persons engaged in these enterprises in East and South districts are Nepalese. At least hundred years ago the picture was completely different when the state's economy was mainly based on agriculture.

So, it may be said that though the Nepalese of Sikkim are not autochthonous to Sikkim, but due to their hardy cheerful and industrious nature they have been able to keep their position prominent in the state's economy, and administration. But in the course of their adjustment periods, they have preferred some particular environmental conditions which suit them the most. For example, during survey, most of the Nepali villages were found below 1,220 m. This is mainly due to paddy cultivated in the region which is the main cereal taken by the Nepalese. But, at places Hindu Nepalese, for their long co-existence with the *Lepchas,* have now acquired a culture, which can neither be called Hindus or Buddhists, but the elements of both religions are found in their beliefs. Moreover, the Nepalese of Sikkim, especially in the urban areas, lead a near cosmopolitan life in which their expenses are in tune with their income and so they do not attract envy of their non-Nepalese neighbours because they rarely add to an affluence. But, inspite of their significant contribution to the total economy they maintain a submerged identity. They are politically aware, and culturally conscious of their status in the state.

REFERENCES

Chie, Nakane (1966), 'A plural society in Sikkim' in C. Von Furer Haimendrof (ed.) *Caste & Kin in Nepal, India & Ceylone*, Asia Publishing House, Bombay.

Das, B.S. (1983), *Sikkim Saga*, Vikas Publications, New Delhi.

Department of Land Revenue (1975), *Report of the Committee on Land Reforms*, Government of Sikkim.

Jha, S.K. & Mishra, S.N. (1984), *Sikkim Govt. and Politics*, Ess Ess Publications, New Delhi.

Karan, P.P. & Jenkins, W.H. (1960), *Nepal — A Cultural & Physical Geography*, University of Kentucky Press, Lexington Ky.

Sinha, A.C., (1973), The feudal polity and political development in Sikkim", *Indian Anthropologists*, Vol. 3(2).

Statesman (1960), Sikkim, *A Statesman Supplement*, June.

14

THREAT TO HIMALAYAN ECOSYSTEM: A Case of Uttarakhand

J. SINGH

The onslaughts of technocratic and industrial society have endangered to many natural ecosystems but it has been more cruel towards mountain ecosystem. In spite of the fact that our hills in the present-day economy (in terms of mineral deposits, hydro-electric potential, perennial source of water, forest wealth, tourist resorts etc) decide the destinies of atleast 40 per cent people living in the foothills and basins of the rivers, the man by his own negligence and activities are facing the wrath of Himalayas — a major mountain system of the world. To assess the magnitude of the problem, Uttarakhand — the crown of Uttar Pradesh has been taken for case study. It comprises the eight Himalayan districts, i.e. Almora, Nainital, Pithoragarh, Dehradun, Pauri Garhwal, Tehri Garhwal, Chamoli and Uttarkashi.

It is doubtless that developmental activities e.g. construction of roads, new buildings and tourist complexes, quarrying and mining, cultivation, industrial establishments, reservoirs and dams etc. have brought changes in the overall socio-economic conditions of the people. On the contrary, these activities have jeopardized the mountain ecosystem on account of loss in vegetal cover, soil erosion, population growth, tourist influx, drying of springs, landslides etc. This is not only the people in downstream also face the consequences like floods, shortage of fuel and timber etc.

With this backdrop, an attempt has been made here to study how the developmental activities and human negligence are endangering this very Himalayan ecosystem and the people in downstream. What we do in Himalayan region today will have considerable impact in the future. Keeping this fact in view, some meaningful and feasible suggestions and steps have been put forward to save this vital ecosystem which is functioning as reserve for the nation in general and the people of hills in particular.

The Study Area

The study region embraces an area of 51,125 km^2 and is inhabited by 4.84 million persons in 1981. The region is larger in area than many states of India, viz. Punjab, Haryana, Kerala, Meghalaya, Manipur, Nagaland and Tripura.

The study area has three parallel ranges extending east-west. They are known as Greater Himalaya, Lesser Himalaya and Siwalik in succession from north to south. It has developed as a result of upheaval of entire material of the Tethys geosyncline in Tertiary era. Greater Himalaya is always clad with snow excepting the lower valleys. Its altitude varies between 3000m to above 7000m. It has several towering mountain peaks, e.g. Nanda Devi (7,815m), Kamet (7,756m), Badrinath (7,138m), Trishul (7,120m), Dronagiri (7,066m), Kedarnath (6,939m) and Nandkot (6,816m). Glacial landforms can be observed in fully developed forms above an elevation of 4000m. Some of the mighty glaciers of this region are Gangotri, Milam, Pindari, Kamet, Kedarnath and Jamnotri. Lesser Himalaya varies in elevation from 1000-3000m. In this part many important hill stations are situated such as Mussoorie (2042m), Nainital (1934m), Ranikhet (1798m), Almora (1,676m), etc. The height of the Siwalik range varies between 300-700m. Structurally, the region as a whole is made up of weak deformed sedimentary rocks. The general pattern of drainage is transverse to the ranges cutting deep valleys and gorges. Soils of the region in general are quite thin-layered; stoney and poor in fertility. Faced further with the problem of soil erosion, the economic value of these soils is considerably too less.

There are a number of components which have threatened to the Himalayan ecosystem but here only following three aspects have been discussed in sequel.

Deforestation

According to official statistics, Uttar Pradesh possesses only 17.41 per cent area under forest in 1981-82 while the results of landsat imageries taken during the period 1980-82 show only 7.14 per cent area under forest. Moreover, the good forest accounts for no more than 6.16 per cent of the state's total geographical area. Between 1972-75 and 1980-82, the state has lost 4,847 km^2 forested tract. According to another statistics, in the last three decades approximately 2,450 km^2 forest land had been cleared off for various purposes (Table 14.1). Besides, between 1978-79 to 1981-82, about 1,562 km^2 forest land was damaged due to forest fires. It is notable that Uttarakhand possesses altogether 34,399 km^2 forested tract which shares 67.28 per cent of the total forested area of the state. It means that the axe of man has fallen mostly on the forests of this region. The landsat imageries taken out in 1973 indicated that only 28.7 per cent area is now forested. Among this, only 4.4 per cent area has a forest with greater than 60 per cent crown density (Table 14.2).

TABLE 14.1: Purpose of Deforestation in Uttar Pradesh and the Area Accrued (in ha)

Period	River Valley projects	Ag. purposes	Roads	Industries	Misc. purposes	Total
1951-52 to 1972-73	93,200	79,700	3,400	19,300	19,990	2,15,500
1973-74 to 1979-80	9,087	4,371	1,571	4,125	10,383	29,537
Total	102,287	84,071	4,971	23,425	30,373	2,45,037
%	41.74	34.30	2.02	9.55	12.39	100.00

Source: Forest Statistics, Uttar Pradesh, 1983.

TABLE 14.2: Distribution of Forest Area (1981-82)

District	% of forest area to total Area[1]	Result of satellite Imageries (in %)[2]			
		Forest	Good forest	Snow	Meadows
Almora	73.25				
Pithoragarh	37.30	34.10	4.8	14.2	3.5
Nainital	59.26				
Dehradun	72.06				
Pauri Garhwal	83.75				
Chamoli	57.74	24.90	4.1	16.6	3.7
Tehri Garhwal	89.84				
Uttarkashi	88.61				
Total	67.28	28.70	4.4	15.6	3.6

Source:1. Forest Statistics, Uttar Pradesh, 1983.
2. Singh, J.S. et al., 1984.

According to one statistics, still today household sector consumes as much as 62.27 per cent energy from firewood. Heavy consumption will certainly lead to reckless cutting of trees. With the growth of population, the pressure on the surviving woodlands is increasing. As it vanishes, stress on the remaining woodland increases. Apart from this, the requirement of firewood increases with increase in altitude, due to lower and lower temperatures. Pandey and Singh (1984) have observed in their random checks that in villages situated at altitudes of around 1,900m, the actual fuel expenditure is about 2.5 kg per day per head as against the average of 1.5 kg. The excessive fuel needs with variation in altitude will automatically result in legal and illegal fellings.

Besides, to meet the fodder requirements during winter and summer seasons as well as for night feeding, frequent and repeated lopping of the broad leafed trees are done which has resulted to their gradual disappearance (Pandey and Singh, 1984). Owing to reckless deforestation on various counts, the situation has become so acute that for collection of firewood and fodder, people have to walk several kilometres everyday. Negi has also made observation that each village of 500 people in the U.P. hills consumes 250 tonnes of firewood each year plus 100 mature trees for housing, 75 for ploughs and 3 lost to fire. In addition, 1.5 ha of forest land is lost to agricultural extension and quarrying while 2 ha are lost to poor irrigation. Even if it is counterbalanced by regeneration and replantation, the hills will continuously lose a large forested area every year and will face dire consequences.

Cultivation and Over Grazing on Hill Slopes

In spite of all hue and cry, the agriculture is the mainstay of economy as 80.78 per cent of total working force are engaged in agricultural practices. With the growth of population which is increasing at a rate of 2.11 per cent/annum in rural areas, the pressure on land is bound to increase. This pressure is driving the people up the slopes, leading to encroachment of forests. Shah (1982) has reported that in the last 15 years the area under agriculture in the hill districts has been increasing at an exponential rate of 1.5 per cent per annum, all at the cost of forests.

The hills have a very harsh environment. Here agricultural land is scarce and scattered. In almost all the districts, agriculture is practised on a very meagre percentage of land. So naturally it will have a excessive human and animal interference to meet the requirement. Physiological density is much more higher than the arithmetic density (Table 14.3). Besides the pressure of population, unplanned and unscientific methods of agriculture on unsuitable steeper slopes has augmented the process of soil erosion. Ultimately it reduces fertility of the soil. Under these circumstances, the hill people have a desperate need for forest products to enhance their income. In any way they are losers.

TABLE 14.3: Rural Population Density (1981) and Percentage of Cultivated Area

District	Population Density per km^2		Percentage of cultivated land to total area, 1980-81
	Arithmetic density	Physiological density	
Almora	132.7	624.8	15.64
Nainital	122.8	402.0	29.30
Pithoragarh	52.4	654.7	11.41
Chamoli	37.0	713.1	5.46
Uttarkashi	22.2	328.8	4.18
Tehri Garhwal	109.1	631.1	13.15
Pauri Garhwal	107.0	562.2	14.20
Dehradun	133.6	690.6	17.46
Total	78.1	561.0	13.22

Source: Census of U.P., 1981 and Season & Crop Report, U.P., 1980-81.

Besides the extension of cultivation on upslope and grazing on hill-slopes have also contributed much to soil erosion. Overgrazing not only enhance the erosion intensity, but also destroys the saplings and adversely affects the regenerative capacity of seedlings. The increase in the number of livestock (3.57 million in 1972 to 4 million in 1982) coupled with reckless cutting of trees, pressure on grazing land has markedly increased. Due to poor productivity, the crop residues could not fulfil the total requirement, so open grazing is most prevalent. Table 14.4 shows the incidence of grazing in all the eight districts. Excepting Uttarkashi and Chamoli, the pressure is nearly 2 to 3 times higher than the carrying capacity of their respective forest floors.

TABLE 14.4: Grazing Pressure in Uttarakhand (Carrying Capacity is 2 Cattle Units per ha)

Cattle	Cattle unit per animal	Grazing Pressure (Cattle Unit per ha)								
		Almora	Nainital	Pithora-garh	Chamoli	Pauri Garhwal	Tehri Garhwal	Uttar-kashi	Dehradun	Uttra-khand
Cow	4	3.508	2.701	3.710	1.818	3.346	1.062	0.564	2.858	2.221
Buffaloes	6	2.169	2.863	1.896	0.703	0.937	1.628	0.301	1.397	1.338
Sheep	1	0.093	0.009	0.173	0.185	0.107	0.086	0.135	0.148	0.118
Goats	1.5	0.655	0.296	0.542	0.290	0.589	0.351	0.149	0.570	0.392
Total	–	6.425	5.869	6.321	2.996	4.979	3.127	1.149	6.122	4.069

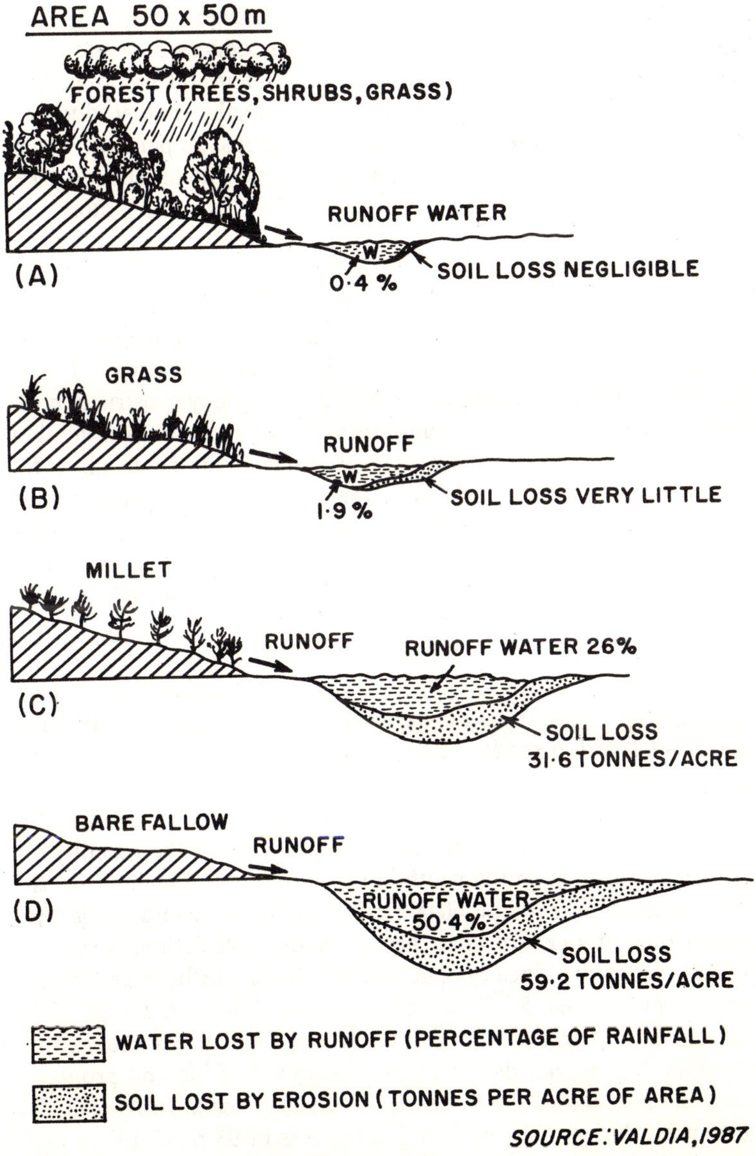

FIG. 14.1: Intensity of Surface Runoff and Soil Loss in Different Protective Cover

In the process of overgrazing and cultivation on upslope, a large area become deforested and deforested slopes yield more water and more sediment quicker than forested slopes (Haigh, 1984). Fig. 14.1 illustrates well that how the protective cover reduces the water loss through surface run-off and as a result reduction in soil loss as well. It also exemplifies how much vegetation is protective and productive. Even in the case of agricultural practices of crops, 26 per cent water is lost through surface run-off and in the process 31.6 tonnes soil acre is lost. On the bare fallow land, the intensity increases much more.

Road Construction

In near absence of railways and water transport in the region, roads are the only means of transport. Though roads are considered best channel for socio-economic development, yet speedy construction of roads particularly after independence without properly assessing the ill-effects on the ecosystem have caused visible and invisible harms to the nature. Among all, landslide is the most glaring example. It has been estimated that the construction of just one kilometre long road requires removal of 40,000 to 80,000 m^3 of debris and on an average about 30 per cent of the debris produced as a result of cutting hill sides is consumed during construction itself, however, about 70 per cent debris are thrown on the sides of the hills (Singh and Singh, 1988). It expands to atleast 20 per cent of its original *in situ* volume. On an average, 550 m^3 per km per year of debris is generated on the already constructed roads. After the roads are built, they are recurrently ravaged by landslides and rockfalls during rains. In the area under study there are altogether 9,483 km surface roads (1981-82). In 1979-80 the total road length was 8,061 km. Though the quantity of debris also depends on the width of the road, even though if it is assumed that on an average 60,000 m^3 of debris is generated on the construction of one km road based on this calculation, between 1979-80 and 1981-82 nearly 8,532X10^4 m^3 debris has been generated and of which about 5,972X10^4 m^3 has slide down the slopes. Ultimately, it destroys plants completely and some times chokes the mountain stream. Besides, it takes a heavy toll of life and property and sometime causes flood menace at a bigger scale. In Alaknanda valley itself during 1971-79, it registered loss of 93 human lives and 1,073 cattle and property worth of Rs. 21.84 million (Pathak, 1983).

Apart from cutting down a number of trees in the way, roads

necessarily involve some cutting and blasting of hill sides at times which results in slope failures. Since the seventies more than 1,000 people and thousands of animals have lost their lives on account of slope instability.

Consequences

Deforestation, cultivation and overgrazing on hill slopes and construction of roads in the mountains have made irrepairable damages in the hills as well as in plains. To mention a few, drying of springs, siltation and recurrent floods in plains are enough.

(a) In hill areas the most alarming consequence is the damage caused to spring environment. Deforestation adversely affects the quality, quantity and periodicity of springs. Singh and Rawat (1985) have reported that in the whole of Kumaun and Garhwal region, out of 120 springs, 22 have become extinct, 40 seasonal and in the remaining there is a reduction in the discharge. They have also reported that gradual depletion of oak forests has adversely thwarted the periodicity of springs. This is not only, excepting Pauri and Tehri Garhwal, the percentage of problem villages in other districts is more than 50 per cent (Table 14.5).

(b) After the evaluation of sediment data of twenty one selected reservoirs, National Commission on Agriculture (1976) has come to a conclusion that on account of deforestation, the rate of siltation has been several times more than the assumed in the project design. It has not only reduced the life span of these reservoirs but also their storage capacity which ultimately will affect the power and irrigation potentials. In the case of Ramganga, the observed rate of annual sedimentation is 18.19 ha m/100 km^2 as against the 4.29 ha m/100 km^2. As a result, the age of Ramganga Dam Project has been reduced to 76 years against the previous estimated age of 100 years (Mithal et.al., 1984).

Valdia (1985) has reported that the river Ganga originating from the Central Himalaya, carries 340X10^6 metric tonnes of sediment each year. This is not only, on account of forest destruction, the present rate of erosion in the catchment area of the Himalayan rivers (100 cm/1000 years) is five times higher than the rate prevailing in the past 40 million years (21 cm/1000 years) (Menard, 1963). On account of siltation, the beds of river rise which in turn increases the intensity of floods.

(c) The National Flood Commission (1980) has opined that among the various causative factors leading to floods in the Gangetic plains, deforestation and erosion is most alarming. In Uttar Pradesh alone the total damages caused by floods during recent years (1976-78) is more than five times than that of earlier twelve year span (1953-65) (Table 14.6). The Ganga river system covers approximately two-thirds area of the state which often experience normal floods and atleast one major flood in each decade.

TABLE 14.5: Problem Villages, 1981

District	% of total inhabited villages
Almora	50.92
Nainital	93.05
Pithoragarh	66.99
Uttarkashi	54.15
Chamoli	53.15
Tehri Garhwal	42.39
Pauri Garhwal	26.63
Dehradun	58.67

Source: Uttar Pradesh Jal Nigam, 1981.

TABLE 14.6: Major Trends in Area Affected and Damage by Floods (Area in Lakh ha and Damage in Lakh Rupees at 1952-53 Prices)

	Period			
Items	1953-65	1966-70	1971-75	1976-78
Total Area affected	17.44	20.12	23.46	40.90
Cropped area	7.93	10.42	13.22	22.33
Non-cropped area	9.51	9.70	10.24	18.57
Total damages	1270.6	1637.2	2728.0	7388.6
Crop damages	1055.6	1296.3	1748.4	5249.2
Non-crop damages	215.0	340.9	979.7	2139.4

Source: National Commission on Floods, 1980.

Remedial Measures

It is encouraging to note that the hill development has lately been accorded greater importance realizing the magnitude and man-nature relationship. But, we have to go a long way. In order to achieve this, following suggestions and steps should be taken into account which certainly will mitigate the problems and pressures. Then this unique ecosystem will prove benefits for our future generations in a better way.

(i) Priority based afforestation.
(ii) Complete prohibition on diversion of forest land to non-forest use.
(iii) People's co-operation and involvement, if necessary with certain incentives in raising the social and agro-forestry.
(iv) National Policy on Forest declared in 1952 should be strictly adhered to.
(v) Like agricultural crops, forest should be treated as crop.
(vi) Ban on practice of mono-culture in afforestation.
(vii) Inculcation of awareness among the people concerned.
(viii) Environmental Impact Assessment before execution of any development plan.
(ix) Continuous monitoring after their execution.
(x) Hazard zone mapping and complete ban on developmental activities in sensitive zone.
(xi) Landuse planning according to slope.
(xii) There should be obligation that the number of trees cut or damaged should be replaced with local species.
(xiii) Each establishment should compensate in equal amount, if the forest tract is cleared off.
(xiv) In construction of roads, debris generated should be arrested by stacking the large pieces of rocks or by making fences.
(xv) Establishment of Forest Ecology Commission under the chairmanship of leading ecologists to decide the future course of development and forestry.

REFERENCES

Haigh, M. (1984), Deforestation and Disaster in Northern India, *Landuse Policy*, July, p. 193.

Mithal, R.S. *et al.* (1984), Environmental impacts of the Ramganga Dam Project, *The National Geographical Journal of India,* Vol. 30(2) p.85.

Menard, H.W. (1963), Some Rates of Regional Erosion, *Journal of Geology*, Vol. 69, pp. 154-161.

Negi, S.S., Need for a Conservation based rural development strategy in the Himalayas, *Journal of Rural Development,* Vol. 2. No.1, pp. 130-43.

Pandey, U. and J.S. Singh (1984), Energy-flow relationships between Agro-and Forest Eco-systems in Central Himalaya, *Environmental Conservation,* Vol. 11, No. 1. p. 51.

Pathak, C.C. (1983), *Pahar*, Nainital, p. 126.

Singh, A.K. and Rawat D.S. (1985), Depletion of Oak Forest Threatening springs: A Exploratory Survey, The *National Geographical Journal of India,* Vol.31(1), pp. 44-48.

Singh, D.S. and A.K. Singh (1988), Development and Environmental Crisis in Kumaon Hills, Paper submitted for *Frontiers in Environmental Geography*, eds. Singh Onkar et al.

Singh, J.S. et al (1984), Man and Forests : A Central Himalayan case study, *Ambio,* Vol. 13(2) pp. 82.

Shah, S.L. (1982), Ecological degradation and future of agriculture in the Himalayas, *Indian Jl. Agr. Econ.,* 37(1) pp. 1-22.

Valdia, K.S. (1985), Accelerated Erosion and Landslide Prone Zones in the Central Himalayan Region, in J.S. Singh (ed.). *Environmental Regeneration in Himalaya.* CHEA, Gyanodaya Prakashan, Nainital, p. 24.

Valdia, K.S. (1987), *Environmental Geology : Indian Context,*, Tata McGraw-Hill Publishing Col. Ltd, New Delhi, pp. 187-89.

15

SOIL EROSION AND EFFECTS OF DEGRADATION OF RURAL ENVIRONMENTS : A Case Study

DEBASHIS DAS

Land and soils are the two important resources of the nature, which should be utilised and preserved for the future generations. Population pressure of present days forces the land to be utilised intensively for the production of both, foodgrain and industrial crops. This is often practised unscientifically so that adverse pressures are created on land to degrade it. Nature's recuperative processes are limited to the productivity of the region. So, proper reclamation measures are needed with the help of appropriate technology in order to regain productivity value of soil and ecological balance. Due to negligence in making reclamation processes by the farmers, mainly due to poverty, the quality of land is further degraded. Besides, flood, drought, waterlogging conditions etc. also deteriorate the quality of land through the intensive soil erosion.

OBJECTIVES

The study has following objectives:

(a) To find out the causes of soil erosion in the study area.
(b) To find out the effects of soil erosion on agricultural landscape of the study area.
(c) To suggest policy measures to alleviate such problem.

DATA AND ANALYTICAL PROCEDURE

Data have been collected from fifteen randomly selected villages in the study area. From these fifteen villages, 360 households were selected by using simple random sampling technique.

Net value of crops have been computed after multiplying production of different crops by prices of respective crops to get gross value of crops. Then gross expenditure is substracted from gross value of crops to get net value of crops.

The relationship between proportion of soil eroded land and other variables has been computed with the help of correlation analysis, and casual relation between net value of crops per acre of GCA and proportion of soil eroded land has been computed by regression analysis. Concentration of land holdings has been computed by Gini's land concentration ratio which is as follows:

$$G = \frac{1}{100 \times 100} \left| \left(\sum_{i=1}^{n} X_i Y_i + 1 \right) - \left(\sum_{i=1}^{n} X_i + 1Y_i \right) \right|$$

Location and Geographical Appraisal of the Study Area

The Ajoy-Kopai inter riverine tract of Birbhum district is located at the southern part of the district and western part of West Bengal. It is bordered on the north and south by two perennial rivers. The Ajoy makes its southern boundary whereas the Kopai forms its northern, boundary.

The average height of the region is 60 m. The highest part (98.5 m) is located at the west whereas the lowest part (21 m) is on the east, implies the slope of the region from west to east. The average slope of the region is 1.15m per kilometre.

The western elevated part is covered by red and laterite soils, whereas as older alluvium soils are predominant in the east. The entire region is covered by sandy loam, sandy clay, loam and clay soils. These soils are fertile and become highly productive if irrigation and other inputs are applied in an appropriate manner.

The region has both the surface and groundwater resources. Surface water resources are utilised through canal and tank irrigation and the groundwater resources are tapped by tubewell and well irrigation. Besides,. river water is also utilised by means of lifting devices. The region has 1,93,593 acres of land under irrigation, which is 58.11 per cent of the net sown area. The same figure for Birbhum is 58.97 per cent. Canal irrigates an extensive area, without being facilitated by field channels. Tank irrigation is available in entire region, whereas canal and tubewell irrigation predominate in central and eastern parts only. Although huge amount of water is available from different sources, but due to the absence of proper management practices, these water resources are not fully utilised for the development of agriculture in this region.

The region receives 78 per cent of the total rainfall during monsoon period. In between 1975-82 the eastern part received more rainfall than western part during winter season, whereas both the western and eastern parts receive more rainfall than the central part during pre-monsoon and monsoon seasons.

Agriculture is the major occupation of this region. During monsoon season, traditional *Aus* and *Aman* varieties of paddy are mainly cultivated. Besides, HYV paddy is grown where irrigation is assured. During winter season potato, vegetables, sugarcane, wheat, mustard etc. crops are grown and in summer season paddy is cultivated where irrigation is assured, besides, till, pulses etc. are grown in other areas.

Causes of Soil Erosion

There are several causes responsible for the erosion of soil in the study area. These causes are grouped into two parts. These are: (a) physical problems, and (b) social problems.

Physical Problems

The surface run off that results from lowered intake of rain water by soil due to inherent characteristics of the soil, adverse soil conditions caused from bad cultivation and cropping practices or sudden and heavy downpour, remove top soil through erosion and expose the less fertile sub-surface soil (Rajan and Rao, 1978, 205).

The western and central parts of the study area (Khoyrasol, Dubrajpur and Illambazar blocks) are more or less dry, where crops are rainfed. The nature and magnitude of soil erosion vary from area to area subject to the availability of moisture that arrive either from rainfall or from irrigation. The moisture holding capacity within the soil zone is relatively less in rainfed areas than the controlled irrigated areas. There exists greater possibility to retain more moisture in the soil zones in the irrigated areas where as small amount of moisture is retained in the soil of the rainfed area. So, the sudden and intensive downpours cause serious soil erosion specially in the western part of the area where the slope is relatively steeper.

Larger number of gulleys and bad land topography are found there. The case is different in the well saturated tract of Nanoor and Labhpur blocks on the east where the rainfall, surplus irrigation water and drained water together flow over the land for some days together. Ballal and Deshpande (1960, 37-41) referred that the run off and soil losses in the saturated condition of soil are generally higher than those caused by rains from air dry soil, The effect of increase in slope is more pronounced in soil losses as compared to run off. The relative density of run off material increases as the slope increases.

Social Problems

Cultivation does not take place on the cultivable waste land or on fallow lands. These lands are liable to erosion. The permanent pasture and other grazing land represent some of the worst eroded areas. These are characterised by unchecked misuse. The grazing is by and large unrestricted and widely practised in the forest areas both sides of the river banks, canals banks, the livestock feeding on undergrowth and trees pass into the young regenerated areas. This type of indiscriminate grazing has caused great damage to forestry ultimately encourages erosive agents to work on the unprotected soil to erode down it (N.C.A., 1976, 232-33).

Soil erosion is acute on unlevelled and undulating land where bunding in the agricultural field is not done carefully and properly. Although the farmers are aware of the fact that bunding may reduce the problem of soil erosion, but the non-cooperative attitude of the farmers and weak coordination that exist between different sister administrative departments of the area restrict the construction of bunds. Further, sometimes facilities are provided to conserve soil and moisture in the field, but the poor farmers and the farmers of relatively backwards areas do not always get such facilities.

Field investigation proves that the areas where the functional relation between landlord and share crop holder (*bargadar*) is unhealthy, there the landlord is not interested to invest money in maintenance work and poor *bargadar* is unable to invest money to upkeep that land, so the reclamation works become piecemeal. On the other hand, the poor farmers with meagre resources cannot up-keep their land rather have to work on the land of landlord due to poverty. So, their own lands become neglected, which are further deteriorated.

Improper management of irrigation water resources, unscientific provision and supply of water for irrigation from dam, improper construction and alignment of side embankments, raise or river bed, etc. factors directly on indirectly encourage to occur floods, which not only cause erosion of soil, but also deposit sand on fertile land and cause water-logging. Sometimes surplus water from canal, drained water from surrounding elevated land and excessive rainfall unite and flow over the land, causes soil erosion. The poor farmers construct cross earthen embankments across the streams and rivulets of the area for catching fish, which derelict it and often overflow to cause soil erosion.

Soil Erosion and Affects of Degradation of Rural Environment

In the present study the rural environment is indicated by agricultural landscape. The objective of this part of study is to assess the causal relationship between proportion of soil eroded land, waterlogged land, flood affected land and land affected due to uncontrolled irrigation on one hand and to find out the proportion of soil eroded land according to the farmers belonging to different socio-economic classes on the other.

The volume of over land flow, soil characteristics and nature of slope together determine the magnitude of soil erosion in the region. The land of the western part of the area are relatively less affected by erosion. Although, the slope of this part is steeper (about 2m per km) than eastern part (0.4 m per km) but the limited flow of drained water and rainfall does not cause acute soil erosion there. On the other hand, the slope of the land of central and eastern parts are comparatively gentler. The surplus water of canal irrigation, rainfall, and drainage water combine together and flow, causing soil erosion intensively in those parts (Appendix I). For example 39.80 per cent of the cultivated area of Jalundi Mouza in Nanoor block is affected by soil erosion, where the proportion of water-logged land (53.07%), and proportion of damaged land due to uncontrolled irrigation (23.16%) are also considerably high. These facts also hold good in Brahman para village of the study area (Appendix I).

The relationship between proportion of soil eroded land and proportion of flood affected land, net value of crops per acre of the Gross Cropped Area (G.C.A.), cropping intensity and land concentration ratio are found out with the help of correlations co-efficients. (Table 15.1)

TABLE 15.1: Relationship Between Soil Eroded Land and Other Variables

Variable I	Variable II	r value	Significant at the level
Percentage of soil eroded land	Percentage of flood affected land	0.75	0.01
	Net value of crops per acre of GCA	(-)0.29	0.50
	Cropping Intensity	0.60	0.50
	Land concentration ratio	(-)0.24	0.50

N = 15

Table 15.1 indicates that flood causes soil erosion, where both the fertility and productivity values of soil are very low, therefore, in those villages the net value of crops per acre of gross cropped area also becomes considerably low. But the cropping intensity values are quite high, because the farmers have to depend upon the production of crops grown at other than kharif seasons for their survival, which may not indicate development of the farmers, but reveal their magnitude of struggle for existence. Due to the above-mentioned poor agricultural situation the propensity to concentrate more land holding by the farmers is at a low ebb, indicating inverse ratio between these two variables.

Regression analysis also has been done in order to find out cause and effect relationship between net value of crops per acre of G.C.A. and proportion of soil eroded land. The regression co-efficient is (-)1.65. Soil erosion degrades the quality of land and causes a decline in the productivity of crops, hence the net value of crops per acre of G.C.A. will also tend to be low. That is why proportion of soil eroded area has been considered as independent variable (X). The equation stands as Y = 998.16 + (-1.65) X. This shows that for per unit increase of soil eroded land (X) in the village, about 1.65 unit of net value of crops (Y) will be decreased in the same village. The significant test proves that, the computed value is more than tabulated value, which means that the null hypothesis, "the proportion of soil eroded land (X) and net value of crops per acre of GCA (Y) are independent of each other" is rejected and the alternative hypothesis is accepted. Intercept (a) shows minimum average net value of crops per acre of GCA. In this case it is 998.16, which means the net value of crops per acre of GCA is on an average Rs. 998.16 and the regression co-efficient shows variation above or below this constant figure. This indicates that soil erosion exerts adverse influence on agricultural productivity and lower down the net value of crops in the villages.

The co-efficient of determination (r^2) is 0.08, which shows that 8 per cent of the total variation in Y is being explained by X.

An attempt has been made to compute the extent of loss that are incurred in agriculture for soil erosion to farmers belonging to different socio-economic classes of the villages.

The study reveals that the major proportion of land-holdings of the weaker sections of the villages are located near the erosive agents. Therefore, a large amount of their land-holdings are affected by the soil erosion, and they have to invest a large amount of money to control it. Side by side of these poor farmers some portions of the land-holdings of the large farmers are also affected by soil erosion. (Tables 15.2 and 15.3)

TABLE 15.2: Land-holding Classes and Nature of Soil Erosion

Land holdings in acres	Total No. of holdings	% of affected holdings to the total holdings	% of affected land of total land	Value of loss per acre of affected land in Rs.
More than 10 (Big)	42	83.64	15.56	77.62
7.5 - 10.0 (medium)	38	83.33	16.30	74.69
5.0 - 7.5 (semi medium)	49	75.32	22.87	77.91
2.5 - 5.0 (small)	60	68.76	20.54	75.69
Less than 2.5 (marginal)	50	73.00	26.66	76.19
Non-recorded share cropper	28	79.98	17.94	75.59
Recorded share cropper	39	80.00	21.06	79.99

TABLE 15.3: Social Classes (Castes) and Nature of Soil Erosion

Caste/Religion	Total No. of holdings	% of affected holdings to the total holdings	% of affected land to total land	Value of loss per acre of affected land in Rs.
High caste Hindu	27	80.77	17.34	79.09
Middle caste Hindu	171	76.73	21.02	79.48
Scheduled caste	70	49.42	17.26	76.74
Scheduled tribe	6	48.33	12.99	75.00
Muslim	32	61.08	28.95	71.88

Note: Percentages are calculated from the total of individual category and not from the total of the region.

These poor farmers do not have adequate money and other relevant infrastructures to tackle the soil erosion problem. Moreover, they cannot upkeep the damaged land in right time, because poverty compels them to work on the land of rich farmers of the villages either as share cropper or as agricultural labourers.

Set against this situation, another social factor causes soil erosion. There are some landlords who are not satisfied with the techniques of cultivation that are performed by the recorded share cropholders. According to them, the recorded share cropholders do not use the entire amount of money that is invested by the land lord on his own land, rather some portions of the invested money of the land lord are used to upkeep the affected land of the recorded share cropholders. This fact discourages the landlord to invest money further for the cultivation even on his land. So, the poor recorded share cropholders can not protect that land from soil erosion.

The study further reveals that the big farmers often encroach upon or divert the alignment of drainage channels and make it more serpentine in appearance. These channels cannot drain away the water, rather surplus water overflows these channels and cause soil erosion. In this way, the soil looses its fertility and productivity value considerably, the level land becomes uneven and then the large sized plots are required to be reduced into many small pieces to avoid the problem of unevenness. This not only causes a decline of the productivity of crops but also increases the cost of production.

The cropping pattern and productivity of principal crops according to different sizes of farms have been studied which can be explained in the Table 15.4.

TABLE 15.4: Cropping Pattern and Productivity of Crops (kg./acre) in Soil Eroded and Non-soil Eroded Areas

Land holding in acre	Soil eroded area			Non-soil eroded area			
More than							
10	Paddy,	Potato	(900)	Paddy,	Wheat,	Mustard,	Potato
	(2400)	Mustard	(300)	(1200)	(600),	(375)	(5400)
				Onion	(1800),	Sugarcane.	
7.5-10	Paddy,	Potato		Paddy,	Wheat,	Mustard,	Potato
	(900)	(2400)		(1200)	(750)	(375)	(6000)
	Mustard	(300)		Sugarcane.			
5.0-7.5	Paddy,	Mustard		Paddy,	Wheat,	Mustard,	Potato
	(600)	(300)		(975)	(750)	(300)	(3300)
				Onion	(1500),	Sugarcane	
2.5-5.0	Paddy,	Potato		Paddy,	Wheat,	Mustard,	Potato
	(675)	(2400)		(1275)	(510)	(255)	(4500)
				Onion	(1500),	Sugarcane	
Less than	Paddy,	Potato		Paddy,	Wheat,	Mustard,	Potato
2.5	(660)	(2300)		(870),	(825)	(195)	(2400)
	Mustard,	Onion		Onion	(1500)		
	(90)	(1200)					

The Table 15.4 reflects thc fact that the cropping pattern is more diversified irrespective of farm size on non-soil eroded land than on soil eroded land and the productivity values of principal crops are also higher in case of non-soil eroded land than in case of soil eroded land.

Suggestion to Control Soil Erosion

The above discussion ncccssarily proves that soil erosion is a problem of agriculture, which may always distort the ecological balance in the ecosystem. Some suggestions may be considered as remedial measures to solve the problems which are as follows:

(a) Provision of controlled irrigation

Planning should be made to extend the irrigation net work to different corners of the study area, and accordingly field channels (preferably lined) should be aligned, so that the irrigation is expected to reach every part almost uniformly. The provision of irrigation should be made systematically and scientifically after considering merits of *Barabandhi* system. The objective is to increase cropping intensity and moisture content in the soil in order to check erosion and maintain ecological balance.

(b) Provision of drainage facilities

Increase of soil moisture is expected to control soil erosion but excess of water becomes hostile to the rural environment. Drainage channels should be made and maintained in order to pass away the surplus water, and the farmers of different socio-economic interests should work together to consider the benefit of the community and not for their own. Besides, engineering structures are to be located at proper places, so that each structure should not get excessive pressure of water and ultimately get collapsed. Water management techniques should be implemented to clear the bed and confluences of the rivers and rivulets, so that overflow of them can be controlled.

(c) Control of indiscriminate grazing

Indiscriminate grazing and unscientific use of land cause soil erosion. Therefore, grazing fields should be located at every locality or at proper places so that haphazard grazing can be controlled. Simultaneously, grass culture should be performed. Grazing operation should be practiced systematically. So that one piece of land should be grazed then grass culture will take place there and grazing will be shifted to another place which is ofcourse not very far from the community. Besides, cultivation of forage and production of animal foods should be increased in order to release pressure of cattle on land. It is, however, accepted fact that the above practise is not very easy process. It requires large unutilized area which the society may not have.

(d) Proper levelling and bunding

Levelling of land and soil erosion is inversely correlated. So, the foremost task of the farmer is to level the land and to make strong bunding around the plot. To control soil erosion, the entire land of the region should be levelled. In a region there are many plots of land of different farmers, so all the farmers should decide together that for the interest of the community the land should be levelled. Besides, adequate funds should be made available to the concerned *gram panchayats* to invest money properly for the purpose. Along with it, tiny drainage channels should be made to drain out surplus water and bunding is necessary for containing moisture in the soil.

(e) Control of flood and waterlogging condition

Due to improper management of water resources for canal irrigation in Canada and Hinglo dams, the beds of the rivers rise and canals become derelict. As the dams are also silted up, so during last phase of monsoon season when excessive rainfall occurs these dams can not catch hold of it, so the surplus water is released through rivers, rivulets and canals. These surplus water becomes voluminous and overflows the banks of rivers and canals. The side embankments are not constructed perfectly every where and the weak points are breached to cause flood. Flood water can not always find passage to be drained out quickly, causing waterlogging. These are the problems of water management that the study area is facing. These problems should be tackled carefully to reduce flood and waterlogging condition.

(f) Incentives to poor farmers and creation of healthy relation between landlord and share cropper

Poor farmers must get some incentives from block or Panchayat Office to reclaim land. The vested land should be reclaimed before these are redistributed. Share cropholders should not misuse money which is given for the reclamation works of landlord's land. Along with the above, modern agricultural method should be applied through the cooperative farming system and further research innovations should be made in order to confront soil erosion.

Conclusion

Soil erosion is a serious problem that causes degradation of rural environment *vis-a-vis* agricultural landscape and distort the ecological balance in the ecosystem. The effects of soil erosion are widely spread in the study area.

Some physical and social parameters cause soil erosion. These problems, in an integrated fashion, cause decline in fertility and productivity values of soil, hence the productivity value of crops becomes low and cropping pattern becomes less diversified. Generally flood, water logging condition, overland flow, etc. are the prime factors that cause soil erosion.

The study reveals that soil erosion has caused decline in the net value of crops per acre of gross cropped area in the region, although cropping intensity values are quite high. This, however, does not indicate development in agricultural landscape, but reveal the magnitude of struggle for existence with which the farmers of soil erosion prone areas thrive. This kind of intensification of crops rarely direct to produce marketable surplus. Soil erosion has discouraged farmers to accumulate land, as that land may not yield much profit as it may be expected.

The study reveals that this kind of degraded quality of land are vested to the government, which sometimes are reclaimed and redistributed to the poor. Sometimes the reclamation works are piecemeal and haphazard. In that case erosive agents further cause soil erosion. Thus the poor farmers suffer much from the soil erosion. Some policy measures are suggested which may yield effective results to control soil erosion and degradation of rural environment.

REFERENCES

Ballal, D.K. and Deshpande (1960), A preliminary investigation in some of the physical properties affecting erosion of Madhya Pradesh soils, *Journal of Indian Society of Soil Sciences*, pp 37-41.

Government of India (1975), *National Commission on Agriculture*, volume V, pp. 232-33.

Ranjan, S.V.G. and Rao, H.G.G. (1978), *Studies on Soils in India*, p. 205.

APPENDIX I

Distribution of Soil eroded land and other variables

Name of Mouza (Block)	% soil eroded Land	% water logged land	% flood affected land	% of land affected due to uncontrolled irrigation	Net value of crops/acre GCA in Rs.	Cropping intensity	Land concentration ratio
1	2	3	4	5	6	7	8
Brahmanpara (Labhpur)	31.67	36.39	70.83	5.76	1142.86	153.81	0.59
Jalundi (Nanoor)	30.80	53.07	53.80	23.16	1013.50	124.79	0.64
Gomai (Nanoor)	24.47	15.69	14.22	13.38	911.78	121.51	0.33
Uchhakaran (Nanoor)	21.64	18.55	—	37.83	850.74	117.55	0.78
Baitara (Nanoor)	30.15	30.61	32.19	26.64	978.19	151.89	0.11

Contd.

APPENDIX I *Contd.*

1	2	3	4	5	6	7	8
Kapastikuri (Bolpur)	28.73	36.56	36.56	31.71	878.19	121.49	0.80
Raipur (Bolpur)	16.70	13.41	19.73	21.93	1275.11	119.70	0.60
Jahanabad (Bolpur)	38.77	15.77	71.63	14.70	696.39	161.45	0.41
Baidyanathpur (Bolpur)	12.66	9.57	–	5.00	896.39	123.15	0.75
Ushar (Illambazar)	6.66	18.90	36.75	–	1154.30	127.34	0.23
Maheswarpur (Illambazar)	13.66	9.34	13.89	15.06	889.78	124.57	0.58
Khagra (Dubrajpur)	–	3.85	–	1.94	1077.56	126.97	0.73
Purbasibpur (Khoyrasol)	5.86	–	–	–	1018.91	103.23	0.66
Pansuri (Khoyrasol)	17.48	2.73	14.20	28.44	755.21	114.92	0.54
Beledanga (Khoyrasol)	10.13	–	17.48	–	784.49	100.33	0.78

Source: Field investigation.

16

ENVIRONMENTAL CRISIS IN SOUTH MIRZAPUR

V.K. KUMRA and J. SINGH

Since the Vedic period we have been very close to nature and the basic philosophy of life has been to live in harmony with the nature. But uncontrolled population growth and thereby an ever increasing demand for resources compounded with careless application of technology in developmental activities during the span of about 150 years have reversed the earlier concept. Today mankind, both poor and rich, either on account of need or greed, conflicts with the nature. In the process, the quantum and scale of environmental degradation are increasing day by day. In this respect our fragile ecosystem of hills and plateaus are worst affected. In many cases in these regions, development has led to deforestation, and the tall chimneys, a sign of progress, add to the pollution of land, water and the atmosphere. Invariably all these cumulatively affect the human well being. Khoshoo (1988) has rightly pointed out that in association with any development, there is always some amount of environmental degradation. But the scientific analysis of the problems may provide a ray of hope for environmental management by paving a ground for planned action and concerted efforts for a better tomorrow. Keeping all these facts in cognizance, pollution and eco-degradation, South Mirzapur has been taken here for case study.

Prior to independence, this region was mainly inhabited by aboriginals who had remained here in isolation for a pretty long time owing to rocky terrain coupled with thick forest cover and restricted transport facilities. But with the construction of rail and road bridges over the river Son at Chopan, the region witnessed considerable progress and gradual transformation owing to commencement of a variety of developmental activities. Further, consequent upon the completion of Rihand hydro-electric power project in the early sixties industrial units like HINDALCO and Kanoria Chemicals at Renukoot, thermal power plant at Obra, cement factory at Dala and Renusagar thermal power plant at Renusagar grew upon in the study area (Fig. 16.1). Later the availability of abundant coal in Singrauli area prompted the state government as well as Central Government to take up thermal power projects at Anpara, Kota (Shaktinagar) and Bijpur — all around Rihand reservoir. Now it is fast emerging as the biggest energy complex of the country. With the successive intrusion of these activities, the demand of land for construction of roads, townships etc. have increased, which have ultimately engulfed the forestland of the region. Though these activities were designed to enhance the economy of the people and the region both but it could not remain free from its biased ill effects. At present the variegated problems of the region are mainly associated with the depletion of forest, soil erosion, road construction, industrial pollution in and around industrial units. Progress and pollution simultaneously have affected agricultural, residential and forest resources to a considerable extent. The greed and need of local and non-local people, industrialists, contractors, government agencies etc. are steadily destroying the environment of the study area. Keeping all these facts in view, a few problems have been undertaken in sequel herewith for discussion and analysis in the present chapter.

Deforestation

The area under study consisting of hills and plateaus was mostly covered by forest since long. Perhaps inaccessibility of the region had contributed much in saving the forest from reckless cutting for a long time. But like other parts of the country, the people, firstly in need and thereafter in greed, have adversely affected the natural forest ecosystem. Way back in 1883, Jackson and Fisher in their report on Dudhi forest mentioned, "at present there is great destruction of the existing forests by the system of girdling to procure resin and clearing large patches by felling the timber and then burning it, for cultivation" (SAP, 1990, p. 16). Though commercial users have been placed more responsible for the destruction of forest, yet slash and burn practice of cultivation has also been majorly responsible in depleting vegetal cover and thereby paved the way for washing away fine particles of top soil after a few rains. In recent years the increased

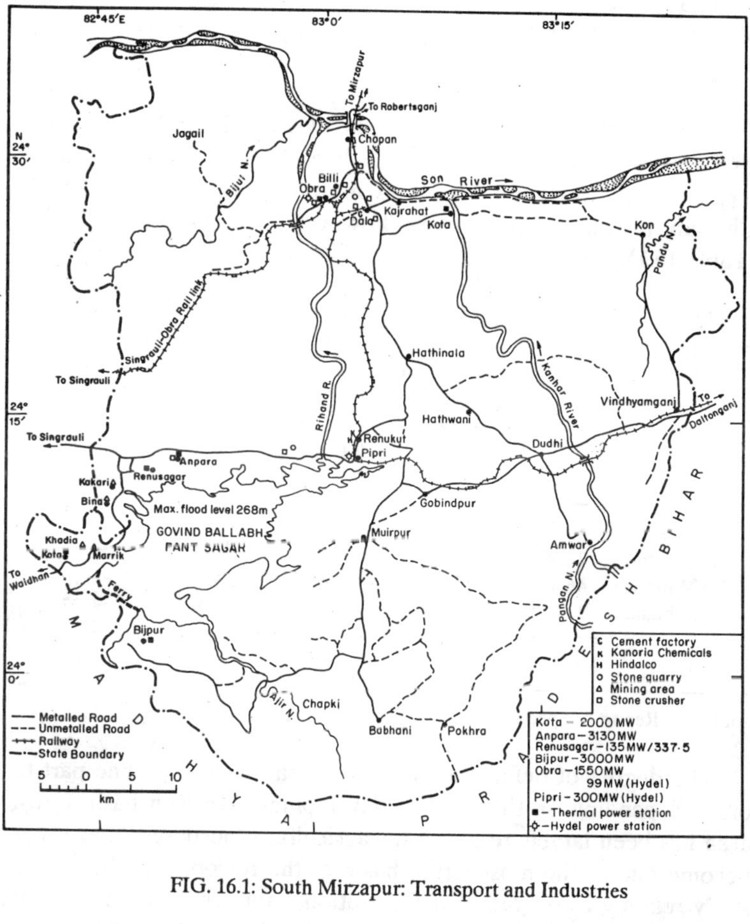

FIG. 16.1: South Mirzapur: Transport and Industries

human activities associated with rapid industrialization have resulted in large scale deforestation. At many places man and cattle have helped in converting the natural ecosystems into Savana and marginal croplands. The story does not end here. The converted ecosystems are also under immense biotic stresses such as lopping for fodder and fuelwood, grazing etc.

Thus, deforestation has caused immense damage to the ecology of study area. At present (1988) only 40.20 per cent of the total area is under forest mostly of tropical dry deciduous type. The forest cover was more extensive earlier. Its coverage was 65.41 per cent in 1954 which came down to 51.80 per cent in 1967 (Yadav, 1984). Between 1977-88 there has been a further decline of about 3 per cent in the forest cover. Maximum variation in forest area is noticeable in Chopan block. Other blocks however exhibit little variation (Table 16.1).

TABLE 16.1: Change in Forest Area (1978-88)

Sl. No.	Block	Forest area (in ha)			Variation (1978-88)	
		1977	1988	Per cent of total geographical area	Absolute (in ha)	In per cent
1.	Chopan	55,176	51,730	37.32	3,446	–6.24
2.	Duddhi	16,099	16,076	36.02	23	–0.14
3.	Muirpur	46,035	45,691	48.65	344	–0.75
4.	Babhani	13,842	13,790	34.92	52	–0.37
	Total	1,31,152	1,27,287	40.20	3,865	–2.95

Source: Report of Special Assistance Programme, 1990, p. 68.

The degree of deforestation in the area varies from one part to other depending on the level of development. In Kon basin forest area has been largely removed for agricultural purposes. Now it has become one of the most fertile basin of the region. According to a study relating to degree of deforestation, more than 60 per cent of the area of Son-Kanhar interfluve is severely damaged, while it is alarming in the north-western part of the Son-Rihand interfluve along with various patches stretching over south-eastern part of the region (Yadav, 1984). It has been observed that forests provide livelihood and income to locals especially the tribals. Thus, deforestation or any reduction in forest cover directly hits their habitat and economy. Thus, in no way the traditional forest dwellers and users may be responsible for the destruction of forest but the commercial users of forests in form of timber-cutter, charcoal-burner and others have played havoc. Usually the contractors destroy more forest area than the auctioned. It, further, aggravates the situation. Besides, 40 thousand evacuees belonging to 8 thousand families of 108 villages of the most fertile Singrauli basin were resettled on Babhani and Muirpur platforms. Such rehabilitation has damaged the ecology of the area in two ways: (a) the destruction of flora and fauna after the submergence of the Singrauli basin, and (b) in the resettlement process, an area of 68.82 km^2 has been cleared of forests (Kayastha and Yadav, 1982). Thus, the ongoing development projects have caused ecological damage in two ways — firstly on the spot and secondly at the resettlement site. A number of projects are also destroying the ecology of the area by way of deforestation. Thus, it may be concluded that no one components of the environment has been possibly so much under stress as the natural vegetation.

In spite of all efforts made so far, the region has been able to maintain only 40.20 per cent of the total area under forest which is much below the national norm of 60 per cent laid in the National Forest Policy of 1952 for the mountainous and plateaus regions. Even in the forested area the vegetation density is not very high. It has been found that the crown density in Hathi Nala and its surrounding has decreased by 15 to 20 per cent during last ten years. It is evinced by the fact that Hathi Nala forest, age old abode of the lions and elephants has no trace of them now. Severe deforestation took place in Dudhi, Kon and Bijul basins. As a result of severe deforestation, droughts have become a frequent phenomenon. At times, the level of reservoir could not reach to maximum.

Surface coal mining in Singrauli area has caused extensive damage to the natural ecosystems with growing dumps of overburden (Table 16.2). According to a statistics Singrauli had acquired 9,508 ha forest land in a total acquisition of 18,383 ha. Three coalfields namely Bina, Kakari and Khadia are part of it. So it must have swallowed the forest area of this region also. In addition, legal and illegal stone quarries in Dala, Billi and Anpara ranges have been destroying the forest and landscape at a fast rate. Whatever remains today, is also under immense biotic stresses.

TABLE 16.2: Amount of coal reserves and overburden to be excavated dumped for certain coal mines of Singrauli coalfield.

Sl. No.	Coal Mines	Coal reserves (million tons)	Overburden (in million m^3)
1.	Bina	128.70	225.4
2.	Kakari	71.93	177.0
3.	Khadia	299.20	1110.0

Source : Singh, J.S. *et al.*, 1990.

Soil Erosion

Soil erosion has been identified as another serious problem in the study region. Excavation for mining, roads, canals, buildings etc. have added greatly to the mass movements and intensification of erosion in hill terrain. Due to limited agricultural land (31.05% in 1988) and poor irrigational facilities (1.43% in 1988), farmers of the study area always face serious shortage of fodder. To meet the fodder shortage, farmers depend on forested area. Lopping of young *shorea* foliage are used as fodder. In addition, heavy grazing by cattles (9.45 cattle units per ha) has destroyed the vegetative cover, paving way to soil erosion. The larger the grazing stock, the more intense would be the severity of erosion. The population of the cattle in the region is growing explosively (Table 16.3). Using the conversion factors,[1] the total number of animal units in the study region comes to about 1.40 million.[2] An ever-increasing pressure of grazing (9.45 animal units per ha) is wholly responsible for severe damage to the soil. Moreover, the soils of the overgrazed area are left with only 8-36 per cent clay content as compared to 50-69 per cent in the unaffected soils. The stability is impaired owing to the deficiency of binding materials (Haigh, 1982). The nutritional deficiency in clay diminishes the productivity of soils.

It has been observed that 25 per cent of the total area under Rihand-Kanhar interfluve has been severely affected from soil erosion. Similarly about 40 per cent of the area lying to the west of Rihand-Son confluence is being affected from moderate to severe

1. Conversion factors : Cow = 4, Buffaloes = 6, Sheep = 1, goat = 1.5

2. Carrying capacity : 2 animal units per hectare

degree of soil erosion. The eastern side soils of Kon basin is also severely eroded.

TABLE 16.3: Grazing animals

Animals	1951-52	1982	Animal units after conversion	
			1951-52	1982
Buffaloes	4,848	42,644	29,088	2,55,864
Cows and bullocks	54,964	2,43,260	2,19,856	9,73,040
Sheeps and goats	12,289	1,27,270	17,819	1,86,017
Total number of animal units			2,66,763	14,14,921

Source: Statistical Diary, District Mirzapur, 1988 (in Hindi)

Besides, numerous development activities were initiated after the construction of Rihand dam at Pipri in 1952. Pant Sagar alone engulfed 344 km^2 area which includes good cultivable land, forest area and settlements. Similarly, other projects also joined hands in playing havoc. Levelling of undulating land, cutting of trees, construction of roads, blasting etc. all added to the magnitude of soil erosion.

From the foregoing it appears that before execution the development process has not taken into account the environmental and socio-economic costs and hence suffers from a serious planning lacunae. Therefore, all development must be well managed and should be made accountable in the local and regional perspective.

Road Construction

Dendy (1968) stated that the amounts of sediments eroded away from sites of road construction is ten times greater than the amount derived from agricultural land, and about 200 times greater than from grassy area and 2,000 times greater from the forested area. Construction of roads disturbs the stability of the hillside, inflicts serious damage to the hydrologic system and removes the protective cover from the vulnerable slopes. In the study region the total length of metalled road in 1986-87 was about 352 km (Table 16.4), which is likely to increase in future with the development of the region. It has engulfed thousands of trees so far and in this process many areas have faced slope failures. Only within a stretch of 14 km of the road constructed from Banvashi Ashram to Renukoot about 1,500 trees were removed which resulted into ecological damage of varied dimensions in the area.

TABLE 16.4: Length of Metalled Road, 1986-87

Sl. No.	Development block	Length (in km)
1.	Chopan	97
2.	Muirpur	114
3.	Dudhi	83
4.	Babhani	58
	Total	352

Source: Statistical Diary, 1988, District Mirzapur.

The excavation of one km stretch of road requires removal of 40 to 80 thousand m^3 (average 60,000m^3) of debris (Singh and Singh, 1988). Therefore, the construction of about 352 km metalled road in the region generated about 21.12 million m^3 of debris which after sliding down not only destroys vegetation but also clogs water course in the down slope. Besides, the construction of unmetalled road has also generated huge volume of debris in the region. To reduce the severity caused by road construction, plantation along the roads and slopes should be made along with bunding of roadside walls.

Environmental Pollution

In the area under study, main sources of pollution are Dala Cement factory, thermal power plants, Kanoria chemicals, HINDALCO and stone crushers which emit gaseous and particulate matters. Besides, liquid effluents discharged into Pant Sagar and Rihand river have also affected the water quality which in turn endangers the biological life.

There has been well-marked pollution in the area around Dala Cement factory and thermal power plants. The deposition of cement dust in soil and vegetation has encrusted the soil with an impervious layer and damaged foliage and reduced transpiration and photosynthesis processes. Soil is gradually being alkaline. The problem of air pollution is felt to such an extent that even the rural inhabitants living within a radius of 10 kms around the Dala Cement factory continue to suffer. Once they demonstrated at the factory gate against uncontrolled emittance of cement dust from chimneys of the dry process cement producing plant installed about eight years back (Singh, 1988). The gravity of the problem can be judged by the fact that during the rainy season, the condition of the ambient air with respect to SPM was found a little disappointing in the residential areas of Dala Cement factory (282-394 $\mu g/m^3$). Further, the survey reveals that the cement dust moves upto Kon which lies more than 20 km away. It has been found that cement dust badly affects Rabi crops. As there is no rainfall during winter, the deposits are not washed away.

HINDALCO, aluminium factory, causes noticeable fluoride pollution which has toxic affects on human, animal and plant life. The maximum and minimum fluoride levels are observed as 423.2 and 20.5 ppm respectively in the north-east at the distances of 0.5 and 15 km from the factory (Pal, 1974).

Kanoria chemicals manufacture gamaxine, bleaching powder and hydrochloric acid. Its effluents create deep foul and abnoxious smell which make the living of workers in nearby areas quite difficult.

Similarly, the thermal power generating plants located at Obra, Renusagar, Shaktinagar, Anpara and Bijpur are adversely affecting the air quality as these units use Singrauli coal which is characterised by high ash content (30%). Though the large height of chimneys and installation of electrostatic precipitators have lessened/arrested the gaseous air pollutants yet the expansion of these units are likely to increase the quantum of CO_2 in the air in near future. Due to high ash contents, large quantity of fly ash is sluiced daily from each plant and hence their dumping grounds create a big problem of air pollution in the form of suspended particulate matter especially during the summer season when the ash is transported farther away by fast blowing winds.

Mining and quarrying have also contributed to the pollution load in the region. The opencast mining at Bina, the first coal mine of Uttar Pradesh, is causing immense air pollution. In a recent air pollution study (Singh, 1990) it was found that the gaseous pollution was much below the risk level but the level of SPM in the residential (summer : 477.57 $\mu g/m^3$ and winter : 442.85) and mining areas (summer : 928.86 $\mu g/m^3$ and winter : 789.20 $\mu g/m^3$) were much higher than the recommended standard of 200 $\mu g/m^3$ and 500 $\mu g/m^3$

of ambient air for the above mentioned areas. Average figures of SPM are found much higher in summer than in winter due to high wind velocity and virtually non-existent natural vegetation. In addition, a large number of stone crushers operating in Billi, Jungail, Dala and Anpara areas are also causing air and noise pollution in a big way.

Singh *et al*. (1990) have come across the conclusion that the industrial development in Obra-Renukoot and Singrauli area has adversely affected the air quality, plant expression and soil characteristics within 8-10 km in prevailing wind direction whereas within 2-4 km in other directions from the source of emission. In their view the fallout of pollutants is mainly governed by the prevailing wind direction.

Remedial Measures

This study, thus reveals that the South Mirzapur region is developing rapidly in the field of power generation and coal mining which will cause ecological damage to the plateau's, environment. Therefore, urgent remedy is the ecological conservation and management programmes. The whole area may be divided into ecological risk zones on the basis of physiography, natural vegetation cover, land use pattern, population, severity of soil erosion, rate of deforestation, rate of air and water pollution and other related factors. It will help in taking decisions for improving the ecological conditions of the area. Afforestation programmes and social forestry can maintain and enhance the forest cover. It will inevitably strengthen the economy of local population. Large barren tracts can be brought under plantation schemes through organised efforts. It will also check soil erosion to a greater extent. Despite so many problems, the perception of villagers regarding environmental and forest degradation is vague and poor due to differences in literacy, resourcefulness and exposure.

Thus, it is imperative to take appropriate steps to reverse current development styles and philosophy of conflict with nature, otherwise no one can save the region from the multiplying environmental damage. Therefore, policy of economic growth should be such that it may cause least ecological harm. To achieve this goal, continuous monitoring of different components at certain interval and assessment of results of already ongoing works is a must Khoshoo(1988, p.4) has rightly cautioned that unless concerted and unified efforts are made now, tomorrow may be too late. Therefore, at all levels of society environmental education and awareness must be imparted and inculcated.

REFERENCES

Dendy, E.E. (1968), Sedimentation in Nation's Reservoirs, *Journal of Soil and Water Conservation*, 23 (4), pp. 135-37.

Haigh M.J. (1982), Soil Erosion in India, *Discussion Paper in Geography*, Polytechnic, Oxford, 17, pp. 1-124.

Kayastha, S.L. and R.S. Yadava (1982), South Mirzapur Region : A Geographical Analysis of the Development Process. *The National Geographical Journal of India*, Vol. 23 (3 and 4), pp. 113-26.

Khoshoo, T.N. (1988), *Environmental Concerns and Strategies*, Ashish Publishing House, New Delhi, p. 9.

Pal, D. (1974), Effects of Fluoride Pollution on Plants and Cattle, *Unpublished Ph. D. Thesis*, Department of Botany, B.H.U., Varanasi.

SAP (1990), Progress Report of the U.G.C. Special Assistance Programme (August, 1988 to February, 1990) in Department of Geography, p.16 (Quoted from Jackson and Fisher, 1883. *Statistical, Descriptive, Historical Accounts of North-West Province*, Vol. 4 (Part 2), Mirzapur, p. 367.

Singh, J.S. *et al.* (1990), *Environmental Degradation of Obra — Renukoot-Singrauli Area and its impact on Natural and Derived Ecosystems*. Centre of Advanced Study in Botany, B.H.U., Varanasi, p. 19.

Singh, Onkar (1988), Pollution Survey of Obra-Dala Area in South Mirzapur Uplands. Abstract published in National seminar on *Environmental Management in Hills and Tribal Areas*, NEHU, Shillong, 14-17 Oct.

Singh, Onkar (1990), The Nature of Industrial Pollution in Sonbhadra District, U.P. Paper presented in National Seminar on *Fast Industrialising Areas : Impacts, Issues & Challenges*, March 21-23, Department of Geography, B.H.U., Varanasi.

Yadav, R.S. (1984), Environmental Problems and Eco-Development Sub-Regions in South Mirzapur, *The National Geographical Journal of India*, Vol. 30 (4), pp. 259-68.

Singh, D.S. and A.K. Singh (1988), Development and Environmental Crisis in Kumaon Hills (Unpublished).

Singh, H.S. and A.K. Singh (1989), [illegible] (unpublished).

17

ECOLOGICAL PROFILE OF KHETRI COPPER COMPLEX : A Study of Environmental Crisis

INDRA PAL

With scientific and technological advancement copper has become more important and valuable metal in modern times. Its demand and consumption is increasing for making wires, alloys, auto ancillary, electricals and electronics and others items (coinage, defence, railways, chemicals, man-made fibres, die casting, etc.). Copper is obtained from Rajasthan and Bihar which are the main producing states in India. The former has more than 22 per cent of the total estimated reserves in the country, but during the five years ending 1980, it had accounted for more than 46.7 per cent of the gross national production.

The copper are having chalcopyrite as the principal copper mineral, is of too low a grade to be smelted directly and, therefore, they are concentrated by floating process to separate the copper minerals from the waste rock. The enriched concentrates are smelted to produce copper which is further refined by electrolysis to get unfinished product as blister copper about 90 per cent pure.

History

Copper mining in India may be traced to the Mauryan period. It is also referred in Ain-i-Akabari. Even during the British times, two mints had been working at Singhania and Khetri, both in Rajasthan. These were closed in 1869. Though regular mining ceased in 1872, the sporadic activity continued until 1910. After independence, new sources of copper were searched for to meet the steadily rising demand, so prospecting and exploratory activities began in the fifties. In 1967 the Government of India set up the Hindustan Copper Ltd. for the development of a copper complex in Khetri. It started with copper mining in 1970. Then came a concentrator plant on stream in 1973, followed by a smelter in 1974. To utilise its waste product, sulphur dioxide (SO_2), a sulphuric acid plant was installed, and in order to consume the sulphuric acid, a fertilizer plant was established, commencing its production in 1975. Thus the copper ore mining, copper smelting, electrolytic refining plants, sulphuric acid and fertilizer plants came up in succession, one after the other along with the auxiliary plants, namely, the rock crushing, rock grinding, mixing, granulation and roasting plants and casting moulds, as well as, the infrastructural captive power plants and stream generation plants. Thus the Khetri Copper Complex (KCC) was set up in the Jhunjhunu district of Rajasthan, about 11 km northeast of Khetri town, 190 km southwest of Delhi and 180 km north of Jaipur. The complex is 370m above the mean sea level.

Geology

The rocks in the Khetri area belong to the Delhi Super group of the Pre-Cambrian age, sub-divided into Older Alwar and Ajabgarh groups (Fig. 17.1). All these formations are metamorphosed to quartzites, schists and phylites. Some intrusive dolerite dykes are present and veins of quartz and carbonates are common.

The general strike of the formation is NNE-SSW. The strike length of the Khetri copper deposit is about 3.6 km while the larger deposit extends in a belt of 76 km from Singhana to Raghunathnagar. The estimated ore reserves are about 47 million tonnes at 0.9 per cent copper grade. The Kharkhara river flows in a valley immediately in the hanging wall of the mineralised rocks in the Khetri mine area.

Plant Capacity

The Khetri copper complex has a design capacity of 31,000 t/year but it is producing only 21,000 t/year electrolytic copper. The sulphuric acid plant has a design capacity of 577 TPD of sulphuric acid and the fertilizer plant is designed to produce 9,000 TPD single/triple super-phosphate. The consumption of basic feed materials in different manufacturing processes is as follows:

	Manufacturing units	Consumption of feed materials
1.	Copper manufacturing plant :	
	Copper ore	8,330 TPD
2.	Sulphuric Acid Plant :	
	Sulphur bearing gases	90,000 NM^3/hr. (as per design)
3.	Fertilizer Plant :	
	Sulphuric acid	577 TPD
	Rock phosphate	900 TPD
4.	Captive power generation :	
	Diesel (HSD)	100 KLD
5.	Boiler plant :	
	Furnace oil	5.6 T/Litre

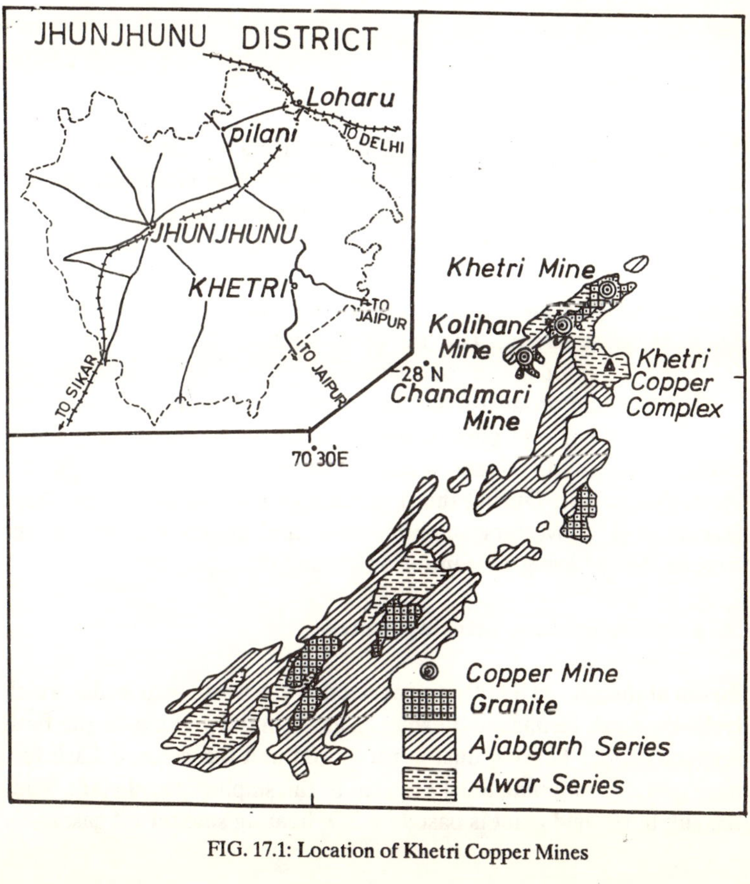

FIG. 17.1: Location of Khetri Copper Mines

Mining

There are four mines in the Khetri copper complex. The mines at Khetri and Kolihan with maximum capacities of 5,000 TPD and 3,000 TPD respectively, are the two of the largest underground mines in the country. Khetri mines has been reported as pioneer in massive underground ANFO blasting to the extent of 14 tonnes of explosive at a time, functioning automatically and operated by remote control. The Kolihan mine has the richest copper deposit so far known in this belt with 1.52 per cent average copper content. The underground mine at Dariba in Alwar district is a small mine with a capacity of 100 TPD. The fourth mine with a capacity of 1000 TPD is an open pit mine at Chandmari situated at about 2.2 km south of the Kolihan mine. The pit is 500 m long, 200 m wide and has a maximum depth of 148 m.

Copper Manufacturing Processes

The ore is primarily crushed to 150 mm size and then to 20 mm size through secondary and tertiary crushers and after classification, sent to the flotation section. The final concentrates are filtered and then final tail after reclaiming the water are sent to the tailing ponds. Blended concentrates contain 16 to 22 per cent copper and are fed into the flash furnace. Here the molten 'matte', slag and gaseous products are separated. The 'matte' contains 45 to 50 per cent copper (Cu) and is treated in the converters.

The finished product from the converter is known as blister copper about 98 per cent pure. It is sent to the anode furnace for anode casting. The anode copper is refined to 99.9 per cent copper in the electrolytic plant. During the treatment of concentrates in the flash furnace and converters sulphur burns and comes out as sulphur dioxide (SO_2) having concentration around three per cent.

Sulphuric Acid Manufacturing Process

Steam at the rate of 33 t/h is generated in the waste heat boiler which cools the flash furnace and exhausts gases in the process. In the flash furnace copper bearing dust is recovered and the cleaned flash furnace gases (containing SO_2) are sent to the sulphuric acid plant. Thus the sulphuric acid plant is based on SO_2 bearing smelter off gases.

Fertilizer Manufacturing Process

In the fertilizer plant phosphoric acid and super phosphate are manufactured, both are based on sulphuric acid as their raw materials. The rock phosphate of required fineness and dryness and the sulphuric acid of required intensity are reacted in the mixing plant. And the mixed product is allowed to cure for a period of 7 to 14 days and subsequently changed into granular product of size — 6 + 14 mesh, confirming to the quality as per the Fertilizer Control Act.

Waste Generation

There is hardly any activity which does not create waste, and waste is therefore invariably associated with all kinds of industrial operations and processes. Thus ore processing is no exception. Since the K.C.C. has grown into a large diversified industry, it generates varied kinds of wastes, liquid, solid, and gaseous wastes and introduces them into the environment.

Liquid Wastes

Large quantities of water is generated in deep mines by seepages. On an average about 97 m^3 of water is pumped out daily in the Kolihan mine, about 4,000 m^3 from the Khetri mine and some quantity from the Dariba mine. A part of this is used in mining operations for drilling bore holes to facilitate blasting, concentration and beneficiation of ores, and the rest overflowing the tank, is discharges as waste into the Kharkhara river. In the Chandmari mine where open caste mining is practiced, the problem of seepage is not encountered. Besides mining, waste water is generated in all the manufacturing units. It is generated on account of furnace cooling, heat exchanger, turbine pump cooling, boiler blow down, cooling towers, slag cooling, etc. and after various operations flows into the main drain. Although this water is relatively free from metal contaminants, but contributes load in term of suspended solids, dissolved solids and sulphates.

The waste water from the wire bar plant on account of furnace cooling, as well as, from the nearby deminerlisation plant is combinedly discharged into the drain. It is acidic in nature and relatively free from metallic contaminants. The combined flow from these plants amounts to 15-20 m^3/hr. Likewise, the sulphuric acid and the fertilizer plants also contribute significant amount of waste water by spillages from cooling tower, acid coolers, heat exchangers, acid sporadic floor washes, etc. It is also acidic in nature and contains suspended impurities. The quantity of waste product from the sulphuric acid plant alone is approximately 200 to 250 m^3/day.

The phosphoric acid plant produces scrubbed liquor. During the reaction of sulphuric acid with rock phosphate fluoride is evolved. The quantity of liquid waste from this source is about 250-340 m^3/day.

During the granulation of the fertilizer production, very fine particles get diffused in the air and these are trapped in a scrubber. The scrubbed liquor mainly contains suspended solids, phosphates, and fluoride. The quantity of this waste is estimated to be 400 m^3/day.

Solid Wastes

Mining operations also generate solid wastes. The ore is subjected to primary crushing in the mines and it is further crushed into fine particles in the crushers. Here very fine particles are diffused into air but as they pass through the rotoclone, these particles are trapped with water. During normal course when there is no power tripping, this water along with trapped ore particles is collected into a pit and thereafter pumped to the milling section. But during power trippings there is huge backflow from the milling section, which is collected in the pit. Owing to insufficient capacity of the pit, the entire slurry overflows into the main drain. The quantity of this flow per tripping is 75 m^3. It is estimated that a total loss of copper per day due to this overflow is about 450 kg. The milled ore is further subjected to floatation where ore concentrates float at the top and the tailings settle in the bottom. The total amount of tailings entering the main drain is about 195 m^3/tripping. The result of analysis indicate that in the rotoclone underflow, metallic impurities like copper and nickel are predominant whereas the tailings backflow contains only solids carrying as high concentration as 11 per cent.

The anode copper is electrolytically refined in the refinery plant where copper sulphate and sulphuric acid are used as electrolyte. The copper from the anode is deposited on cathodes and the impurities nobler than copper remain in the solution. When the strength of the electrolyte is depleted this is rejected as waste. The maximum quantity of the spent electrolyte is 30 m^3/day. This waste contains copper and nickel which are highly acidic in nature. It is estimated that about

220 kg of copper and 6 kg of nickel is wasted per day in this effluent.

The phosphoric acid plant produces gypsum as solid waste to the order of 1,000mt/day. It is on account of the reaction of the sulphuric acid with rock phosphate resulting in the evolution of phospho-gypsum slurry. This slurry is centrifuged to recover phosphoric acid and produce gypsum.

Waste Management and Pollution Control

It is now amply clear that almost all the industrial processes in all the units of the KCC produce tremendous amount of solid, liquid and gaseous waste materials incessantly day in and day out. Most of them create environmental crisis for the poor people of the surrounding area where agriculture is the only occupation. In order to maintain good quality of life, industrial waste management and environmental pollution control has become the concern of the modern society. Waste management does not exclusively relate to the pollution control aspects but also explores the possibilities of minimising the waste at source by process optimisation, conservation measures like recovery of useful elements and recirculation of wastes. With rapid depletion of natural resources, several scientific advances have brought about many methods for alternative uses and recovery of byproducts from industrial wastes.

The quantity of sewerage generated in the complex is of the order of 4 million litres per day. It is collected in a sump and treated in oxidation ponds located about one km from the township. The overflow from the ponds is used by farmers for irrigation.

In the underground mines dust is generated in spitc of wet collaring and drilling. In order to overcome this problem ventilation must be provided to keep dust concentration down to safe level. It has been reported that in most instances power ventilation has become necessary to remove nitrogen oxide and carbon monoxide generated from blasting and also to control heat and humidity. The industry has already provided two blowers on the reclaiming conveyor terminal. It has also provided three rotoclones to control the dust pollution from the screen crusher conveying system, one at the angle station and one on the fired or bunkers. The use of water in drilling and wetting the ore in open pit mining and over waste piles will go a long way in controlling the pollution. Water should be sprinkled before loading the material into trucks and railway cars for transport from the mines. Roads may be sprinkled to reduce dustiness, and ores should be wetted while crushing.

Eighty-five per cent of the ore fed to the milling circuit is rejected as tailings which contains 30-50 per cent solids and traces of metals like copper and iron. This waste is being used for road and building constructions and to fill up the low lying areas. It should be advantageously used to fill up the mined out areas, though this kind of solid waste does not pose significant environmental repercussions. Plantation of trees and shrubs and cultivation on ground cover may be adopted to act as wind breaks.

The entire copper complex contributes about 8.4 ml/d waste water mainly arising out of the concentrator, smelter and refinery. The character of the waste water corresponds with the flow rates which is affected due to power cuts. In such circumstances the average flow of about 350 m^3/hr shoots to as high as 1,000 m^3/hr mainly due to tailing thickness and pump tripping. Suspended solids, copper and iron, are the main contaminants with traces of nickel, zinc and lead. The greenish colour is imparted to the effluent by refining waste which discharges spent electrolyte solution.

It has been seen that in the absence of adequate storage facilities in the copper manufacturing section, large quantities of concentrated slurry is wasted when backflow takes place due to power failure and pump trippings. This kind of wastage can be curtailed by managing the supply of power without cuts on the one hand and by providing adequate storages on the other.

The quantity of waste in the refinery though comparatively less, presents both the alternatives of recovery of useful elements as well as recycling by purification of the electrolytes as is being practised all over the world but not in the Khetri copper complex. The rejected electrolyte is very rich in copper, nickel and iron, and therefore, justifies their recovery. Even otherwise, the cyclones and electrostatic precipitators can be used fo: the collection of copper dust and fumes generated from roasting, smelting and converting.

The atmospheric pollution in the industrial complex environs creates the most hazardous situation. Its cleaning problem is equally challenging. This problem can be realised from the fact that the mass of air handled daily in a smelter is far greater than the mass of solid materials. The latter may be several thousand tons per day, in addition to the potential for industrial emissions of particulate matter, some of which could be toxic. There is a concomitant high production and emission of sulphur dioxide.

The copper concentrates contain as much as 30 per cent sulphur.

When they are treated in the flash furnace to provide 'matte' (40-50% Cu) for further refinement progressively in converters and anode furnaces, sulphur burns and comes out as sulphur dioxide (SO_2) as a result of oxidation. Thus sulphur is a mixed blessing. Its oxidation provides valuable process heat and conserves energy. It is also an useful raw material for sulphuric acid production in areas where acid can be used economically. But it is an onerous problem because it causes extensive vegetation damage nearby. Hence its maximum proportion must be consumed to produce sulphuric acid and to reduce air pollution. But acid plants entail heavy capital investment and substantial operating cost. In the United States, the Clean Air Act of 1970 forced many smelters to install acid plants regardless of economic consequences. Electric smelting reduces sulphur oxides emission, but the region suffers from power deficits. At the top of it if the sulphuric acid plant is not in operation, the entire 90,000 m^3/hr of sulphur dioxide has to be bled off through a 120 m tall stack. As per original design of the smelter plant the surplus SO_2 to the tune of 60,000 m^3/hr is discharged untreated from the main smelter stack into the environment.

In a couple of newly developed processes sulphur is converted either to elemental form or to sulphate, and hence an important advantage is that sulphuric acid disposal is avoided. Such processes avoid air pollution by sulphur dioxide but require careful evaluation with regard to liquid effluents and solid disposals.

In Hayden and Arizona in U.S.A. tall stacks up to heights of 305m have been provided to enable sufficient dispersion thereby keeping the ground level concentration to a minimum. In order to control air pollution, all the three manufacturing units in the Khetri complex have been provided with several cyclones, scrubbers, bag houses, electrostatic precipitators and 39 tall stacks which minimise pollution effects in one form or the other. Fifteen of the stacks have been provided with the diesel generating sets. There are two stacks in the smelter plant and one in sulphuric acid plant.

One stack is provided for the removal of combustion gases from the preheating furnace, which is mainly CO_2, produced on account of the heating of the furnace with the help of the burners. The second stack is used to take out off gases from the arc smelting furnace. In this furnace a wet scrubber is provided between the stack and the arc furnace to collect copper oxide and copper dust which comes out of the furnace. A tall stack has been provided to emit 2 per cent dust amounting to 36 kg/hr generated in the grinding plant untrapped by dust collectors.

From the processes described earlier it may be understood that fluoride contamination occurs in a wide range particularly in the manufacturing process of phosphoric acid and super phosphate. During the manufacturing of phosphoric acid fluorine is liberated in the gaseous form which passes through the wet scrubber to form $H_2S^i F_6$ solution. However, the unscrubbed fluorine is vented through exhaust fan at a 45 m high stack. The main consideration for reducing pollution due to this waste are acidity and suspended solids. The acidity of the waste can be neutralised with lime and the suspended solids can be settled in a well designed classifier. Use of coagulants are recommended where colloidal impurities are predominant.

The main consideration in the wastes produced from these operations is the reduction of fluorides to the tolerance limit. The concentration of fluorides for phosphatic fertilizer has been fixed at a concentration of 15 mg/l and the generalised limits for all other industries is 2 mg/l for discharge into waters. It has been seen in the monitoring results that fluoride concentration in the ground water is persistent even though most of the results are within permissible limits of standards for drinking water. The persistence of fluorides in the ground water can largely be ascribed to the discharge from the fertilizer plant particularly of those monitoring stations which are located downstream from the final discharge point. For the reduction of fluorides, the most widely used method is precipitation of fluorides. The neutralisation of hydrofluorosilicis acid with lime fluoride is precipitated as calcium fluoride. The precipitation of fluoride with alum is used to remove more fluorides ensuring better quality of effluent. The aluminium sulphate binds the aluminium and fluoride ions as a complex. It is therefore reported that activated aluminia can be used successfully to reduce the fluoride concentration to 1 to 2 mg/l.

The wet process for phosphoric acid production generates about 4.8 tonnes of phospho-gypsum per ton of $P_2 O_5$ produced. Its total production is estimated to be 1,000 tonnes per day for the installed capacity of phosphoric acid plant at Khetri. One can get rid off this wastage by using it advantageously for the following purposes:

(a) As soil conditioner
(b) As an additive to cements for producing super sulphated cement.
(c) To produce ammonium sulphate by reacting it with am-

monium carbonate.

(d) To produce phospho-gypsum as hydrite plaster as building material.

(e) It can alternatively be used for reclaiming lagoons.

Environmental Impact Assessment

Environmental impact assessment has become an increasingly important sector for curbing pollution and minimising environmental disruption. But the environment is unbelievably complex. It encompasses not only the more obvious areas of air and water quality and ecology but less tangible areas as well, such as sociology, economics, land use and aesthetics. No individual possesses the skill and the range and depth of knowledge to adequately address all aspects of the environment in an environmental impact analysis. Environmental impact analysis is a study of probable changes in the various socio-economic and bio-physical characteristics of the environment which may result from a proposed or impending action. It can be assessed with the assistance of experts from varied disciplines and technologies.

Environment consists of both natural and man-made attributes (elements), interacting in complex relationships among the attributes. Changes in the attributes of the environment and inter-relationships are defined as impacts. These attributes are : (a) air, (b) water, (c) land, (d) ecology, (e) sound, (f) human aspects, (g) economy, and (h) resources.

In a sense, the set of attributes must provide a model for the prediction of all impacts. In order to determine the impact on environment the following steps have to be undertaken:

— Identification of impacts on attributes.
— Measurement of impact on attributes.
— Aggregation of impacts on attributes to reflect impact on the environment.

It has been found that since the copper industry at Khetri grew into a large complex, the total industrial processes started drawing various inputs from the environment and introducing some outputs to the environment. However, the copper plant is still working at its two-third designed capacity and it is proposed to expand it to its full capacity. Hence the organization has sought the permission of the

Rajasthan Board for Prevention and Control of Pollution. But the government, after the 1984-Bhopal tragedy, has made more stringent rules for allowing the expansion of the existing units or installation of new ones which may prove hazardous for the society. The Board now insists on a feasibility report to evaluate and assess the totality of the attributes. It includes: (a) Landuse Survey within 20 km radius surrounding the industrial complex encompassing socio-economic aspects, flora and fauna, human habitation, employment opportunities in the vicinity, national monuments, historical and religious places, industries and industrial estates, ground and surface water hydrology, agriculture and soil quality, land use map for the 20 km radius; and (b) mathematical modelling: environmental impact assessment with a view to predict ambient air quality, monitoring under average and worst meteorological conditions. The model should consider the shock pollutional loads, leakages, accidental emissions due to unforeseen events, also dynamic effects, micro-meteorological conditions, and topography, etc. The stability classes of environment should be worked out on the basis of practical observations of vertical temperature profile and stack monitoring and analysis. It is only after being assured that the development and expansion of the industry would not bring about pollution beyond recognised standard limits, the permission is issued.

The KCC has its own research and development department which takes care of the several aspects of industrial pollution and regularly monitors solid, liquid and gaseous polluting attributes. In addition, the organisation had sought the services of the National Environmental Engineering Research Institute (NEERI) in 1984 to monitor emissions from the 10 of its 39 stacks for one year, collect the data and give an analytical report. The report had suggested both long term and short term measures to control environmental pollution and also to install additional equipment wherever necessary to prevent the same.

As referred to earlier the entire waste from the KCC is discharged into the Kharkhara river which is a non-perennial river having virtually no surface flow except during the monsoon period. The emphasis of assessments, therefore, has to be concentrated on the ground water quality. For monitoring purposes 8 stations (wells) were selected at strategic locations to assess the impact on water quality. One station is selected nearby the KCC and a few locations down stream at certain intervals and the last one where the effects of effluents are completely diminished

The results of monitoring of these wells over a period of time indicate that the first station located above the effluent discharge point, does not show any contamination and the water conforms to tolerance limits prescribed for the drinking water supply. Well number two indicates abrupt increase in the dissolved solids, total hardness, chlorides, and sulphates which do not conforms to the acceptable limits. The wells feeding supplies to the KCC are relatively free from contaminants even though the total dissolved solids and hardness are not within acceptable limits, but these still can be tolerated in the absence of a better source.

The impact studies carried out so far indicate one striking feature that none of the water sources has been contaminated by metallic pollutants. Even though the metals like zinc and copper are reported but the contaminants fall within the limits. The impurities are in insoluble form and tend to settle in the drainage system or in such zones of the system where the flow conditions are quiescent. Hence the river bed is the most sensitive zone inflicted with metals pollution and in due course of time the adverse effects in the ground water quality, which is already showing signs of deterioration, are very imminent. This zone structures over a distance of about 8 km from the final point of discharge at the KCC.

The Hindustan Copper Ltd. is undertaking all the short term and long term measures with regard to the environment's potential degradation. The installation of a Rs. 3.64 crores treatment plant is a step in the same direction. It would also act over the feasibility report regarding the environmental impact assessment including the land use survey and the mathematical modelling. Thus it would achieve its objective of functioning at its full designed capacity and avoiding any environmental crisis.

18

AGRICULTURE AND ENVIRONMENTAL POLLUTION

LALLAN RAI

The civilised world is at present passing through an anti-intellectual or rather an anti-scientific phase. It has willingly accepted with pleasure the technological renovations and social benefits of modern scientific break-through, but it has not been willing to accept the fact that we are going through a revolution which is bringing changes not less stringent than the first industrial revolution about 100 years ago. Moreover, the world is not willing to accept the fact that we are not always able to predict the changes that will or may occur in such a situation.

Now there is a more serious argument which is used against science and scientist, namely, that the order of discovery of scientists and the greed of the industrialist are destroying the environment in which we are living, that the development of science and technology has disregarded the side-effects which are now threatening environment and the very existence of humanity. We are told that the discoveries in the field of atomic energy have increased the radioactive background in atmosphere. In our food, the antibiotics have led to the development of resistant bacteria against which we are helpless. The automobile industry and the industries of plastic materials and packaging are creating unmanageable problems of waste disposal. The pesticides have not so much controlled the pests as they have destroyed other forms of wildlife and entered the food web of living organism. The use of chemical fertilizers is turning many cultivable lands deficient in some vital trace minerals and brought in undesirable physico-chemical changes, and so on. The chemical fertilizers have been polluting water.

This argument disregards one fundamental fact that there is no such thing as a defined environment, a natural ecological equilibrium, at least since the human race appeared on this planet. Everything it did, every invention it made and every step of its progress have changed the human environment by its ecological impact. It is extremely difficult to assume that any change would be wholly good or wholly bad. It is the duty of the scientist and planner to weigh the good and the bad against each other, to the extent they are capable of. We have to recognise that this unfortunate limitation exists, and it undoubtedly, puts on the scientist and the industrialist the heavy responsibility of updating constantly their knowledge on this point. Perhaps, it would not be right to draw the conclusion that the best we could do in order to preserve the environment is to stop completely scientific injury and technological development. It would also not be right thing to say that 50 years ago our environment was natural and pure and now it has deteriorated through the scientific renovations.

There can not be a better example to demonstrate these points than the field of pesticides and chemical fertilizers, which are generally called agricultural chemicals. Pesticides, because they are toxic substances, are potentially hazardous to human and wildlife, yet very essential so as to protect agricultural production. Likewise, fertilizers, though not toxic chemicals but have some adverse ecological impact on the soil and its micro-organisms.

The annual report of the United National Population Fund (UNPF) states: "The earth is capable of supporting the increasing needs of the next century" but only if the industrialized and developing countries cooperate effectively to reverse the environmental destruction. The UNPF's Executive Director said, "All countries must face the fact that the combination of population and industrial growth could destroy the soil, water and air on which everything else depends." It is obvious that in order to provide adequate food, fibre and timber and to attain greater stability of agricultural production men have been utilizing natural resources (soil-water-air) erratically and the resources are becoming more seriously out of balance.

In the semi-arid and tropics (SAT) of the world the country like India whose majority of population is dependent on agriculture, has no alternative than to exploit the limited natural resources, especially, soil, water, air, vegetation, etc. The extremely low per capita land available in our country has resulted in cropping pressure on soil, use of huge amount of agricultural chemicals, exploitation of irrigation resources, etc.

The soil and water are two major components of environment, hence any activity which is reducing usefulness of soil and water, in other words, is polluting them. If we look to the reality of the environmental pollution, human activity is polluting the agricultural components and on the other hand agriculture is polluting environment which we can say as "pollution to agriculture and pollution by agriculture". Here, let us analyse the problems related to the pollution of environment by agriculture. The main agricultural factors which are responsible for pollution of environment are as follows:

(1) Agricultural chemicals

- (i) Pesticides — Insecticides, fungicides, herbicide, bactericide, rodenticides, etc.
- (ii) Fertilizers — NPK fertilizers
- (iii) Soil amendments and conditioners carrying heavy metals.

(2) Irrigation system : Canals, reservoirs, tubewells (Salinity hazards of soils)

(3) Multiple cropping, monoculture, etc.

Ecological Impact of Pesticides

Let us examine the ecological impact of pesticides. Since biblical times, harms caused by insects, such as locusts have been a curse to mankind. Today, the annual worldwide damage to agriculture, by insects, fungi and noxious weeds, together with the losses caused by insects, affecting man and animal health probably runs into hundreds of millions of rupees. Of all the pests that attack agro-horticultural crops, insect pests are the most notorious and harmful.

For sustaining and increasing the agricultural productivity and the maintenance of health of man and domesticated animals, it is imperative that the insect pests are controlled. Although some crop pests have been artificially controlled in intensive agriculture by cultural techniques and by growing resistant crops during the last three decades, they have been controlled mainly by widespread use of pesticides. A recent FAO report indicated that, on the global basis, pesticides have contributed to over 47 per cent increased crop yields. In our country too an increase of crop yield to over 20 to 50 per cent has resulted due to the adoption of chemical pest control measures. Insect vectors of human disease have also been effectively controlled by pesticides, so that millions of lives have been saved and many more serious illnesses avoided.

In 1974, it was estimated, that over 6 lakh tons of technical grade pesticides were used in the world. In India the use of pesticides per year at present is estimated at about 75 thousand tons. This quantity is not large, considering the vast and extensive area of land under cultivation in the country.

Organochlorine pesticides such as DDT, BHC, dieldrin and heptachlor; and some other pesticides like mercurials, arsenicals, malleic hydrazide 2, 4, 5-T etc., are persistent and the use of some of these cause long-term ecological hazard in the environment.

Since the pesticides are toxic in nature they pose problems even during manufacturing, storage, transportation, handling and application. During the manufacturing process various types of toxic fumes, solid and liquid effluents are produced. If not properly controlled, they will pollute the surrounding atmosphere, soil and water ecosystems.

Extensive researches have shown that the persistent pesticides upset the ecological balance. Also they reach the body of human and domesticated animals either directly or through the simple or multiple "food-chain" links. Deposits of DDT, for instance, have been found in fatty tissues of human and animals, milk, eggs, meat etc. But DDT still continues to be the key insecticide in malaria eradication project throughout the world because of its killing efficiency of the anopheles mosquitoes which transmit the malaria disease and is available at low cost.

Rapid discoveries have been made in the last decade to develop better pesticides with selective action for killing the target pests, but not the beneficial parasites and predators which are the natural enemies of the pest. These selective pesticides are designed in such a way that they have very little or no side-effects on the environment. Many pesticides now available in the market are of short residual action. After use, they kill the pest quickly and get degraded or broken down into nontoxic ingredients within a few days of their application.

Fertilizer Use and Environmental Pollution

Undoubtedly chemical fertilizers have been acknowledged as vital input in both developed and developing countries to achieve the higher food production. The increased level of fertilizer consumption has, however, alarmed some of the environmentalists with the fear of environmental pollution. Fertilizers when applied to soils meet the nutrient needs of the crops, and there remains some residual quantity in soils. If we analyse the situation of input and output of nutrients in arable soils it is apparent that there are following three "leaks" in the cycle :

(1) Export of food from the farm to population (NPK, Zn)
(2) Leaching to the ground water (NPK) and removal through surface runoff.
(3) Volatilization to atmosphere.

Figure 18.1, shows nutrient cycle and 1, 2, 3 leaks of nutrients responsible for environmental pollution. Nitrogen is an abundantly used fertilizer in tropics and semi-arid tropics and their losses from soils by leaching and/or surface runoff are the reasons of environmental pollution. The leachates pollute groundwater and surface runoff is responsible for pollution of surface waters such as lakes, ponds, reservoirs and rivers. Nitrogen loss to groundwater deteriorates drinking quality of water due to nitrates ($NO_{3\text{-}N}$) which causes diseases in babies and possibly cancer in digestive tracts of older people. When $NO_{3\text{-}N}$ exceeds the amount-45 ppm in drinking water it is dangerous for health (Singh and Biswas, 1982). The N losses due to denitrification affects ozone layer by forming nitrogen oxide gas. The studies conducted in Punjab (1980-81) wherein N consumption was 115 kg/ha have shown that 90 per cent water samples from wells and tubewells were under upper safe limit. The phosphorus does not leak up to groundwater but it is lost through surface runoff and its accumulation in ponds, lakes, tanks, etc. may result in nuisance growth of algae. A concentration of 0.05 ppm of P supports abundant algae growth. A very small quantity of potassium can be lost by leaching to groundwaters, however, its surface runoff may be more and soil erosion only causes major K loss.

Fertilizers and Soil Amendments as Source of Heavy Metals

Lead, cadmium and arsenic are added to soils as incidental impurities. Some of the fertilizers such as calcium ammonium nitrate, superphosphate, rock phosphate, muriate of potash, DAP and

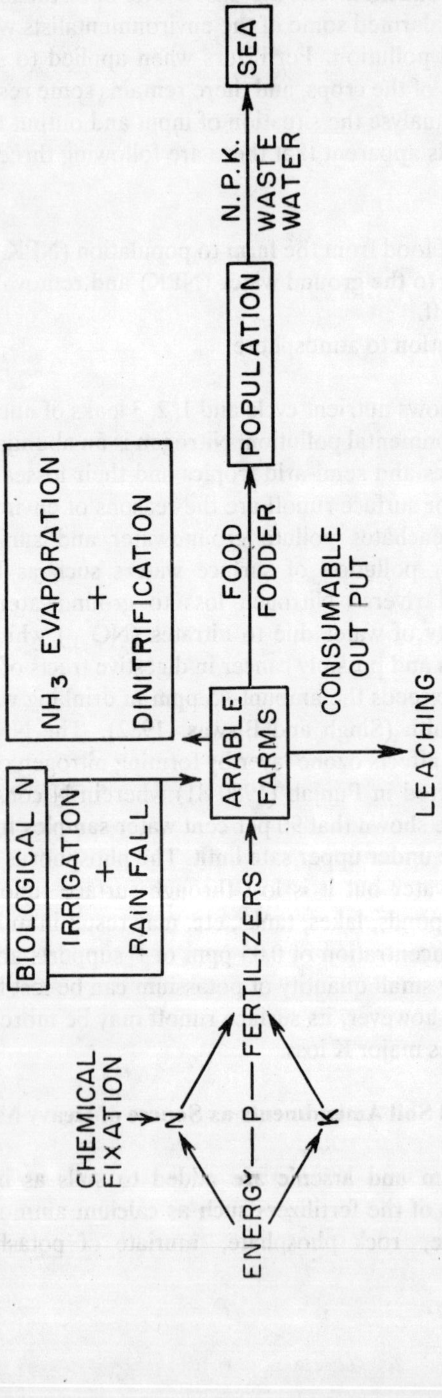

FIG. 18.1: The Nutrient Cycle of an Arable Farm (Source — Kolenbrander, 1982)

suphala have been reported to contain lead between 116 to 962 ppm and cadmium 1 to 300 ppm, which may contribute over a period of time to significant quantity of heavy metals which would while entering to food chain through soil, prove toxic to human beings and animals. At present the addition of small amounts of heavy metals have not caused much concern as the soil due to its dilution effects overcomes the hazards.

In our country gypsum and phospho-gypsum (bi-product of phospheric acid industry) are being used for the reclamation of sodic soils. They contain fluorine, some times to toxic levels. This may cause toxicity in plants and diseases in human beings and animals.

Apart from above mentioned factors more and more use of irrigation waters has resulted in world's about half of irrigated land in danger due to salinization. Irrigation has degraded the soils and soils are moving towards desertification.

One of the most important natural resources and a component of environment as well agriculture . . . the soil (base of agriculture) is being polluted by addition of contaminants nevertheless its usefulness is being reduced (pollution) due to its degradation by erosion and compaction. The recent data have revealed that about 53 per cent of the total geographical area of our country which comes to approximately 175 million ha has been degraded due to erosion. The annual plant nutrient loss comes to about 8.5 million tonnes which is making soils infertile and polluting the environment due to surface runoff. The deforestation is one of the major cause of hazardous soil erosion, silting of rivers and reservoirs and formation of deltas resulting in drought and floods.

Due to the erratic use of heavy field machines and shallow ploughing of soils, the soils are becoming compact, hard and structureless. Thus, the compaction results in poor crop production, less infiltration of rain water, less recharge of ground water and more surface runoff and erosion. That is why now the soil scientists are pressed hard to give the slogan *save soil* rather *save oil,* an earlier slogan of energy conservation.

Conclusion

Coming to conclusion it would be appropriate to say even though the majority of developing countries, such as India, are now trying to reduce population growth, their needs are increasing and their attempts to find new land and increased food production, are destroying, the environment at a rapid rate. There is a need to check the hazards of those agricultural activities which are disturbing and destructing the environment.

In order to safeguard our future, the menace of environmental pollution occuring due to agriculture could be reduced if emphasis is given to the followings :

(1) Integrated crop nutrition approach by —

  (a) balanced, judicious and efficient use of chemical fertilizers,
  (b) use of organic and green manures,
  (c) recycle of agricultural and human wastes, and
  (d) use of bio fertilizers and incorporation of legumes in crop rotation to reduce the use of chemical fertilizers.

(2) Efficient and judicious use of pesticides.
(3) Efficient soil water conservation practices —

  (a) Efficient use of irrigation water to reduce salinization of soils and rise in watertable,
    (i) lining of canals, and
    (ii) avoidance to excess use of irrigation water,
  (b) Soil conservation (to check erosion)
    (i) land levelling, contour, bunding, individual bunding of plots, gully plugging, crop structures, etc., and
    (ii) afforestation, grass ways development.

(4) Use of soil amendments and conditioners on the basis of soil test.

Although it is difficult but possible to act fully on above agricultural practices. This needs farmers education to create awareness in them so as to make them conscious of environmental pollution due to agricultural activities.

On global scale, United Nations, through its agencies such as Food and Agriculture Organization, World Health Organization, Man and the Biosphere Programme, United Nations Environmental Programme, Global Atmospheric Research Programme, etc. are trying to evolve strict safety codes for agricultural chemicals to ensure safe environment for our living with a long range objective.

An objective study of the situation will prove that we cannot do without chemical fertilisers and pesticides in order to sustain and augment the growing requirements of food, fibre and other agricultural produce in the world. Similarly, for helping the humanity in its constant fight against insect vectors of dreaded disease such as malaria, filariasis, typhus fever, plague, cholera, bubonic fever, sleeping sickness etc. use of insecticides is a must. A victory of the insects would change the human environment much more than any use, even any misuse, of chemicals. Biological or other non chemical control is at present impractical for a majority of pests. Thus, pesticides are the only principal means to control insects, pests, plant diseases and weeds.

However, in order to minimise the adverse ecological impact of chemical fertilisers and pesticides strict safety measures and code of conduct for their safe manufacture, transport and use will have to be evolved, applied and constantly reviewed. In this endeavour the scientists, industrialists, administrators, regulatory and enforcement authorities, farmers, traders and the rest associated with the work will have to work in a logical and coherent manner with the sole objective of ensuring the cleanest possible environment.

REFERENCES

Kolenbrander, G.J., (1982), *Fertilizer and Pollution whither Soil Research,* Panel discussion papers, 12th International Congress of Soil Science, New Delhi, 1982.

Singh R. and B.C. Biswas (1982), *Fertilizer use and environmental quality. Whither soil research,* Panel Discussion papers, 12th International Congress of Soil Science, New Delhi, 1982.

19

FLOOD HAZARD : A Study of Natural Hazards in India

N.K. DE and H.K. SARKAR

Floods are a natural environmental phenomenon of riverine tracts. When the rivers overflow their banks and flood the countryside, such a phenomenon is a blessing in disguise and helps in building upland, in bringing natural nutrients in the plains and in conservation of natural drainage systems and cleansing the environment polluted by living creatures. But the rivers become disastrous when they cause environmental degradation such as, landslides, overbank flood, massive erosion, shifting of river course, silting on riverbed, deposition of unwanted sands, etc. which all pose a serious threat to the human society.

Flood may be defined as 'a body of water overflowing land not usually thus covered'. The distribution of rainfall during the year is not, however, uniform in India. Most of the precipitation here occurs during the period of monsoon. But even then the distribution of precipitation is erratic. When there is torrential rain in an area, the rivers are unable to carry that runoff water, banks overflow due to excess of water causing floods. Then flood submerges countrysides including valuable croplands and destroys life and property.

Floods occurred even before the advent of man, but there is no record. Hundreds of years ago our forefathers established their houses on the banks of the rivers which have witnesses of our ancient civilisation. Many of these rivers have acquired religious sanctity. There is also a darkside to this picture as these rivers can be agents of sorrow. It is still believed by the people that flood is a divine punishment to the misdeeds of the people. The present study attempts to assess the environmental degradation (physical) brought down by floods and their control in India.

Significance of Flood

Water evaporates from the sea, part of which comes down as rain which fills drainage systems and rest flows to the sea. While speed of rain water is slowed down, a part of it infiltrates into the soil. These processes depend according to the pattern and frequency of rainfall, topography, soil characteristics, etc. When the amount of rainfall exceeds the drainage capacity and the subsoil storage capacity, the surplus causes problems. Sudden accumulation of huge amount of water overflows the river banks causing floods. The virulence of flood is dependent on the quickness of accumulation in comparison to the means of disposal and outflow. Flood, a blessing in disguise, brings fertile soil materials and deposits on the croplands.

Cause of Flood

We generally experience two types of floods. Their origins are different, but their impact in respect of devastation is of the same magnitude. Sometimes during a cyclonic storm in the Bay of Bengal, sea waves of upto 18 m in height are formed. They overflow the embankment and inundate large areas in the coasts of Andhra Pradesh, Orissa and West Bengal. The other type of flood is caused by heavy and intense rainfall in the catchment areas of rivers which overflow their banks. The riverine flood has become more frequent in recent times.

The factors causing floods are as follows :

(a) Overflow of natural banks.
(b) Meandering of rivers.
(c) Siltation of riverbeds reducing water carrying capacity.
(d) Heavy rains in areas having poor drainage system.
(e) Poor river outlet into the sea.
(f) High sea tides while rivers are in spate.
(g) Cyclonic storms along the sea coast.

Effect of Rainfall

India lies in monsoon land. It receives bulk of her rainfall from the southwest monsoon between June and October every year. Only the southeastern part of India receives most of its rainfall between

November and December from northeast monsoon (Rao, 1975). The belt of heavy rainfall comprises Assam, Tripura, North Bengal, North Bihar, while the central region and the Ganga plain lie in the belt of moderate rainfall. The north Deccan and the adjoining areas receive heavy rainfall in September.

Heavy rain in the Himalayas is the cause of floods in the rivers of Uttar Pradesh, Bihar, West Bengal and Assam, while in the central and the southern regions rivers like the Narmada, Tapi, Sabarmati, Mahanadi, Godavari, Krishna and Cauvery get into flood as a result of heavy rain. Heavy rainfall in the Western Ghats causes floods in the rivers.

Effect of Siltation

Rivers rising from high hills and mountains flow very fast in the upper reaches, carry large load of sediments and deposit in their beds in the lower reaches where the rivers begin to change their courses (Valdiya, 1987). These silts reduce water carrying capacity of the rivers causing inundation. Teesta is an example of a river shifting its course because of heavy load of sediments. The silt accruing from soil erosion is carried from upper reaches and deposited in lower reaches where the flow is slower (Chaturvedi and Mohan, 1983). Simultaneous flood in the main rivers as well as in its tributaries is a source of danger as well. As a result, the confluence gets congested and the water spreads into the tributaries. In the sea coasts, similar congestion takes place on account of high tides which do not allow flood water to flow into the sea causing river bank inundation.

Effect of Earthquakes and Landslides

Earthquakes and landslides frequently occur in the seismic areas. Landslides can also be caused by soil erosion. These natural calamities aggregate the flood problem. The Brahmaputra and other rivers of Assam lying in seismic zone tend to change their courses. The landslides create natural dam across the rivers in upper areas and flood water change the river courses leaving a trail of destruction.

Floods in Different Regions

Brahmaputra river system in Assam represents an area of about one-fourth of the total area. All the rivers under this river system cause devastation in Assam during heavy rains every year. According to old records, a part of Dibrugarh town was washed away in 1880 by the Dibru river. Since 1913 the Brahmaputra and its tributaries have been changing their courses. Severe flood occurred in 1931 and flood water rose over one metre. above railway line. In 1955 flood water of the Brahmaputra inundated all the lowlying areas of Lakhimpur, Sibsagar, Nowgong, and Goalpara districts. Devastating floods occurred again in 1956. River Dihing overflowed its banks inundating Larua mouza. The Umtru, the Kopili and the Barpani rivers submerged vast areas in Nowgong district. In 1956 flood water inundated lowlying areas of Tripura damaging crops and houses. In the northeastern region recurring flood dismay the people every year. Devastating flood in 1988 has caused a serious threat to the civilisation in this region.

Almost in every year during monsoon widespread floods occur in West Bengal causing loss of life and property. In 1956 about 2 million people were made homeless and damages amounted to over Rs. 20 million. In 1959 more than half of West Bengal was severely affected by floods. About 1.2 million people were made homeless, crops of one million acres were damaged and about 100 people lost their lives. In 1956 Howrah, the 24-Parganas and Medinipur districts were worst sufferers as a result of devastating flood. During May-June under the influence of cyclone tidal waters inundate the Sundarban region. In 1978 out of the state's 335 development blocks about 130 were affected severely. About 15.3 million people were affected in an area of about 30,000 km^2 spread over the catchment areas of the Mayurakshi, Damodar, Kangsabati, Subarnarekha, Padma and Bhagirathi rivers with enormous loss of life and property which were greater than ever before (Fig. 19.1).

The Himalayan rivers like the Teesta, the Jaldhaka, the Torsa, the Raidak, the Sonkosh, etc. pour tremendous amount of water into the piedmont plain of North Bengal during rainy season. Here the floods are due to heavy rainfall, landslides and ruthless felling of trees, etc. in catchment region. In 1988 disastrous flood due to unprecedented rainfall has caused widespread damage in the districts of North Bengal.

In the central and the eastern parts of West Bengal the rivers being cutoff from their headwaters can not carry sediments, but instead, deposit them in their beds. In the active delta the Muriganga, the Saptamukhi, the Thakuran, the Matla, the Bidyadhari, etc., deposit silt in their beds because they bring in more silt during high tides. In the western part of West Bengal the Brahmani-Dwarka and the Mayurakshi system, the Damodar, the Rupnarayan and the Kangsabati system and the Subarnarekha river are active in creating floods in their lower reaches (Fig. 19.1).

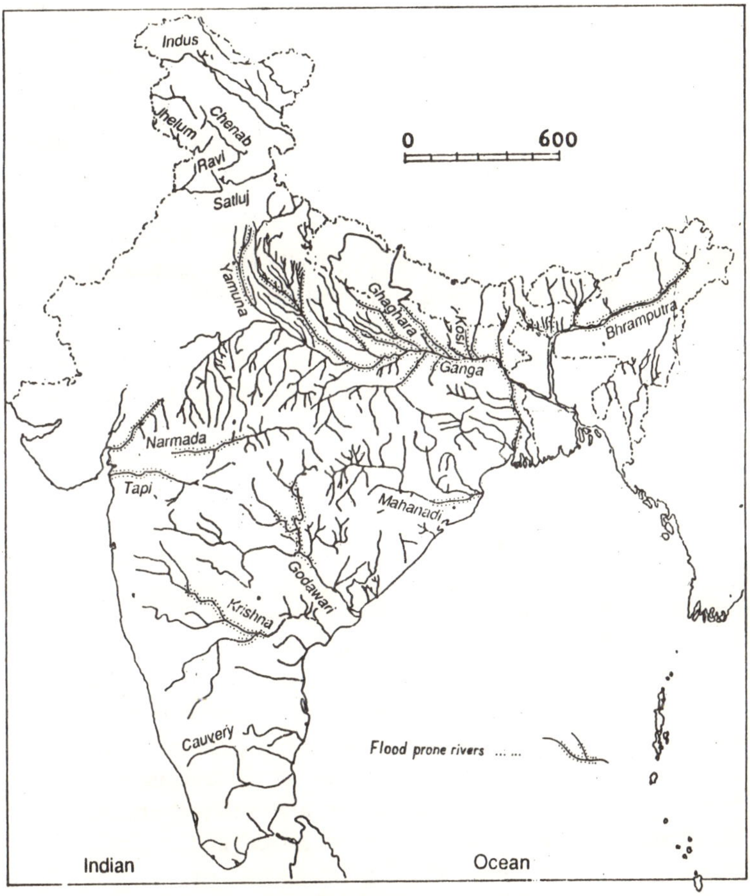

FIG. 19.1: Floodprone Areas of India

In Bihar, Kamla, Kosi, Burhi Gandak and Bagmati rivers ravage the northern parts almost every year and frequently change their courses. Disastrous floods occurred in north Bihar in 1954. Unprecedented floods occurred in Burhi Ganga and Ganga in September 1956 and in subsequent years including 1988. In south Bihar, Patna, Shahabad, and the Santhal Parganas were badly affected. Recurring flood is a yearly episode in this region.

Uttar Pradesh is affected by flood especially in the northern region. All the rivers in Uttar Pradesh drain into the Ganga which itself is subject to heavy floods. Its tributaries including the Yamuna are no less destructive. Rapti and Ghaghra are notorious for their devastating floods. In 1955 extensive areas in Basti, Gorakhpur and other districts were submerged. The southern rivers like the Chambal, Dhasan, Sindh, Betwa, Ken and Tons, tributaries of the Yamuna, overflowed their banks. The riverine tracts of Uttar Pradesh suffered from loss of life and property due to heavy floods occurred in recent years as well as in 1988.

Delhi experienced severe floods in 1924, 1950, 1955 and 1956. In 1955 Shahdara town and outskirts of Delhi were inundated and the traffic system was disrupted. In 1988 severe floods also caused widespread damage in Delhi region.

The northwestern region of India comprising the Punjab, Himachal Pradesh and Jammu and Kashmir is less affected by floods compared to the above regions. Unprecedented floods occurred in the rivers of the Punjab in 1947. In 1955 floods exceeding those of 1947 caused breaching of the embankments and submergence of entire area of Ambala division. Floods in 1988 exceeded those of previous years. In Jammu and Kashmir flood caused considerable damage in 1954, 1956, 1957 and 1988.

Flood is a problem in the coastal areas of Orissa and Andhra Pradesh. In Orissa, floods in Mahanadi, Brahmani, Baitarani, Salandi and Subarnarekha rivers cause damages in Baleswer, Puri and Cuttack districts. The disaster is accentuated if the floods coincide with high tides. In Andhra Pradesh floods in the Godavari, the Krishna and the Penner rivers cause considerable damage in Vijayawada, Nellore, Eluru, Rajamahendravaram and other areas. In Maharashtra unprecedented floods occurred in the Krishna, the Bhima, the Chandrabhaga and the Ghataprava rivers in 1956 affecting Pune, Sholapur, Bijapur and South Satara districts.

All the states of India suffer more or less due to floods but the devastation is most frequent and heaviest in Assam, West Bengal, Bihar and Uttar Pradesh drained mainly by the Brahmaputra, the Ganga and their tributaries.

Preventive Measures

The following measures have been adopted to control floods in different parts of the country:

(a) *Multipurpose projects* — Under the Five-Year Plans some multipurpose projects have been launched in some major rivers in different parts of the country. The objective of these projects is to provide water for irrigation, generate hydel power and regulate flood water.

The setting up of the Damodar Valley Corporation is to harness the Damodar, the river of sorrow. There are several other multipurpose projects in the country which help to control floods in the lower reaches of the concerned rivers.

(b) *Soil conservation* — It is an important method for reducing floods indirectly. This method plays significant role in controlling the basic factors that result in floods. It conserves soils and preserves their nutrients thereby reducing silt load in the rivers. The Planning Commission made a specific provision for soil conservation works in First Five Year Plan. In 1953 the Central Soil Conservation Board was established. The Board is interlinked with the Central Water and Power Commission, Central Flood Control Board, River Commission and similar other organisations.

(c) *Artificial embankments* — Construction of embankments is a popular method for flood control all over the world. Embankments are the most extensive measures taken to control flood and reduce damages. They are economic and quick, but they have their limitations.

(d) *Improvement of river channel* — The capacity of river channel is enlarged by widening and deepening it. This method in combination with construction of embankments on either bank is very effective in flood control measures. This method is rarely used in India as the maintenance is costly and difficult.

(e) *Diversion of flood water* — The diversion of flood water is merely a method of returning of the river a portion of its own flood water after making relatively harmless.

(f) *Reservoirs* — The method of constructing embankments, improving channel capacity and diverting flood waters are applicable in lower reaches where rivers carry enormous quantities of silt laden waters. In the upper reaches method of reservoir is suitable, especially for minimizing violence of floods and for hydel power generation.

(g) *Irrigation planning* — Irrigation planning on the basis of watershed management is now an accepted means of maximising wateruse efficiency. Of the factors of watershed management reduction of silt-laden water is necessary for flood control measures. These processes connected with watershed management should be so designed as to reduce soil erosion wherever possible.

Flood Control Works

During the British regime some relief was given to the affected areas, but no considerable attempt was made to minimise the floods. Local embankments were occasionally constructed by the landlords and some sporadic efforts were made by the then provincial governments. These were all temporary measures. A great change, however, has taken place after Independence. A brief account of flood control works carried out in various regions of India are as follows:

(i) In Assam embankments have been constructed along Brahmaputra, Barak rivers and their tributaries. Shifting of Palasbari town and protection of other small towns have been done.

(ii) In Manipur protective works have been done on the banks of the Imphal river for storage reservoirs and protection of towns. In Tripura works have been done for protection of Agartala, Durgapur, Somasura and other towns.

(iii) In West Bengal flood protection works in existence consist of the embankments in the Sundarban region where dilapidated embankments have been reconstructed and many new embankments have been erected. Flood control works providing protection to the towns of Jalpaiguri, Cooch Behar, Alipur Duar, Mathabhanga, Siliguri and Toofanganj have been done. The construction of embankments for protecting of extensive areas from the Teesta river near Barnes and Domohani has been done. The Damodar Valley Corporation (DVC) has also taken up measures for flood control of this state.

(iv) In Bihar besides the existing embankments along the Kosi, the

Bagmati, the Burhi Gandak and the Ghaghara new embankments have been constructed along the Kosi, the Kamla, the Burhi Gandak, the Bagmati, the Mahi, the Khirohi and the Adhwara rivers.

(v) In Uttar Pradesh some protective works have been done along the Rapti, the Tons, the Ghaghara and the Gandak rivers. Protective works include embankments along the rivers, protection of Varanasi, Rajapur, etc. and raising of the plinth of lowlying villages above flood level. Lowlying villages of Bahraich, Faizabad, Ballia, Basti, Deoria, Gonda, Gorakhpur, Ghazipur, Jaunpur etc. have been provided proper protection.

(vi) In Delhi area embankment to save Shahdara town has been constructed along the Yamuna river. The old Mughal Bund and other old bunds have been raised and strengthened.

(vii) Protective measures including desilting and improvement of waterways as flood control works on the Ravi, the Beas and their tributaries have been taken up. In Jammu and Kashmir past constructions consist of embankments and a supplementary spill-channel of the Jhelum. Major works of the Kashmir Valley Flood Protection Scheme have been carried out. Some minor schemes have also been taken up.

(viii) In Orissa repairs of breached embankments have been done. Several schemes for constructing new embankments, improving existing embankments and other protective works have been completed. The schemes of raising of lowlying villages and raising and strengthening of bunds on the Mahanadi and the other rivers in the delta region have been taken up.

(ix) In Andhra Pradesh flood discharge of the Budameru river into the Krishna river has been done through a diversion channel. The canalisation of the Budameru in its certain reaches and improvement of the Tungabhadra drainage have been done. The construction of detention reservoirs on Penner, Erranpalli, Yerrakalva, Byneru, Pulivagu and Yeleru and the embankments on Tammileru, Godavari Svarnamukhi, etc. have been done.

Various flood control methods applied in other countries are being adopted in India with modifications to suit local conditions. We have made considerable progress within a comparatively short period in respect of flood control, in spite of lack of meteorological, hydrological and topographical data relevant to the flood problem of the major river basins.

Conclusion

Flood hazard has become a serious problem because it causes enormous destruction of life and property. Steps must be taken to moderate the intensity of flooding and devastating effect of uncontrolled fury of nature to a tolerable level. It must be always remembered that neither it is possible to eliminate fully the devastating nature of flood nor it is desirable to shut out the beneficial effects of floods on our environment. The most important consideration is that the floods do not bring in misery only, they bring in blessings as well.

It should be borne in mind that final selection of flood control methods does not depend on its technical merit alone, but on its economic justifications and other conditions such as, the availability of capital, labour and materials. These economic considerations are, therefore, of vital importance in flood control methods. Besides these, the social conditions of the people in and around flood prone areas are required to be analysed scientifically in the sphere of flood, its occurrences, remedies and control.

In India the flood problem has been taken up only recently on a national plan. It received little attention so long as India was under Foreign domination. The achievement of independence in 1947 ushered in the concept of welfare state in which flood control is now being tackled on a national footing.

REFERENCES

Chaturvedi, R.S. and Mohan, R. (1983), Delineating Flood affected areas of south U.P. using satellite remote sensing techniques, *Proc. Nat. Symp. Remote Sensing in Development and Management of Water Resources*, Indian Institute of Photointerpretation and Remote Sensing, Dehradun, pp. 175-86.

Coates, D.R. (1976), *Geomorphology and Engineering*, Dowden, Hutchinson and Ross, Stroudsburg, p. 360.

Rao, K.L. (1975), *India's Water Wealth*, Orient Longman, New Delhi, p. 255.

Seth, S.L. Das, D.C. and Gupta, G.P. (1982), *Floods in Arid and Semi Arid Areas — Rajasthan*, National Flood Commission Report, Vol.1, Chapter 10.

Valdiya, K.S. (1987), *Environmental Geology* Tata McGraw Hill Publishing Company, New Delhi, pp. 329-51.

20

ECOLOGY OF THE RIVER GANGA

B.D. TRIPATHI

Ganga is the river of life for every Indian. It is regarded as the most pious river of the world by Hindus and worshipped like a goddess. Considering the total length, it ranks 15th in Asia and 39th in the world. Ganga basin receives water from an area equal to about one quarter of the total area of India, i.e. 8,61,404 km^2. River Ganga has two main sources in the upper reaches of Himalaya. They are Bhagirathi originating from the Gangotri glacier at Gomukh and the Alaknanda from glacial spout of the Alkapuri glacier. After passing through the mountainous region of more than 300 km, Ganga enters into plains at Hardwar, and then flows through the thickly populated and intensively cultivated vast stretch of the most fertile land, i.e Ganga Plain and ultimately ends her journey in the Bay of Bengal covering a length of 2,525 km. About 350 million people live in the Ganga basin area, and 58 million ha of land is under cultivation.

A number of rivers from Himalaya and Vindhya join Ganga in its catchment area. Yamuna which originates from the Yamunotri glacier, joins Ganga at Allahabad. The main rivers joining Ganga or Yamuna are Chambal, Betwa, Ken, Sind, Ram Ganga, Gomti and Ghaghara in Uttar Pradesh, Gandak, Burhi Gandak, Son, Kosi and Bagmati in Bihar and Mahananda in West Bengal. After Farakka Ganga bifurcates and a number of tributaries join from left and right sides. The sacred river Ganga, whose water is regarded as purifier of human body and soul, poses a unique challenge to the people in India in view of the accumulated input of untreated pollutants. From the recent past, researches conducted on the water quality criteria and Composite Index (CI) calculation based on the Biochemical Oxygen Demand (BOD), Dissolved Oxygen (DO) and total coliform bacteria, it is revealed that the Ganga is highly polluted at Kanpur (CI = 12–17), followed by Dakshineshwar (CI = 20–31), Patna

(CI = 22–26) and Varanasi (CI = 28–34). However, the water quality at Hardwar is reasonable good (CI = 44.53). Low volume of water, specially during summer months, is associated with the construction of a number of dams and canals on Ganga.

Research Work Done on River Ganga

A number of investigations have been made on river Ganga from time to time. Biswas (1942) studied the algal community of Hooghly (river Ganga) to assess its impact on the drinking water quality. Roy (1959) has estimated various physico-chemical factors and studied phytoplankton periodicity and ecology in West Bengal. He concluded that Diatoms were responsible for the chocking and clogging of the filter beds of the Calcutta water supply system. At Varanasi, Singh (1962), Lakshminarayana (1965), Rai (1977), Tripathi (1982), Tiwari (1983) and Sikandar and Tripathi (1984) have investigated the hydrobiological features of River Ganga. Pahwa and Mehrotra (1966) studied the phytoplankton fluctuation and abundance in relation to hydrological conditions. Saxena *et al.* (1969) assessed the pollution level by surveying different regions receiving domestic and textile wastes and reported that the self purifying capacity of river was moderate to high at different localities. Further, Ray and David (1966) assessed the effects of textile and tannery effluents on fish life and chemical and biological composition of water.

Agrawal *et al* (1976 a & b) studied the bacteriological characteristics and found that the water quality at different ghats in Varanasi was unsatisfactory owing to heavy sewage pollution. Bilgrami and Siddiqui (1980) reported on the algal flora at Bhagalpur. Chattopadhyay *et al* (1984) have conducted a short term study on pollution status of river Ganga in Kanpur region. Singh and Bhowmick (1985) studied the effect of sewage on the physico-chemical characteristics and bacterial population of river Ganga at Patna. Bhowmick and Singh (1985) investigated the phyto-plankton population of river Ganga at Patna. Bilgrami and Dattamunshi (1979 & 1988) performed extensive work on the hydrobiology of Ganga water from Barauni to Farakka.

Tripathi *et al* (1984) have studied the plant biomass and energy used for the burning of human dead bodies at Varanasi ghats. Pandey and Tripathi (1984) have observed the contamination of Ganga water and its impact on mammals. Sikandar *et al* (1980) have also studied the impact of dead body cremation on the river ecology. Later on Saran (1988) studied physico-chemical properties and population dynamics of zooplankton and macrofauna of river Ganga at Patna. Singh (1988) monitored the water quality at Gangotri and Rishikesh, while in the same year Mathur (1988) studied the physico-chemical properties of water at Roorkee. Extensive researches have been conducted by Ambasht (1988) to study the ecobehaviour of plants growing on the left and right banks of the river from Mirzapur to Ballia. Das (1988) and Samaddar (1988) have reported the presence of DDT residue in river Ganga. Mehrotra (1988), Chaudhary (1988) and Rupenwar (1988) have detected a few heavy metals in the water and mud. Sinha (1988) has reported the physico-chemical and biological properties of water near Rai Bareilly. Recently author has analysed the physico-chemical and biological properties of river Ganga from Mirzapur to Ballia and assessed the impact of major pollutants on quality of river water, primary productivity of the phytoplanktons and population dynamics of micro and macrophytes.

Physico-chemical Properties of Ganga Water

In general the minimum and maximum temperatures are lower in non-polluted water as compared to the polluted water (minimum temperature 8.4°C vs 21.2°C in January; maximum temperature 32.7°C vs. 34.8°C in June). Considerably higher temperature (37.5°C) was recorded near an area where distillery effluents (60-67°C) are discharged (Bilgrami and Dattamunshi, 1979). Transparency values range from 10.7 cm to 19.0 cm at unpolluted sites and 7.9 cm to 17.5 cm at polluted sites. Polluted water has a higher pH (9.6) compared to about neutral at non-polluted sites, as are the electrical conductance (949 μm hos/cm in polluted sites vs 135 μm hos/cm) at unpolluted sites.

Acidity and alkalinity values showed similar trend of variation. The oxygen content was recorded very low (0.2 mg/l) in highly polluted areas specially in the summer season. Whereas in unpolluted areas it was very high (12.8 mg/l) in comparison to other rivers. The decreasing trend of DO content in all the seasons from upper Ganga to the middle and lower Ganga may be associated with the variation in the ambient temperature. The concentration of Biochemical Oxygen Demand reveals very high value (660.7 mg/l) in some of the polluted stretch of the river, whereas it ranged from 0.5 to 2.0 mg/l in the unpolluted water. From the DO and BOD values it is clear that the Ganga water in most of the regions are of good quality. However, in polluted areas it is neither useful for drinking nor for bathing. Chemical Oxygen Demand (COD) values also showed similar trends as BOD. The optimum values of chloride varying from 8.5 mg/l to 106.7 mg/l was reported. Nitrate and phosphate contents were higher, 2.18 mg/l and 1.59 mg/l in polluted areas respectively. Nevertheless, these contents support the growth of aquatic plants. On the other hand unpolluted water indicates very low concentration i.e. 0.2 mg/l and 0.008 mg/l indicating poor condition of nutrients. The concentration of free carbon dioxide varied from a minimum (0.9 mg/l) at unpolluted areas to the maximum (41.5 mg/l) in the highly polluted areas. Higher hardness value (218 mg/l) has been observed at polluted site which decreased to 26.5 mg/l at unpolluted sites. High sulphate content (293 mg/l) was also noted in polluted area, whereas it ranged from 14.5 mg/l to 94.0 mg/l at unpolluted site. The silica content did not show any marked variation in polluted and unpolluted Ganga water. Higher calcium content 193.7 mg/l was noted in polluted area whereas it was normal (18.5 mg/l) in the unpolluted water. Total nitrogen (217.8 mg/l) and Magnesium (24.3 mg/l) contents were also higher in polluted water.

Biological Properties of Ganga Water

The biological assessment of river Ganga made by numerous workers from time to time (Table 20.1) revealed that there are 146 algal species found in Ganga water. Out of the total 82 species belong to chlorophyceae, 20 bacillariophyceae, 40 gyanophyceae, 2 euglenoidae, 1 xanthophyceae and 1 rhodophyceae. Total 98 fungal species are also reported including 15 phycomycetes, 12 ascomycetes, 3 basidiomycetes, 86 deuteromycetes and 2 species of mycelia sterila. About 51 species of macrophytes have also been reported. Zooplankton observation reveals the presence of about 35 species. Out of which 3 species belong to ciliata, 14 rotifera, 10 cladocera, 5 copepoda and 3 species of rhizopoda.

Insects and molluses have also been reported. There are 7 species of aquatic insects, 7 species of gastropoda, and 2 spccics of rivalvia were observed, 4 species of rana, 72 species of fishes, 100 species of turtlcs, 1 specie of snake, 13 species of birds and 1 specie of mammalia, i.e. *platanista gangetica* which has been reported. Recent observations made by the author from Mirzapur to Ballia reveals the following properties of river Ganga.

TABLE 20.1: Biotic Composition of river Ganga

Biological Properties	Name of Species	No. of Species
ALGAE	Chlorophyceae	82
	Bacillariophyceae	20
	Cyanophyceae	40
	Euglenoidae	2
	Xanthophyceae	1
	Rhodophyceae	1
FUNGI	Phycomycetes	15
	Ascomycetes	12
	Basidiomycetes	3
	Deuteromycetes	66
	Mycelia sterila	2
MACROPHYTES		51
ZOOPLANKTON	Ciliata	3
	Rotifera	14
	Cladocera	10
	Copepoda	5
	Rhizopoda	3
INSECTS AND MOLLUSES	Aquatic insects	7
	Gastropoda	7
	Bivalvia	2
AMPHIBIA	Genus Rana	3-4
PISCES	Fishes	72
REPITLIA	Turtles	100
	Snake	1
AVES	Birds	6
MAMMALIA	Platanista gangetica	1

River Ganga at Varanasi

Varanasi, a famous seat of learning and ecological research centre of repute has the distinction of being the oldest living city of the world. Varanasi is famous for its cultural, educational and religious heritages. River Ganga having great religious importance when flows through the city, the religious significance of the place becomes further glorified. People from all over the world come to take bath into the holy river Ganga and carry its water as *elixir* and distribute a spoonful to others as *prasad* (gift of God).

The river Ganga flows along the eastern corner of the city over a distance of about ten kilometres in a crescent shaped river bank. There are eighty four ghats on the left bank of the river (Sikandar and Tripathi, 1984). Most of the ghats are *pucca* (lined by stones or concretes). These ghats are used for bathing, washing, recreation, fishing and drinking purposes. Moreover, two ghats i.e. Harishchandra and Manikarnika are used for cremating human dead bodies. Some ghats are specially used for cleaning domestic animals and their waste disposal (Gai ghat), some for sari printing (Lalita ghat and Hanumangarhi ghat) and for washing clothes by washerman (Baccharaj ghat). At certain points sewage is discharged just near the points where from water is pumped for domestic consumption.

Sources of Pollution

There are three main sources, viz. industrial effluents, domestic sewage and dead body disposal, by which the water of river Ganga at Varanasi is being polluted. During rainy season, agricultural runoff also play important role in polluting river Ganga.

Industries

About one thousand five hundred large, basic and small scale industries (Table 20.2) including manufacturing metal products (27.68%), chemical and chemical products (14.04%), food stuffs (3.59%), textiles (3.53%), transport equipments (2.13%), furnitures and fixtures (2.06%), non-metal mineral products (1.37%), beverages (0.8%), rubber products including tyre and leather, fur product (0.6%) and numerous other products are located in and around the city of Varanasi. Most of them discharge their untreated effluents containing huge amount of mercury, lead, iron, nickel, cobalt, manganese, boron, detergents, paints, dyes, varnishes, oil, grease, kerosene, alcohols, phenols, acetones, chlorinated hydrocarbons, soda, acids, alkalies, carbon and silicon, carbohydrate, protein, fats, aldehydes and fatty acids either directly or through the city sewage are being discharged into river Ganga (Sikandar, 1987).

Dead Body Disposal

Hindus due to religious faith burn their dead bodies and throw ashes into the river. More than 30 thousand human dead bodies annually are brought from distant places and cremated at Harishchandra and Manikarnika ghats. About 40 per cent of total dead bodies cremated every year are brought from outside the Varanasi district. Among those coming from Varanasi district about 60 per cent are brought from outside the Varanasi city. Out of total dead bodies about 83 per cent are cremated at Manikarnika ghat and remaining 17 per cent at Harishchandra ghat. From these burning ghats half-burnt flesh content, unburnt human and animal dead bodies are also thrown into the river. These practices of disposal of dead bodies add another dimension of pollution in river Ganga at Varanasi.

TABLE 20.2: Nature and number of industries located in and around the city of Varanasi (Industrial Directory, 1981)

Name of Industries	Number	Percentage
1. Metal Products (Aluminium, steel, brass, iron and copper)	416	27.7
2. Chemical and Chemical Products (Candles, agarbatti, cosmetics, soap, detergents, paints, varnishes, ayurvedic medicines, ink, liquid phenyl, oils, pharmaceutical chemicals etc.)	211	14.0
3. Electricals	201	13.4
4. Food Products (Ice, cold drink, oil and dall mills, salt stuffs, salt and sweet biscuits, bread etc.)	54	3.6
5. Spinning, Weaving and Finishing of Textiles (Hosiery, printing of sarees and clothes, cotton and thread processing)	53	3.5
6. Transport equipments (Automobile services, truck, bus body, motors, engineering works and motor parts)	32	2.1
7. Furniture and Fixtures	31	2.1
8. Non-metallic Mineral Products Except Petroleum and Coal	26	2.1
9. Wooden and Cork Work Except Manufacturing Furnitures	18	1.2
10. Beverages (Tobacco, zarda, pan massala and surti etc.)	12	0.8
11. Leather Fur, rubber and tyre etc.	9	0.6
12. Printing Press and Paper Products	107	7.1
13. Miscellaneous (Plastic, polythene, elastic, silk and sarees fabrics, metallic yarns, carpets, bricks, cake, brequiting, cement, tiles, surkhi, sport equipments, pens, photostats, sindhur, gulal etc.)	333	22.2
Total	1,503	100

Sewage Disposal

The population of Varanasi city (7.97 lakh in 1981) having water supply facility for 180.83 MLD, generate huge amount of sewage (about 126.58 MLD) which is discharged through several minor and major *nalas* into river Ganga.

Agrawal *et al* (1976. a) pointed out that there are 34 sewage discharge points, but owing to the action programme of connecting several sewers in main sewer lines, only eight sewage discharge points have been located by Sikandar and Tripathi (1986) along the whole 10 km crescent course of the river. These are near Assi, Shiwala, Harishchandra, Chauki, Rajendra Prasad, Gola, Telia and Rajghat. Among these, six discharge points are major (Fig. 20.1) and perennial discharging huge amount of sewage into the river. These points are: (a) Assi ghat, (b) Shiwala ghat, (c) Harishchandra ghat, (d) Chauki ghat, (e) Rajendra Prasad, and (f) Rajghat nala.

Physico-chemical Characteristics of Varanasi Sewage discharged into the River Ganga

According to a study conducted by the author, the temperature of sewage discharged into Ganga varied from a minimum 20.0 $\pm$ 0.6°C in February to a maximum 34.0 $\pm$ 0.6°C in June. The pH ranged from 7.0 $\pm$ 0.1 to 8.1 $\pm$ 0.1. The pH values were usually low during rainy season. The maximum value of acidity i.e. 72.8 $\pm$ 1.0 mg as $CaCO_3$/l was recorded at Rajghat nala and a minimum value 6.0 $\pm$ 0.1 mg $CaCo_3$/l at Shiwala nala in October. The highest value of total alkalinity was recorded as 798 $\pm$ 8 mg $CaCo_3$/l at Rajghat and the lowest 171 $\pm$ 1.2 mg $CaCo_3$/l at Harishchandra ghat.

The DO values ranged from a minimum 0.2 $\pm$ 0.1 mg/l at Rajghat to a maximum 4.4 $\pm$ 0.1 mg/l at Shiwala ghat. The values of BOD (at 20°C for 5 days) were usually higher during the month of summer and lower in rainy season. It ranged from a minimum 20 $\pm$ 2 mg/l to a maximum 500 $\pm$ 40 mg/l at Rajghat. It was usually lower at Assi ghat nala. The minimum value of COD 53 $\pm$ 15 mg/l was obtained at Assi ghat whereas the maximum value 870 $\pm$ 30 mg/l was recorded at Rajghat. The lowest concentration of chloride as 15 $\pm$ 0.5 mg/l was noted at Assi ghat and the highest 100 $\pm$ 1.0 mg/l at Rajghat. The values of electrical conductance ranged from a minimum 175.0 μ mhos/cm to a maximum 915.8 μ mhos/cm. The

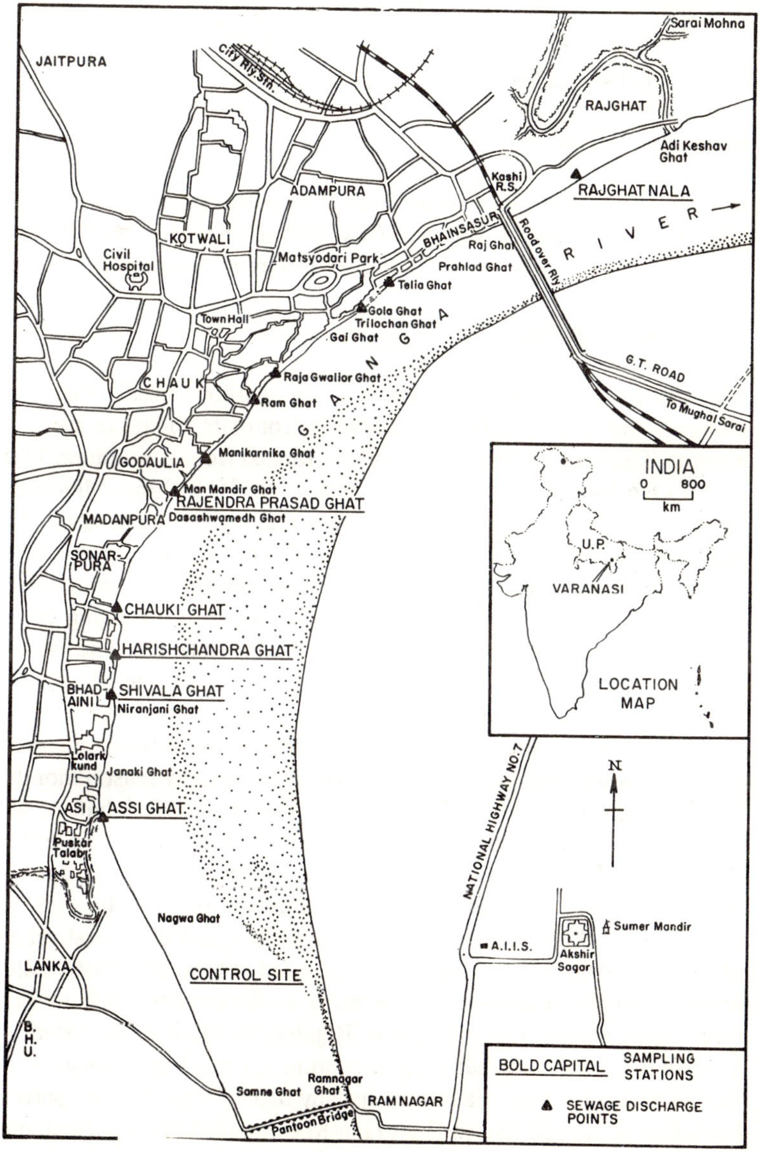

FIG. 20.1: Location Map of Various Sampling Sites

nitrate-nitrogen ranged from a minimum 0.089 mg/l at Assi ghat nala to a maximum 0.98 mg/l at Rajghat nala. The phosphate content varied with a minimum of 0.14 mg/l at Assi ghat to a maximum of 5.9 $\pm$ 0.12 mg/l at Rajghat nala.

Definition of Control Sites

Control site is the place located on the upstream of River Ganga after which it enters the Varanasi City. This site has been used for the collection of the water and soil samples and treated as unpolluted.

Ganga Water Quality

The personal observations made by the author reveals that the Ganga water quality varied significantly from site to site in various months. The temperature of river water ranged from a minimum 18.0 $\pm$ 0.4°C in January to a maximum 34.6 $\pm$ 0.5°C in June. Measurement of water transparency of Ganga water by Secchi disc at all the sites was usually lower than that of control site 8.0 $\pm$ 1.0 cm to 190 $\pm$ 2.0 cm. Moreover, the transparency was very low during rainy season. At Rajghat water showed always lowest transparency ranging from 6.2 to 10.0 cm. At other sites it ranged from 9.0 to 64 cm. The values of pH at all the sites were usually lower than that of control site. At control site it ranged from 7.0 in rainy season to 8.5 in winter season. Very high fluctuation in the values of acidity among various sites was recorded. It was zero at control site except a few rainy season month. The highest value (70.0 $\pm$ 1.4 mg/l) was recorded at Rajghat. At all the sites the alkalinity values were lower than that of control site the maximum values being at Rajghat. The values were lower during rainy season and attained the peak in January at most of the sites. It ranged from 100 mg $CaCO_3$/l to 475 mg $CaCO_3$/l. The highest value of dissolved oxygen (10.0 mg/l) was recorded in March at control site which gradually decreased to 3.6 mg/l in September. The lowest values have always been noticed at Rajghat, where it ranged from a minimum 1.2 mg/l in July to a maximum 2.5 mg/l in August. The BOD values at all the sites were usually higher than that of control site where it ranged from 0.7 mg/l in March to 3.0 mg/l in September. At all the sites, the BOD values decreased during rainy season whereas at control site it increased. Next to control site the lower value of BOD was noticed at Assi ghat (3.7 to 20 mg/l). At other sites it ranged from 16 mg/l to 100 mg/l. The highest BOD values were always recorded at Rajghat (80 mg/l to 277 mg/l). COD values ranged from 1.1 to 6.0 mg/l at control site, 12.0 to 47 mg/l at Assi ghat and 45 to 185 mg/l at all other sites except Rajghat. At Rajghat it was maximum ranging from 180 to 475 mg/l.

The chloride content was usually higher at all the sites than that of control site, the highest value (81.5 $\pm$ 0.5 mg/l) being at Rajghat. The chloride content of Ganga water decreased during rainy season at all the sites, the minimum value being as 8.2 $\pm$ 0.3 mg/l at control site. Similarly, the values of electrical conductance (EC) at all the sites were higher than that of control site, and it has decreased during the rainy season. It ranged from a minimum 175.0 μ mhos/cm during rainy season to a maximum 915.8 μ mhos/cm during winter season. Nitrate-nitrogen at all the sites were higher than that of control site, the maximum value (0.63 to 0.98 mg/l being at Rajghat. It ranged from 0.015 to 0.085 mg/l at control site and from 0.100 to 0.345 mg/l at all the sites except of Rajghat. Moreover, during rainy season nitrate showed a sharp increase at control site and slight decrease at other sites. Similar trend was also obtained for phosphate. At control site it ranged from 0.005 mg/l to 0.120 mg/l, whereas at other sites except Rajghat it varied from 0.120 to 0.990 mg/l. At Rajghat the values were usually higher and ranged from 0.260 to 3.78 mg/l.

Lower values of pH, DO, BOD, COD, acidity, alkalinity, chloride, EC etc. at most of the sites during rainy season may be found due to dilution by rain and flood water. A sharp increase in nitrate and phosphate contents at control site may be due to addition from atmosphere through rain and from the croplands through runoff during rainy season.

Sikandar and Tripathi (1984) have also studied the physico-chemical characteristics of Ganga water at a few bathing ghats (Assi ghat, Janaki ghat, Dashashwamedh ghat) and burning ghats (Harishchandra and Manikarnika ghats). The investigation in the latter ghats revealed higher values of these characteristics. Values of most of the physico-chemical parameters were lower during rainy season months. Temperature ranged from 19.1°C in December to 30°C in May. The highest value of pH (8.9) was recorded in January whereas lowest value (7.7) was recorded in August. The chloride content of water varied from 8.2 mg/l in August to 58 mg/l in February. Transparency ranged from a maximum of 190 cm in April to a minimum 5.1 cm in July. Total alkalinity of Ganga water ranged from 108 mg/l $CaCO_3$ in August to 251 mg/l $CaCo_3$ in December and January. Maximum value of 9.8 mg/l DO was recorded in February and gradually decreased to a minimum 3.0 mg/l in August. The minimum BOD and COD values were obtained as 0.2 and 1.1 mg/l respectively in February and gradually increased to 13.1 mg/l and 19.0 mg/l in July. The values of pH, chloride and alkalinity were higher during winter season and lower during rainy season, whereas the values of BOD, COD and temperature were higher during summer and lower during winter season. At Assi ghat, Janaki ghat, and Dashashwamedh ghat the BOD values were usually higher than that of the permissible limit during summer and rainy season months.

Biological Characteristics of Ganga Water

In Varanasi Lakshminarayana (1965), Rai (1977), Tripathi (1982) and Tiwari (1983) have studied various aspects of phytoplankton dynamics of river Ganga. Recently, Shukla *et al.* (1988) have performed extensive research work on the phytoplankton population, its density, diversity, similarity index and primary productivity. Macrophytic flora of river Ganga at Varanasi have also been studied.

The phytoplankton community of river Ganga in Varanasi includes 42 genera and 90 species (Table 20.3). These forms belong to the members of chlorophyceae (16 genera and 30 species), bacillariophyceae (17 genera and 42 species), cyanophyceae (7 genera and 13 species), and euglenophyceae (2 genera and 5 species). On the basis of frequency and density these *taxa* have been presented in A, B and C denoting three distinct groups. Group A is represented by the plant species present at control site and Assi ghat. These algal forms were either drastically reduced or absent at other sites. The group B species were found at Assi ghat nala and at 5 m distance from Shiwala ghat, Harishchandra ghat, Chauki ghat and Rajendra Prasad nalas. The species of group C were recorded at most of the sewage discharge points.

The most common species of green algae recorded at unpolluted site and Assi ghat are *ankistrodesmus, chlorella, closterium, coelastrum, cosmarium, mougeotia, pediastrum* and *scenedesmus*. The bacillariophyceae get abundance during winter season, the most common species being *gyrosigma, navicula, synedra*, and *nitzschia*. Other species are *amphora, cocconeis, cyclotella, cymbella, gomphonema, pinnularia* and *tabellaria*. Four species i.e. *amphora, navicula, gom-*

phonema and *nitzschia* have been recorded at almost all the polluted sites. The most abundant genus of the cyanophyceae is *oscillatoria*. Other species are *phormidium, lyngbya* and *arthrospira*.

TABLE 20.3: List of Observed Algae in River Ganga at Varanasi

Name of Algae	Unpolluted area (Group A)	Moderately polluted area (Group B)	Highly polluted area (Group C)
1	2	3	4
A. CHLOROPHYCEAE :			
Ankistrodesmus falcatus	+	—	—
Actinastrum hanatzachi	+	—	—
Chlorella vulgaris	+	+	—
C. subsala	+	—	—
Cladophora glomerata	+	+	—
Closterium rostratum	+	—	—
C. incurvatum	—	+	—
C. parvulum	—	+	—
C. paracerocum	—	+	—
Coelastrum sphaericum	+	—	—
C. reticulum	—	+	—
Cosmarium dentiferum	+	—	—
C. denatum	—	+	—
C. undulatum	—	+	—
Hydrodictyon reticulatum	+	—	—
Crucigenia truncata	+	—	—
Mougeotia sp.	+	—	—
Pediastrum simplex	+	—	—
P. boryanum	+	—	—
Scenedesmus acuminatus	+	—	—
S. bijuga	+	—	—
S. bicaudatus	+	—	—
S. falcatus	—	+	—
S. quadricauda	—	—	+
Spirogyra subsala	+	—	—
Chlamydomonas derenberji	—	+	—
C. mirabilis	—	—	+
Pandorina morum	—	+	—
Ulothrix subtilisma	—	+	—
U. zonata	—	+	—

(Contd.)

TABLE 20.3: *(Contd.)*

1	2	3	4
B. BACILLARIOPHYCEAE :			
Amphora ovalis	+	−	−
A. venata	−	−	+
Achnanthes clevei	+	−	−
A. microcephala	+	−	−
A. axigua	+	−	−
Cocconis placentula	+	−	−
Cyclotella comta	+	−	−
C. meneghiniana	+	−	−
C. stelligera	+	−	−
Cymbella offinia	+	−	−
C. termida	−	+	−
C. ventricosa	−	+	−
Fragillaria capucina	+	−	−
Gomphonema constrictum	+	−	−
G. parvulum	+	−	+
G. clevia	−	−	+
Gyrosigma acuminatum	+	−	−
Navicula cryptocephala	+	−	−
N. gracilis	+	−	−
N. mutica	+	−	−
N. cinata	−	+	−
N. minima	−	−	+
N. cuspidata	−	+	−
N. grigaria	−	+	−
Nitzschia gracilis	+	−	−
N. linearis	+	−	−
N. amphibia	−	+	−
N. acicularis	−	+	−
N. filiformia	−	+	−
N. pallia	−	+	−
N. sigmoidea	−	+	−
N. tryblionella	−	+	−
N. communis	−	−	+
N. thermalis	−	−	+
Pinnularia viridis	+	−	−
Surirella elegans	+	−	−
Synedra ulna	+	+	+
S. acus	−	+	−
Tabellaria flocculosa	+	−	−
Diatoma elongatum	+	−	−
Melosira ambigua	+	−	−
Diatoma vulgare	−	+	−

(Contd.)

TABLE 20.3: *(Concld.)*

1	2	3	4
C. CYANOPHYCEAE :			
Anabaena variabilis	–	+	–
A. cylindrica	–	+	–
Phormidium inundatum	+	–	–
Oscillatoria limosa	+	–	–
O. limnetica	+	–	–
O. formosa	–	+	–
O. subbrevis	–	–	+
O. tenuis	–	–	+
Merismopedia minima	+	+	–
M. glauca	–	+	–
Microcystic aeruginosa	–	+	–
Lyngbya limnetica	–	+	–
Spirulina major	–	–	+
D. EUGLENOIDAE :			
Euglena gracilis	–	+	–
E. acus	–	–	+
E. viridis	–	–	+
Phacus acuminatus	–	–	+
P. eusillus	–	–	+

(+) PRESENT; (–) ABSENT

The phytoplankton density (Table 20.4) was higher at control site (81.17 X 10^4 individuals per litre) and decreased gradually as are moved towards down stream. The minimum value was recorded at Rajghat (6.90 X 10^4 individual/l). The diversity index (Table 20.5) was maximum at control site and gradually decreased as the pollution load increased. It ranged from 1.1 at Rajghat to 4.96 at control site. The net and gross primary productivity values were higher at control site and reduced gradually at polluted sites, the minimum values were noted at Rajghat. The NPP ranged from a minimum zero to a maximum 2.035 $gC/m^3/d$ (Fig. 20.2). The GPP varied from a minimum 0.245 to maximum 2.28 $gC/m^3/day$ (Fig. 20.3). The phytoplankton density, diversity index and primary productivity usually decreased during rainy season months. The similarity index for phytoplankton (Fig. 20.4) reveals that Assi ghat is highly similar (s=0.95) and Rajghat is highly dissimilar (d=0.98) to the control site. Rajghat is also least similar (s=0.02 to 0.18) to all other sites. However, Shiwala, Harishchandra, Chauki and Rajendra Prasad ghats are highly similar to each other (s=0.82, to 0.89) and highly dissimilar to the control site and Assi ghat (d=0.85 to 0.94) and Rajghat (d=0.82 to 0.88).

Most common species of macrophytes (Table 20.6) are *aponogeton natans, ceratophyllum, hydrilla verticillata*, and *naja* species. Some species e.g., *azolla pinnata, lemna minor* and *spirodella polyrrhiza* appeared occasionally. *Fichhornia crassipes* was found growing near sewage discharge points on the bank of river Ganga.

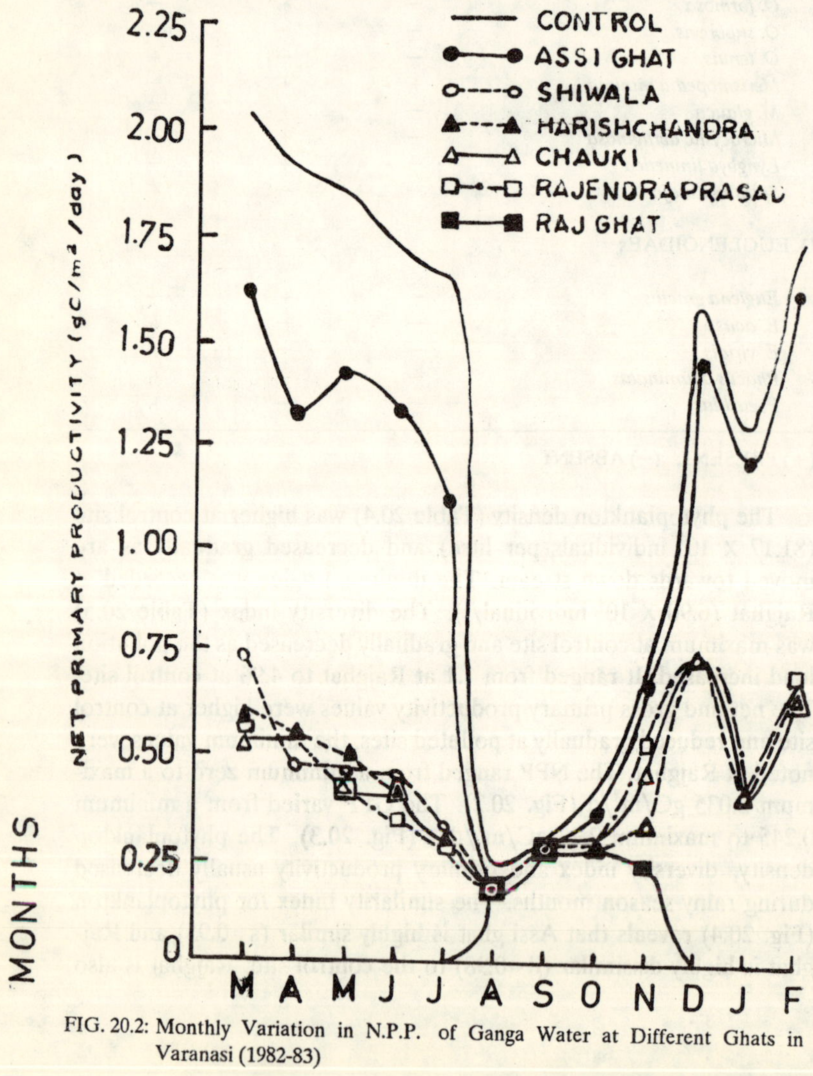

FIG. 20.2: Monthly Variation in N.P.P. of Ganga Water at Different Ghats in Varanasi (1982-83)

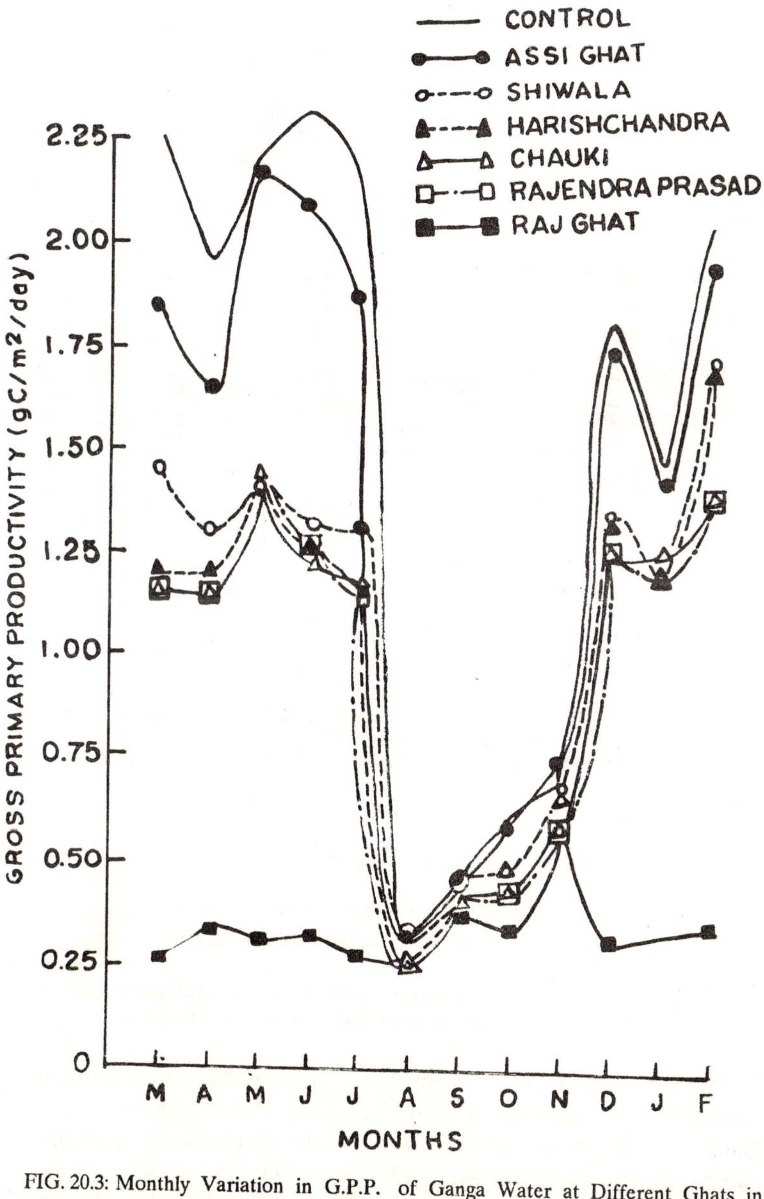

FIG. 20.3: Monthly Variation in G.P.P. of Ganga Water at Different Ghats in Varanasi (1982-83)

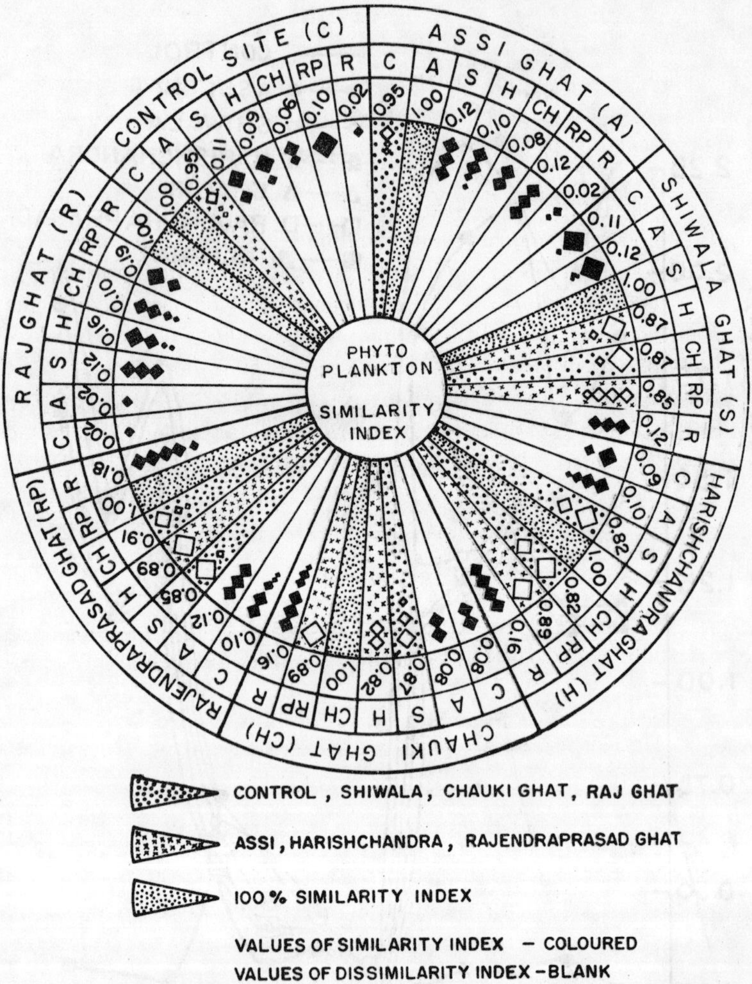

FIG. 20.4: Similarity Index Values Between Various Sites Calculated on the Basis of Phytoplankton community of River Ganga in Varanasi (1982-83)

The highest number of macrophytes (10 species) has been recorded at control site followed by Harishchandra ghat (9 species) and Chauki ghat (8 species). The macrophytic species identified near control site are *aponogeton natams, chara* sp., *naja* sp., *nitella* sp., *hydrilla verticillata, ipomoea aquatica, potamogeton, pectinatus* and *vallisneria spiralis*. Up to a distance of nine metres from the Rajghat

TABLE 20.4: Monthly Variation in Phytoplankton Density* (1×10^4 Individual/L) of River Ganga at Different Ghats in Varanasi (1982-83)

Sites	M O N T H S												Average
	MAR	APR	MAY	JUNE	JULY	AUG	SEPT	OCT	NOV	DEC	JAN	FEB	± S.D.
Control	79.10	82.70	75.60	74.47	49.40	5.70	10.00	62.23	75.56	76.73	78.27	81.17	62.63
site	±1.20	±0.89	±1.16	±1.19	±0.92	±0.06	±0.12	±0.40	±0.94	±1.62	±0.61	±1.17	±26.45
Assi	78.40	81.00	74.70	73.20	48.47	5.70	10.00	62.00	74.37	76.23	77.77	80.30	61.86
ghat	±1.40	±0.58	±1.27	±1.33	±1.32	±0.06	±0.12	±1.15	±1.24	±0.72	±1.22	±1.30	±26.11
Shiwala	53.13	57.50	50.07	48.27	38.20	6.60	9.90	49.13	50.07	51.40	52.33	55.00	43.46
ghat	±1.16	±0.78	±1.73	±0.96	±1.01	±0.06	±0.06	±1.16	±1.09	±0.74	±1.73	±1.32	±16.68
Harischandra	53.40	58.21	50.50	49.03	38.27	7.00	9.93	49.33	50.27	52.13	52.87	56.63	43.96
ghat	±1.31	±1.16	±1.15	±1.16	±0.72	±0.06	±0.09	±1.01	±1.22	±1.16	±1.30	±1.19	±16.84
Chauki	42.63	47.23	40.03	38.60	37.47	6.60	9.90	39.14	40.47	40.50	42.50	44.63	35.81
ghat	±1.39	±1.30	±0.87	±1.73	±1.03	±0.06	±0.06	±1.17	±1.16	±1.15	±1.44	±1.16	±12.87
Rajendra Prasad	52.33	56.70	49.17	48.00	37.70	7.00	10.20	49.13	50.03	50.53	49.90	54.03	42.90
ghat	±1.45	±0.66	±1.17	±1.15	±1.30	±0.06	±0.16	±1.16	±1.16	±0.99	±1.07	±1.18	±16.26
Rajghat	11.63	12.57	11.67	11.50	11.53	6.90	10.07	12.17	12.27	12.07	11.53	12.13	11.33
	±0.59	±0.58	±0.35	±0.29	±0.29	±0.06	±0.14	±0.12	±0.18	±0.18	±1.73	±0.20	±1.57
Average	52.50	56.59	50.25	49.00	37.24	6.50	10.00	46.14	50.43	51.45	52.17	54.93	43.14
± S.D.	±21.91	±22.37	±20.67	±20.28	±11.94	±0.55	±0.19	±16.22	±20.41	±21.10	±21.59	±22.23	

* MEAN ± S.E.

TABLE 20.5: Monthly Variation in Diversity Index (Bits per Individual) of Phytoplanktons of River Ganga at Different sites in Varanasi (1982-83)

Sites	M O N T H S												Average ±
	MAR	APR	MAY	JUNE	JULY	AUG	SEPT	OCT	NOV	DEC	JAN	FEB	S.D.
Control site	4.34	4.96	4.23	3.77	3.20	1.99	2.07	3.17	3.55	3.86	4.27	4.79	3.68
Assi ghat	4.20	4.81	4.10	3.62	3.06	1.90	2.00	3.10	3.45	3.76	4.16	4.69	3.57
Shiwala ghat	3.03	3.72	3.01	2.50	2.19	1.85	1.95	3.01	2.48	2.60	3.03	3.52	2.74
Harischandra ghat	3.18	3.72	3.01	2.49	2.02	1.86	1.98	3.05	2.42	2.55	3.04	3.52	2.74
Chauki ghat	2.82	3.61	2.87	2.42	2.00	1.85	1.93	3.03	2.23	2.41	3.05	3.44	2.64
Rajendra Prasad ghat	2.44	3.10	2.98	2.11	2.12	1.84	1.69	2.82	2.73	2.52	3.13	2.93	2.53
Rajghat	1.10	1.42	1.21	1.11	1.62	1.70	1.64	1.26	1.14	1.24	1.78	1.25	1.37
Average	3.02	3.62	3.06	2.57	2.32	1.86	1.89	2.78	2.57	2.71	3.21	3.45	

TABLE 20.6: Distribution of Macrophytes in River Ganga at Different Sites in Varanasi (1982-83)

Name of the Species	SAMPLING STATIONS						
	Control Site	Assi Ghat	Shiwala Ghat	Harishchandra Ghat	Chauki Ghat	Rajendra Pd. Ghat	Rajghat
Aponogeton natans	+	–	+	+	–	–	–
Azolla pinnata	–	–	–	+	+	–	+
Chara sp.	+	+	–	–	–	–	–
Ceratophyllum demersum	–	–	–	+	–	+	+
Ceratophyllum sp.	+	–	–	–	–	–	–
Cyperus rotundus	+	–	–	–	+	–	–
Eichhornia crassipes	–	+	+	–	+	–	+
Hydrilla verticellata	+	+	+	+	+	–	+
Ipomoea aquatica	+	–	–	+	–	–	–
Lemna minor	–	+	–	–	+	+	+
Najas graminea	+	–	–	+	–	–	–
Nitella sp.	+	–	–	–	+	–	–
Pistia stratoites	–	–	–	–	+	–	+
Potamogeton crispus	–	–	–	+	–	–	–
Potamogeton pectinatus	+	–	+	+	–	–	–
Spirodella polyrrhiza	–	–	–	–	+	–	+
Vallisneria spiralis	+	–	–	+	–	–	–
Number of species occuring	10	4	4	9	8	2	7

sewage discharge point no macrophytic vegetation was noticed. Thick vegetation has been recorded after about 100 m distance down stream. This is comprised of *fichhornia crassipes* near the bank and *hydrilla verticillata* towards slightly deeper region. They were sometimes intermingled by *lemna minor, azolla pinnata, ceratophyllum demersum* etc.

Agrawal *et al* (1976a) have performed a bacteriological study of the Ganga river from bathing ghats and sewage outfalls at Varanasi. Near University ghat and Bhadaini ghat where no sewage is discharged, the mean MPN index/100 ml was 2,408 and 1,054 respectively. The facal coliform was 468 and 595 respectively. At the bathing ghats the average MPN index/100 ml was 9219 and the value of facal coliform was 4598/100 ml. The values increased manifold near the sewage outfalls (MPN index $152.4 \times 10^3/100$ ml and facal coliforms $128.4 \times 10^3/100$ ml). The pathogenic bacteria reported from the sampling stations, are *vibri cholerae, vibrio nag, salmonella* group A, *salmonella* group B and *salmonella* group C, *Shigella boydii* type 1, *shigella flexneri* and *salmonella* E etc.

Statistical Analysis

Author has obtained a positive and significant correlation of BOD with COD, nitrate, temperature and acidity. A similar correlation was exhibited by electrical conductance with transparency, alkalinity, DO and chloride. Dissolved oxygen showed a significant and positive correlation with transparency, alkalinity and phosphate. A significant negative relationship of temperature with transparency, DO, C1, EC and phosphate has also been established. Similar relationships have been shown by acidity with transparency, pH and DO with BOD, COD and NO_3.

Phytoplankton density and diversity index both exhibited significant and positive correlation with dissolved oxygen and transparency of Ganga water. Moreover, a significant inverse relationship between diversity index and temperature, diversity index and nitrate, and a positive and significant correlation between primary productivity and transparency has been recorded by the author.

A significant and positive correlationship between phytoplankton density and dissolved oxygen have been established by Tripathi (1982) and Tiwari (1983) in Ganga water and by Venkateswarlu

(1969b) in river Moosi. However, Lakshminarayana (1965) and Rishi (1983) have reported negative relationship. Author has obtained negative and nonsignificant relationship between phytoplankton density and DO content of water at control site. At polluted sites the relationship was significantly positive. Similar relationship has been established between nitrate and cell density by the author.

Dead Body Cremation at Varanasi

In Hindu traditions, the dead bodies are burnt and ashes are scattered into the rivers or other water bodies. In Varanasi city, which is considered to be the holiest among all Hindu religious places, dead bodies are brought from distant places even out of Varanasi district. The ritual is performed at Harishchandra and Manikarnika ghats on the bank of river Ganga. The dead bodies are usually burnt with the help of fire wood. In open burning system about 95 per cent energy is wasted.

Tripathi *et al.* (1984), Sikandar and Tripathi (1984) and Sikandar (1987) have studied the impacts of dead body cremation on river Ganga environment and reported that at the aforesaid two burning ghats of river Ganga at Varanasi 23,312 dead bodies in the year 1971; 27,479 in 1975 and 28,701 in 1980 were cremated (Table 20.7) using dry wood biomass 8,159 ton (350×10^8 k cal wood energy), 9618 ton (414×10^8 k cal wood energy) and 10045 ton (432×10^8 k cal wood energy) respectively. Using polynomial equation it has been predicted that 32,478 and 35,952 dead bodies will be cremated at Varanasi in 1990 and 2000 A.D., with the help of wood biomass and energy 11,376 ton (488×10^8 k cal) and 12,583 ton (541×10^8 k cal) respectively. They have reported that 51,321 kg nitrogen (N), 31,821 kg potassium (K) and 3,753 kg phosphorus (P) in the year 1971, 60,495 kg N, 37,509 kg K and 4424 kg P in the year 1975, and 63185 kg N, 39,177 kg K and 4,621 kg P in the year 1980 was released into the river. The predicted quantities of minerals to be released are 70,914 kg N, 43,969 kg K, 5,186 kg P in the year 1990 and 77,869 kg N, 46,281 kg K and 5624 kg P in 2000 A.D.

The calculated amount of ash content discharged into the river after burning of huge amount of wood biomass were 117 ton in 1971, 137 ton in 1975, and 144 ton in 1980. The predicted values of ash content to be discharged into the river are 161 ton and 177 ton for the years 1990 and 2000 A.D. respectively. The amount of half-burnt

TABLE 20.7: Dead Body Cremation and its Effect on Ganga River Ecosystem

Year	Number of dead bodies cremated	Wood Biomass required (in ton)	Energy consumption (1×10^8 k cal)	MINERALS (in kg)		
				Nitrogen	Potassium	Phosphorus
1971	23312	8159.2	350.8	51321.4	31820.9	3753.2
1972	31990	11196.5	481.5	70425.9	43666.3	5150.4
1973	25517	8930.9	384.0	56175.7	34830.7	4108.2
1974	25185	8814.8	379.0	55432.8	34377.5	4054.8
1975	27479	9617.6	413.6	60495.0	37508.8	4424.1
1976	25961	9086.4	390.7	57153.0	35436.7	4179.7
1977	27556	9644.6	414.7	60664.5	37613.9	4436.5
1978	28108	9837.8	483.0	61879.7	38367.0	4526.6
1979	30600	10710.0	460.5	67365.9	41769.0	4926.8
1980	28701	10045.4	431.9	63185.2	39176.8	4620.8

flesh thrown into the river after the cremation of dead bodies were 140 ton, 165 ton and 172 ton for the years 1971, 1975 and 1980, respectively. The predicted values of half-burnt flesh to be thrown into the river Ganga at Varanasi in 1990 and 2000 A.D. are 193 ton and 212 ton respectively. About 3,000 human and 6,000 animals unburnt dead bodies are also dumped into river Ganga every year at Varanasi.

With the population growth and availability of transport facilities, an increase in the number of dead bodies cremated at burning ghats of Varanasi has been observed with increased demand of wood biomass. The traditional method (open fire system) of energy extraction from wood is wasteful because of its low efficiency i.e 3 to 7.6 per cent (Bodria and Sangiorgi, 1983). Therefore, the use of open fire system in India for the burning of dead bodies is wasteful. During burning process whole of the nitrogen and about 80 per cent phosphorus are released into the atmosphere. However, other minerals locked up in wood and released into the river.

Most of fuel wool comes from forests. Massive deforestation results problem of soil erosion, sitation, flood and water pollution and even decrease in precipitation which make the soil infertile. Huge amount of ash content released from the cremation ground makes the water turbid, enhances siltation process at the ghats and help the growth of aquatic weeds. Addition of half-burnt flesh content is also responsible for the decrease in DO content and increase in BOD and COD values. The disposal of unburnt animal and human dead bodies into the river may also cause nuisance and pollution problem. It has been calculated that by using electric crematoria for burning dead bodies about 464×10^8 k cal and 515×10^8 k cal of energy could be saved in the years 1990 and 2000 A.D. with saving Rs. 5,986 thousands and Rs. 6,585 thousand respectively.

REFERENCES

Agrawal, D.K., S.D. Gaur, I.C. Tiwari, N. Narayanswami and S.M. Marwah. (1976a), Physico-chemical characteristics of Ganges water at Varanasi. *Indian Journal Environmental Health*, vol. 18(3): 201-6.

Agrawal, D.K., S.D. Gaur, P.C. Sen and S.M. Marwah. (1976b), Bacteriological study of Ganges water at Varanasi. *Indian Journal Medical Research*, 64(3): 373-83.

Ambasht, R.S. (1988), Plant population and community structure and function of riparian and aquatic macrophytes in relation to pollution of river Ganga from Mirzapur to Ballia. *Technical Report*, Department of Environment, Project,

Govt. of India.

Bhowmick, B.N. and A.K. Singh (1985), Phytoplankton population in relation to physico-chemical factors of river Ganga at Patna. *Indian Journal of Ecology*.

Bilgrami, K.S. and J.S. Dattamunshi. (1979), Limnological survey and impact of human activities on the river Ganges. Barauni to Farakka. *Technical Report*, MAB Project, UNESCO.

Bilgrami, K.S. and F.N. Siddiqui. (1980), Effects of industrial pollutants on phytoplankton composition of river Ganga at Barauni. *Indian Botanical Society*, 59: 9-10.

Bilgrami, K.S. and J.S. Dattamunshi. (1988), A multidisciplinary coordinated project on "Study of river Ganga". Munger to Farakka. *Technical Report*, Department of Environment Project, Govt. of India.

Biswas, K. (1942), The role of common algal communities of the river Hooghly on the drinking water of Calcutta. *15th Ann. Vol. Roy. Gdn*. pp. 189-206.

Bordia, L. and F. Sangiorgi. (1983), Wood burning stoves and their efficiency. The case of Kenya, *Renewable sources of Energy*, 1(2): 131-5.

Chattopadhyaya, S.N., T. Routh, V.P. Sharma, H.C. Arora and R.K. Gupta. (1984), A short term study on the pollutional status of river Ganga at Kanpur region. *Indian Journal of Environmental Health*, 26(3): 244-57.

Choudhary, U.K. (1988), Coordinated study of river Ganga Civil Engineering Aspects. Mirzapur to Ballia. *Technical Report*, Department of Environment Project Govt. of India.

Das, T.M. (1988), Survey of Agro-chemicals in the riverine lands and its impact on the level of the pollution of Ganga water. Bally to Banded. *Technical Report*. Department of Environment Project, Govt. of India.

Lakshminarayana, J.S.S. (1965), Studies of the phytoplankton of the river Ganges, Varanasi, India, Part I. Physico-chemical characteristics of river Ganga. *Hydrobiologia*, 25: 119.

Mathur, R.P. (1988), Pollution modelling of upper Ganga basin. *Technical Report*, Department of Environment Project, Govt. of India.

Mehrotra, M.N. (1988), Role of sediments in attenuation of trace metals in polluted aquatic environment of river Ganga. Mirzapur to Ballia. *Technical Report*, Department of Environment Project, Govt. of India.

Pahwa, D.V. and S.N. Mehrotra, (1966), Observation on fluctuations in the abundance of plankton in relation to certain hydrological conditions of river Ganga *Proceedings of the National Academy of Sciences* India, 36B(2): 157-89.

Pandey, B.N. and B.D. Tripathi. (1984), Contamination of Ganga water and its impact on mammals. *Proceedings of National Conference on River Pollution and Human Health*. pp. 29-32.

Rai, L.C. (1977), Ecological studies of algal communities of the Ganges river at Varanasi. *Indian Journal of Ecology*, 5(1): 1-6.

Ray, P. and A. David (1966), Effect of industrial wastes and sewage upon the chemical and biological composition and fisheries of river Ganges at Kanpur, U.P. *Environmental Health*. 8: 307-39.

Rishi, V. (1983), Ecology of stream of Doodh Ganga catchment area. *Ph.D. Thesis*, Kashmir University.

Roy, H.K. (1949), Some potamological aspects of the river Hooghly in relation to calcutta water supply. *Science in Culture*. 14(8): 17-20.

Roy, H.K. (1955), Plankton ecology of the river Hooghly at Patt., West Bengal. *Ecology*, 36(2): 169-75.

Rupenwar, D.C. (1988), Pollution of river Ganga with heavy and toxic metals. Mirzapur to Ballia. *Technical Report*, Department of Environment Project, Govt. of India.

Saran, R.K. (1988), An interdisciplinary study on the problems of water pollution of the river Ganga with special reference to luota. Buxar to Barh. Department of Environment Project, *Technical Report*, Govt. of India.

Samaddar, K.K. (1988), Biotic communities and productivity of river Ganga and its banks as affected by pollution. Dehagram to Naihati. *Technical Report*, Department of Environment Project, Govt. of India.

Saxena, K.L., R.N. Chakrawarty, A.K. Khan, S.N. Chattopadhyaya and Harish Chandra (1966), Pollution studies of the river Ganges near Kanpur, *Environmental Health* 8: 270-85.

Shukla, S.C., Rajani Kant and B.D. Tripathi. (1988), Ecological Investigation on physico-chemical and phytoplankton productivity of river Ganga at Varanasi. *Geobios*. vol. 15.

Sikandar, M. (1987), Ecology of river Ganga in Varanasi with special reference to pollution. *Ph.D. Thesis*. B.H.U., Varanasi.

Sikandar, M. and B.D. Tripathi. (1984), The physico-chemical characteristics of Ganga water at Varanasi. *Proceedings of the National Conference on River Pollution and Human Health*. pp. 53-61.

Sikandar, M., B.D. Tripathi and Suresh C. Shukla. (1988), Possible ecological effects on the dead body cremation on the bank or river Ganga, *Tropical Ecology* (Accepted).

Singh, C.S. (1962), Studies on the growth and distribution of fresh water diatoms in relation to physico-chemical conditions of some ponds and rivers of U.P. *Ph.D. Thesis,* B.H.U. Varanasi.

Singh, H.R. (1988), Pollution study of the upper Ganga and its tributaries. Gangotri to Rishikesh, *Technical Report*, Department of Environment Project, Govt. of India.

Singh, A.K. and B.N. Bhowmick. (1985), Effect of sewage on physico-chemical characteristics and bacterial population of river Ganga at Patna. *Indian Journal of Ecology*, 12(1): 17-9.

Sinha, A.K. (1986), A comprehensive study of Ganga and its dependents. *Technical Report*, Department of Environment Project, Govt. of India.

Tiwari, D. (1983), Pollution phycology of the Varanasi frontage of river Ganges, *PH.D. Thesis*, B.H.U., Varanasi.

Tripathi, C.K.M. (1982), Investigation on Ganga river to determine biological indicators of water quality. *Ph.D. Thesis,* B.H.U., Varanasi.

Tripathi, B.D. (1988), Physico-chemical and biological characterization or river Ganga in relation to pollution — Mirzapur to Ballia. *Technical Report*, Department of Environment Project, Govt. of India.

Tripathi, B.D., M. Sikandar and R.S. Ambasht. (1984), Ecological investigation on the plant biomass and energy used for the burning of human dead bodies at Varanasi ghats, *Proceedings of the National Conference on River Pollution and Human Health*, pp. 62-9.

21

GROUNDWATER WITHDRAWAL : A Cause of Environmental Problems

G.C. CHOWDHARY

VIRAT PURUSHA
After coming out of the "Golden egg" created unpolluted water to live in a pure environment (2/10/9).

— *Sri Bhagavata Purana* by Vyas (200 BC)

Pollution and utilization of groundwater especially for drinking have been points of great concern for people in India since ancient days. In the past great man of the medicine world sage *SUSHRUTA* (200 B.C.) prescribed many properties of groundwater regarding human health and one of them is that, "Water from a well when taken generates bile, which may be bad for health sometimes, but is mostly appetising". At the same time he suggested that impurities in water may be treated with the help of special stones and minerals as well as various types of vegetations (Chowdhary and Agrawal, 1981). Much late during the Christian era (11c A.D.) King Someshwara recommended in his epic *Mansollas* various uses of water from different sources at different times in a year from the point of view of health and preferred *Bhaum Jal* or groundwater to be taken during rainy season as it is free from all pollution. Behind it is attached an important fact that in a country like India where 69 per cent of water received as rains goes from land as run off both surface and subsurface, during all seasons in a year. He prescribed consumption of groundwater for drinking as it is always safe. The idea is to save oneself from water pollution effects. The famous incidence of the jet of water which gushed out from the ground when Arjuna pierced an arrow in it, is quoted by Vyas in *Mahabharata* (200 B.C.). We have such references of ancient period in our old scriptures many of them are about water pollution.

During the modern period groundwater studies first related to its quantitative assessment in principle and now in these days of concern about environmental pollution, its multiple use has been viewed with a great suspicion regarding its quality under the conclusion that for the major part the adverse effects on groundwater quality are the results of man's activity at ground surface, unintentionally by agriculture and industry, unexpectedly by sub-surface disposal of sewage and industrial waste water and by solid waste dumps, illegally by small workshop and unfortunately also by the abstraction of groundwater itself (Huisman, 1981). The last problem is connected by underground activities too i.e. mining. This aspect could be taken as absolutely new, since land subsidence or land surface settling although is much a natural phenomenon occurring all over the world, is now taken marked often as a result of man's activities, such as mining of water, gas, oil from underground sources. But, whether of natural or man made causes, the effects of land subsidence are mostly undesirable, ranging from minor changes in the landscape to major damages to structures or environment itself. Often these deleterious effects do not become evident until long after their cause has been set in motion.

Subsidence of Land

The seriousness of land subsidence was recognised by the UNESCO as long ago as in 1964, when this problem was first included for study under the International Hydrological Decade Programme and later the International Hydrological Programme. Carbognin (1985) has recognised both the natural and unnatural causes among the most common causes which leads to subsidence. According to him the most common manmade subsidence results from:

1. **Removal of fluids from the subsoil**
   - (a) Withdrawal of groundwater, thermal water and gas bearing water, causing a decline of the water head;
   - (b) Pumping of crude oil, natural gas, oil and associated gas, causing underground pressure depletion.

2. **Compaction of fine sediments induced by**
   - (a) *Application of water*: Some unconsolidated dry deposits collapse

when wetted; this process has been referred by him as hydro-compaction or shallow subsidence.

(b) Land drainage and reclamation and biochemical oxidation of heat and organic soils.

(c) Loading by buildings or other engineering structures and vibration of sediments.

3. **Mineral mining**

Pumping of sub-surface fluids causing subsidence was first observed in Galvaston Bay, Texas, U.S.A. in the early 1920s. It was pumping of petroleum causing land to sink and sudden development of cracks in the ground. It is no wonder the heavy extraction of water for industrial, domestic and agricultural uses may result in such catastrophic phenomenon.

Geological conditions play a great part in such phenomenon. Deposits composed of unconsolidated mostly of late Cenozoic and Mesozoic sediments bear high initial porosity. It is observed that almost the subsiding strata tapped are, confined or semi-confined water bearing formations (aquifers) naturally, made up of loose sand or gravel and usually with high permeability and low compressibility, interbedded with layers of clay and/or silt of low vertical permeability and high compressibility, i.e. the aquitard beds. Both aquifers and aquitards are found in varying thickness. So far subsidence by extraction of oil and water are concerned they differ but have the same mechanism. The mechanism has some variables as size of the reservoirs, the lithologic nature of the deposits, their depth and physical characteristics, the effective tension and its increase and the time of subjection to the increase. All these variables determine the nature of the subsidence.

Consequences of tapping of water is very interesting as it is the most common sub-soil fluid for extraction by mining. In this phenomenon structure of sub-soil plays a vital role. Commonly, sand and gravel, silt, clay and peat are found as common to the entire zone of subsidence. Each of them have properties typical of their own. For example Sand and Gravel — differ in size of their particles which are commonly loose and non-coherent. Naturally, their high permeability makes them a good aquifer. Ofcourse, when subjected to loading their behaviour is essentially elastic. Silts are intermediate in size and properties, between bigger sands and smaller grains of clays. Thus they are of medium permeability. Clay are smaller than silt grains. It develops plasticity when limited amount of water is present and naturally the permeability values are very low. Still microscopic size of sediments are that of *Peat*, mostly composed of organic matters while all the above, i.e. sand, gravel, clay and silt, are made up of rock fragments. But it is highly compressible. However, the basic physical characteristics which the behaviour of sub-soil depends for subsidence comes out to be permeability and consolidation. Some well known factors about these two factors related to subsidence can be described here.

Regarding permeability indeed, the coefficient is higher for the coarse material and lower for fine grained sediments. The law of permeability had already been formulated by Darcy as said before as early as 1856 the reaction of an aquifer system of any, more or less rapid, perturbation could only be understood when Karl Terzaghi developed his theory of vertical consolidation in 1923, introducing the fundamental principle of *Effective stress*. Using this principle from a practical point of view looking to the Figs. 21.1, 21.2 and 21.3 one can speak as given under the captions with each figure. So far validity of these phenomenon is concerned it is dependent on consolidation of strata.

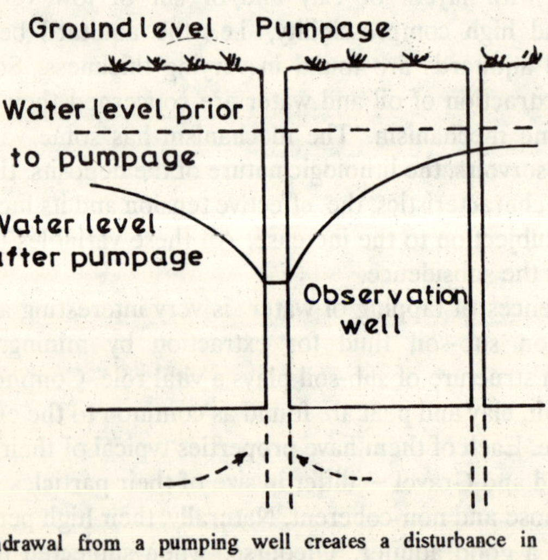

Water withdrawal from a pumping well creates a disturbance in the pre-existing hydrodynamic equilibrium and generates a cone of depression.

FIG. 21.1: Water Withdrawal from Pumping Well

The decrease of the water pressure produces on increase of the effective intrgranular stress under whose influence soils compact

FIG. 21.2: Decrease of the Water Pressure

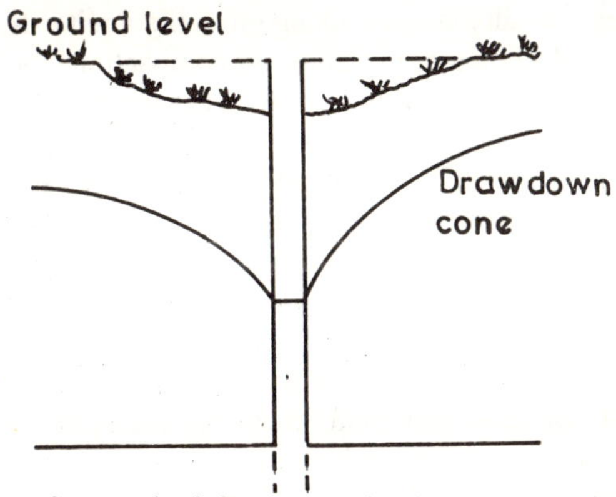

The underground compaction induces a ground surface settlement, which is proportional to the decrease in water pressure and follows the shape of the drawdown cone. (After Carbognin, 1985)

FIG. 21.3: Underground Compaction

It is valid as given in Figs. 21.1, 21.2 and 21.3 for poorly consolidated strata, however, for strata where grains are strongly held together as usually in hard rocks with only exceptions of carbonate rocks with solution cavities where subsidence is only possible due to it, that the rock shall retain its structure after the water has been withdrawn, without any subsidence in loose sediments in concerned, has another story.

Consolidation is the gradual reduction of the soil volume when underground equilibrium conditions are disturbed. Compaction differs with consolidation as a decrease in the volume by increase in load. In a broad sense they are used with the same meaning. It is because in all cases it decreases either by overloading or by percolation of water from pores. Their evaluation by usual odemometric test gives different result. In compaction it occurs that the flow structure partly breaks and a part is absorbed by the work of elastic deformation. However, during consolidation and compaction taking place by increase of hydrostatic pressure, water content and volume increases. In a sand bed where flow structure is absent, greatest part of energy is spent in the permanent breakdown of the flocculation structure. The two contrary responses of clay and sand bring out two important observations as below, made by Carbognin.

Firstly, it is clear that aquifers of sand and gravel compact quite elastically, their compaction is negligible owing to their low compressibility. Clayey and silty layers making often the aquitards, are highly compressible and will undergo significant compaction which is mostly non-recoverable. In hard rocks especially carbonate a very different case is observed.

The carbonate rocks, limestone and dolomite, are responsible for the most wide spread incidence of subsidence related to solution, not because of a high degree of solubility, but because of wide geographic distribution. A great deal of information is available on solution fractures in carbonate rocks. Tolson and Doyle; (1977), Transportation Research board (1976). La Moreaux and others (1975) have listed more than 50 symposia and conferences on hydrology of carbonate rocks held throughout the world in the past 30 years.

The incidence of sinkhole development may be greatly increased when equilibrium conditions are disturbed by man's construction of projects or mining operations particularly those that alter groundwater levels or increase surface infiltration (Allen, 1984). Earlier, Newton (1976) noted that more than 4,000 induced sinkholes, areas of subsidence or other related features have occurred in Alabama since 1900 and that too most of them since 1950.

In Missouri, 97 catastrophic surface failures have been recorded since the 1930's of which 46 were attributed to man's activities (Williams and Vineyards, 1976). Subsidence accelerated by dewatering of under ground mines in carbonate terrain has been described by Foose (1968) two decades ago.

Collapse at the ground surface may appear suddenly, but it is the culmination of a sequence of processes starting with the development of solutions. Passage always develop over geologic time and persist owing to a combination of the slow rate of dissolution of Carbonate rocks and their high compressive strength, which maintains the integrity of the cavity systems. Subsequently, unconsolidated over burden materials may be slowly washed down into bed rock cavity systems. The resulting voids in the overburden may become enlarged until the remaining cover is too thin to support the surface and collapse takes place. In the same decade another case of hardrock mining related to groundwater pollution resulting in environmental pollution has been reported from East Central Scotland by Henton (1981). There arose the problem of water table rebound after mining activity and the water quality was effected.

It is well known that mining activity, open or closed, effects the hydrogeology and also interferes with the natural surface water hydrology. In temperate climatic zones infiltration from surface needs removal of excess groundwater accompanied with mining. They may result depletion of general water level in the mining area and also reduction of the base flow of river as reported in river Ore valley in Scotland. Also deterioration in water quality was marked. However, the fact that, this river recovered as water table rebound eventually restored better base flows but streams in adjoining mining areas elsewhere which do not receive pumping waters could be depleted for many years was observed in the Fourth Annual Report (1966) of River Purification Board (U.K.). The incidence quoted brought out many observations which are useful for other areas as well. It was found that where there is no check on mine water discharge, after the mine is abandoned, a rebound of water table takes place. In mines where ferragineous water is high in dissolved iron salts and contain free sulphonic acid coating of ferric hydroxide on flora and fauna, and bacterial slime are marked. Farmers in the field may complain of water logging. All streams when charged with it may become fishless. In Scotland during 1979 the river Ore became free of best salmon fishes and the mine had to be closed resulting unemployment to 500 persons.

Remedial measures for such incidences are chemical treatment neutralization with lime followed by settlement or simply oxidation. Second method is by continued pumping from suitably sited shafts which could dewater sensitive zones in the same way as active coal field. There is an easy measure for coal-field near a sea. A channel to the sea can be constructed which atleast protects the fresh water streams. Although people are not fussy about iron discharges through groundwater leading to the level of environmental pollution, but no such excuse from man made causes would be in all circumstances be acceptable.

Acknowledgement

This article is due to Prof. B.B. Dhar's Deptt. of Mining, B.H.U. constant encouragement and persuasion to write one for this book.

REFERENCES

Allan, A.S. (1984), *Types of Land subsidence*, UNESCO-Paris Studies and Report in Hydrology, No. 40, pp. 133-38.

Carbognin, L. (1985), *Land Subsidence: A Worldwide Environmental Hazard*, UNESCO, Nature and Resources, Vol. XXI, No. 1, p. 2.

Chaudhary, G.C. (1981), Water: Some facts from Indian Scriptures, *Prajna*, B.H.U., Varanasi, Vol. 26 and 27, No. 10, p. 29.

Foose, R.M. (1968), Surface subsidence and collapse caused by groundwater withdrawal in carbonate rock areas, in the *Proc. 23rd Int. Geol. Congress*, Vol. 12, p. 155.

Henton, M.P. (1976), Predicting the quality of mine water discharges, *Jr. of Effluent and Water Treatment*.

Henton, M.P. (1981), The problem of water table rebound after mining activity and its effect on groundwater quality, *Studies in Environmental Sciences*, Vol. 17, Elsevier Scientific Publishing Company, New Delhi, p. 111.

Huismann, I.L., (1981), Quality of groundwater, *Proceedings of International Symposium*, Elsevier Scientific Publishing Company, New York.

Keller, E.A. (1978), *Environmental Geology* (IInd Edition), Charles E. Merill Publishing Company, London.

La-Moleaux, P.E., Legrand, H.E. and Stringfield, V.T. (1975), *Progress of Knowledge about Hydrology of Carbonate Terrains*, International Association of Hydrogeologists and International Union of Geological Sciences, Series B., No. 3, Paris, pp. 41-51.

Mishra, S.S. (1966), *Mansollas — Ek Sanskritic Adhyayana*, Chaukhambha Sanskrit Series, Varanasi.

Newton, J.G. (1976), *Induced and Natural Sinkholes in Alabama — A Continuing Problem Along Highway Corridors*, Nat. Acad. Sc. Transportation Research Record 612, Part I, pp. 9-16.

Sharma, K.L. (1963), *Sushruta Samhita*, Chaukhambha Sanskritic Series, Varanasi.

Terzaghi, K. (1925), Principles of Soil Mechanics, *Engineering News Record*, Vol. 95, p. 987.

Tolson, J.S. and Doyle, F. Leds. (1967), *Soil Mechanics in Engineering Practices* (II edition), John Wiley and Co., New York.

Vyas, *Mahabharata*, Indian Press Publications, Allahabad, 1920.

22

ASSESSMENT OF WATER QUALITY OF A PERENNIAL STREAM FROM SEMI-ARID MAHARASHTRA FOR HUMAN CONSUMPTION AND IRRIGATION

V.S. KALE, N.J. PAWAR and S.C. GUPTE

In recent decades, the uncontrolled growth of population in all the states of the Indian Union has demanded the rapid expansion of water supplies. In semi-arid areas where natural water resources are not plentiful, even the problem of drinking water is more acute. This is partly due to higher frequency of droughts and partly due to over-exploitation of the available surface and sub-surface water resources (Tarate, 1988). Although, to overcome this problem, the union and state governments have undertaken programmes to build large storages for greater water transference to the main centres of consumption, an unbalanced demand-supply situation continues to prevail in many parts of the country. In the drought-prone areas the sources of drinking water and disposal of sewage are scarcely separated and therefore, they are a cause of increasing concern to the health department. Further, natural waters are relatively dilute aqueous solutions containing varying concentration of organic and inorganic constituents. It is therefore, possible that excess concentrations of certain constituents naturally can make the water unacceptable for human consumption as well as for irrigation purpose.

Studies on river water quality are gaining increasing importance in our country and many exhaustive and comprehensive studies have been carried out (Sivagnanam and Kumarswamy, 1983; Pawar, 1985; Handa *et al,* 1986; Phadtare, 1986 and Subramanyam, 1987). In Maharashtra, the Maharashtra Engineering Research Institute, Nasik (MERI, 1982) undertook a three year study of water quality of rivers and reservoirs. The present study is designed to assess the water quality of a small perennial stream in the severely drought-prone area of the Sangli district of Maharashtra, both for human consumption and irrigation.

The Sukadeo Stream : An Introduction

The stream under review originates on the Khanapur plateau, at an altitude of 903 m ASL and descends down the plateau to 680 m ASL within a short distance of 5 km, giving rise to an impressive waterfall. At the edge of the plateau a perennial spring feeds the stream and below it, the stream is characterised by thick calcareous deposits. (Fig. 22.1).

Geographically, the stream is located in the Man basin in southern part of western upland Maharashtra. Geologically, it is characterised by Cretaceous-Eocene basalts, and geomorphologically, the area reflects a semi-arid (rainfall between 500 to 650 mm) landscape. Lithologically, the Khanapur plateau is underlined by varieties of basalts which include, compact, fine grained and amygdaloidal basalts. Interestingly, although the area is semi-arid, Neogene — Quaternary laterites occur in the source of the stream under review. These highly porous rock formations favour infiltration of rain water leading to unusually high groundwater levels. Further, along the plateau edge, springs are located at the contact of underlying compact basalt and overlying vesicular and pervious basalts.

In the absence of any other source of water for drinking and irrigation purposes, the local population entirely depends upon the perennial Sukadeo stream. Therefore, in order to study the water quality of the stream, the present investigation has been undertaken.

Measurement of Water-Quality Variables

While, the natural quality of stream water depends upon its interaction with the rock and various geochemical reactions taking place within the lithologic framework (Drever, 1982), the pollution and deterioration of water quality is related to the impact of human activity. Hence, to assess the quality of water, analysis of physico-chemical parameters as well as major and minor ions present in the stream waters is imperative. This is achieved by proper selection of sampling sites, collection of water samples and by chemical analysis of water.

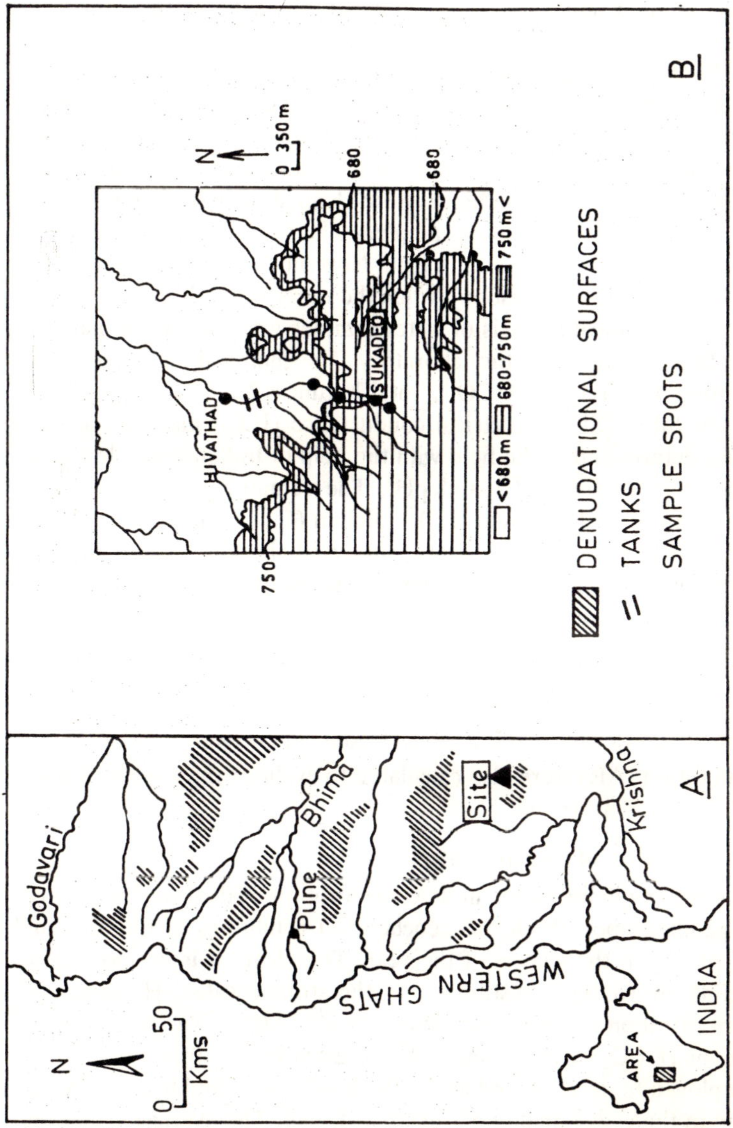

FIG. 22.1: Location Map

The choice of sampling spot, timing and frequency of sampling was based on field survey and the geographical knowledge about the area.

Four sampling spots were identified along the Sukadeo stream, after a detailed dumpylevel survey (Fig. 22.1). Since, water is susceptible to changes in the chemical composition due to its powerful solvent properties, one litre sample was collected in plastic containers from each spot. The containers were cleaned with distilled water and rinsed with the water being sampled. The sampling programme was undertaken once every month from November, 1987 (post-monsoon) to May, 1988 (pre-monsoon), when the water is not diluted.

The chemical analysis of water samples was carried out partly in the field and partly in the laboratory. The chemical parameters such as total alkalinity (as HCO_3), total hardness (as $CaCO_3$), Chloride (Cl–), Silica (SiO_2) electrical conductance (E.C.) and Ca hardness were determined in the field with the help of Indian portable kit, besides temperature (^{O}C) and pH. The remaining constituents viz. Sodium (Na^+), Potassium (K^+), Calcium (Ca^{2+}), magnesium (Mg^{2+}) and Sulphates (SO_4^{2}–) were analysed in the water analysis Laboratory. The flame photometric techniques were used for the determination of Na^+ and K^+. The pH was re-checked by using digital pH meter. The SO_4^{2}– was determined by using turbidity method (HMSO, 1972). In addition to these, Ca^{2+} and Mg^{2+} were analysed by using atomic absorption spectrophotometer (AAS). Further, Kelly's ratio, Ca/Mg ratio, total dissolved solids (TDS) and sodium absorption ratio (SAR) were also calculated to evaluate the quality of water of the Sukadeo stream.

Geochemistry of Stream Water

The water quality varies in its concentration of dissolved substances depending upon the hydro-geochemical conditions of the area and according to the season of the year. Therefore, it is not possible to define an average constituency of the stream water. However, the concentration of abundant cations (Ca^{2+}, Mg^{2+}, Na^+ and K^+) and anions (HCO_3^{2}–, SO_4^{2}– and $C\,1^{1}$–) present in the stream water were employed to describe and assess the longitudinal and temporal variation in the water quality of Sukadeo stream. The average values and the range calculated from physico-chemical analysis of samples collected from the four sample spots for seven months are given in Table 22.1, and are illustrated in Figs. 22.2 and 22.3.

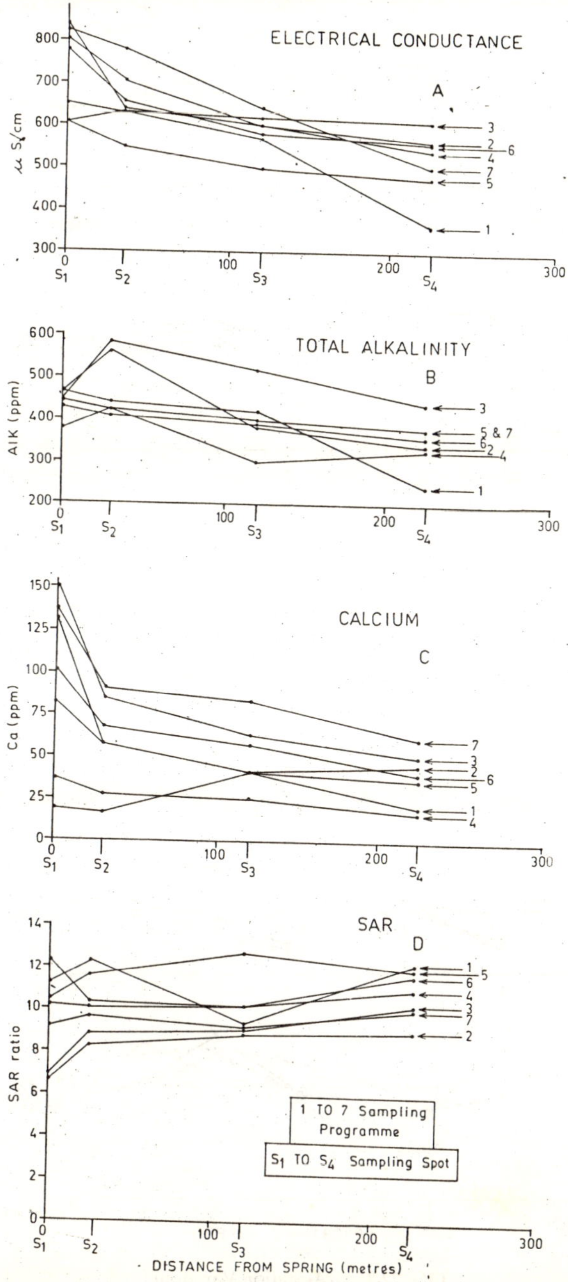

FIG. 22.2: Downstream Variation

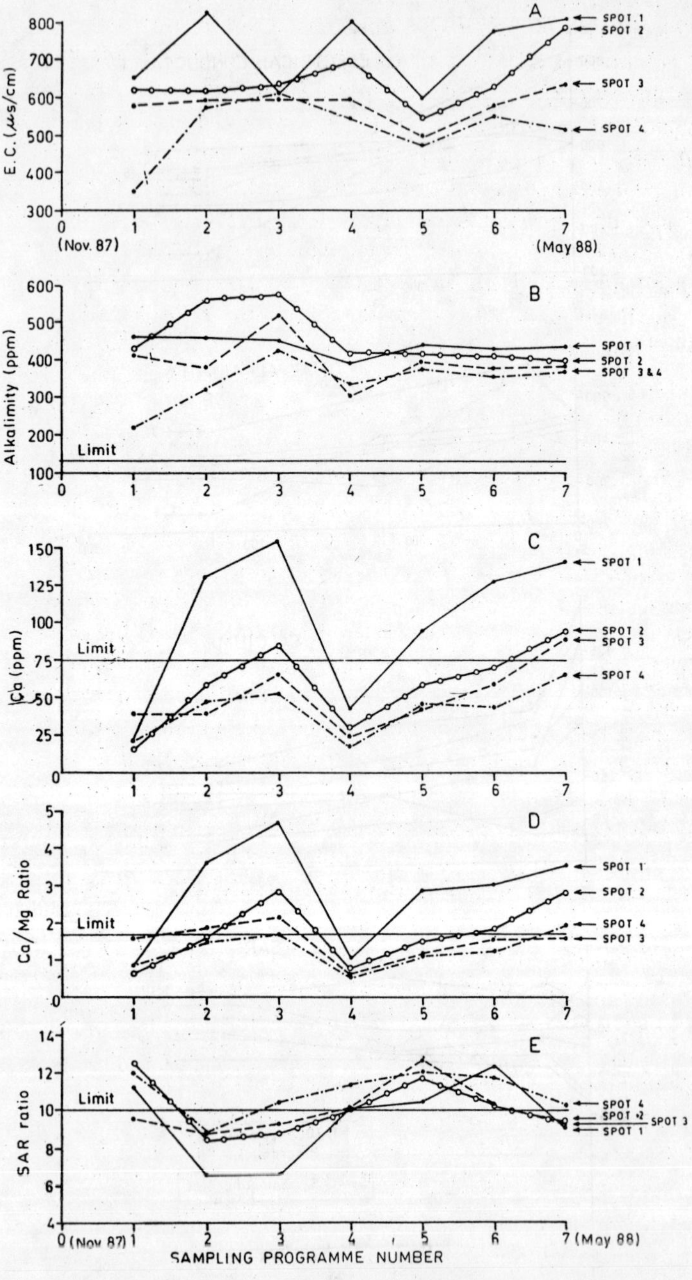

FIG. 22.3: At-A-Station Variation

TABLE 22.1: Geochemical Properties of Sukadeo Stream Water

Variable	Mean	S.D.	Range	C.V.%
pH	8.07	2.03	7.5 - 8	25.15
Temperature °C	21.3	0.66	20 - 22	3.1
E.C.	620.7	105.98	350 - 830	17.09
Ca hardness(ppm)	135.5	38.41	60 - 220	28.35
Total Hardness (ppm)	363.71	81.89	175 - 550	22.51
Silica (ppm)	65.71	11.78	40 - 80	17.92
Chloride (ppm)	35.01	5.67	30 - 50	16.91
Alkalinity (ppm)	412.86	70.29	220 - 580	17.03
Na (ppm)	67.29	9.72	54 - 87	14.45
K (ppm)	0.65	0.32	0.1 - 1.0	49.23
Ca (ppm)	57.59	36.55	8.76 - 155	63.47
Mg (ppm)	32.03	3.94	23.9 - 38.9	12.31
Ca/Mg	1.85	1.01	0.7 - 4.7	54.59
Sulphate (ppm)	26.61	5.65	19 - 40	21.23
Total Carbonate (ppm)	90.17	33.43	41 - 172	37.07
TDS (ppm)	409.86	61.16	312 - 539	14.92
Kelly's ratio	0.81	0.25	0.34 - 1.37	30.86
SAR	10.19	1.57	6.6 - 12.7	15.41
Odour	Unobjectionable			
Taste	Unobjectionable			

S.D. = Standard deviation; C.V. = Coefficient of variation.

Downstream Variation

It is obvious from Fig. 22.2 a that E.C., as a measure of solubility, goes on decreasing from spot 1 to 4. This implies that the mineral content of water is highest when water emerges from the underground and as the water flows downstream the conductance goes on decreasing (Fig. 22.2a). Alkalinity, which is imparted by presence of the bicarbonates, carbonates and hydroxides ranges between 220 and 580 ppm and decreases downstream (Fig. 22.2b). Similarly, Ca (8 - 154 ppm), Mg (22 - 38 ppm), $CaCo_3$ (175 - 550 ppm) and total carbonates (41 - 170 ppm) also decline in the downstream direction. The above results indicate that the groundwater emerging in the form of a spring is highly charged with different species. The water, as it moves down the rapids and falls, precipitates calcium carbonate as tufa (Tarate, 1988) and hence, ionic content goes on decreasing. The remaining constituents such as sodium, potassium, silica, chlorides and sulphates do not show much variation along the stream course.

This can be attributed to the conservative nature of these cations and anions which inhibits their precipitation.

At-A-Station Variation

The at-a-station variations in water quality parameters are large, almost for all the parameters (Fig. 22.3). Nevertheless, the most prominent variation is seen in the case of total hardness and calcium. The total hardness values are highest (550 ppm) during summer and lowest during winter. A more-or-less similar trend was observed in case of calcium. The values of remaining parameters such as SiO_2, Mg, Na etc., are high during winter, perhaps due to enhanced solution activity in response to winter showers and low temperatures (Tarate, 1988).

Geochemical Character of the Stream Waters

For the purpose of comparison and also to evaluate the differences and similarities between the compositions of different samples collected during the study period, data were plotted on Piper's Trilinear diagram (Piper, 1953). It is observed from the Fig. 22.4 that the alkaline earths (Ca + Mg) exceed alkalies (Na + k) except for samples spot 2 in January and May, (Na + K > Ca + Mg) and weak acids (HCO_3 + CO_3) predominate over strong acids (SO_4 + Cl). Further, the secondary alkalinity (CO_3 hardness) exceeds 50 percent. Therefore, the chemical properties of stream water are dominated by alkaline earths (Ca + Mg) and weak acids (HCO_3 + CO_3).

Overview and Discussion

The importance of the knowledge of the geochemistry of drinking water and its relation to water quality and hence to health cannot be underestimated (Amgino, 1983).

Examination of the chemical data clearly reflects that the stream water predominantly carries mineral constituents (derived from rocks) and that some of the constituents are in excess. The water contributes significantly to man's nutritional requirements. However, many times due to peculiar local geo-environmental conditions, the nutritional requirements of human health are not properly fulfilled. In order to assess the standard of water for human consumption,

World Health Organisation (WHO) in 1972 and Indian Standard Institution (ISI) in 1974 have recommended permissible limits for some important constituents. Thus a comparison of the stream water these limits can throw some lights on the potability of water for drinking as well as suitability of water for irrigation.

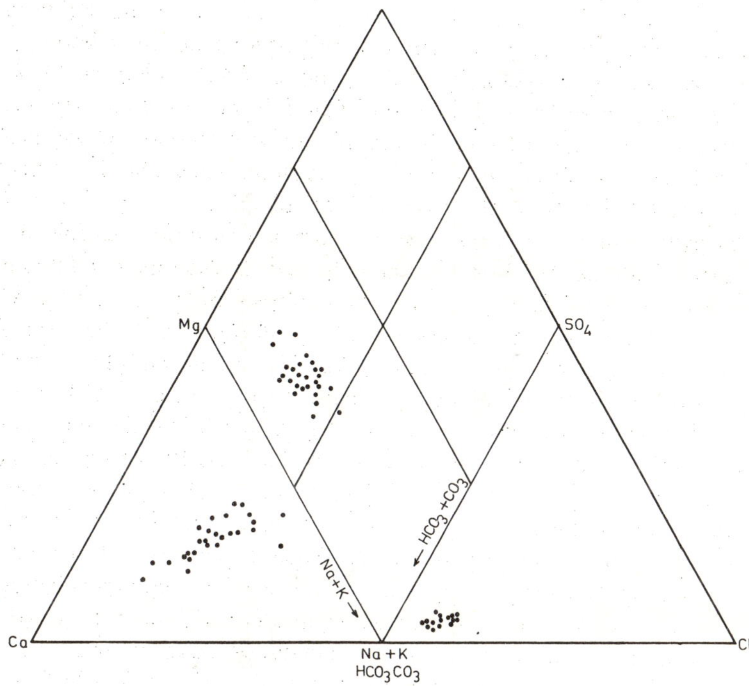

FIG. 22.4: Piper's Trilinear Diagram

The results obtained during present investigations reveal that total alkalinity, total hardness, Ca, Mg and Ca/Mg ratio are in excess at-a-station as well as downstream (Table 22.2). The Mg content remain high for most part of the non-rainy season, except during November and February. Further, the values of Ca and Ca/Mg ratio are higher than the permissible limits (75 and 1.7 respectively), from February onwards and reach their maximum in May (pre-monsoon). The total alkalinity also shows an analogous trend. However, due to precipitation of calcium carbonate the water quality improves comparatively downstream.

TABLE 22.2: Excess and Deficiency Chart

	Nov.	Dec.	Jan.	Feb.	Mar.	Apr.	May
Excess (Above Permissible limits)	HAR^{1,2,3}	HAR	HAR	HAR	HAR	HAR	HAR
	ALK	ALK	ALK	ALK	ALK	ALK	ALK
	—	Ca^{1}	Ca^{1,2}	—	Ca^{1}	Ca^{1}	Ca^{1,2,3}
	—	Mg	Mg^{1,2}	Mg^{1,2,3}	Mg	Mg	Mg
	KELLY^{1,2,4}	—	—	KELLY^{4}	KELLY^{3,4}	KELY^{4}	—
	SAR^{1,2,4}	—	SAR^{4}	SAR	SAR	SAR	SAR^{4}
Deficiency (Below Permissible limits).	HAR^{4}	—	—	—	—	—	—
	Ca	Ca^{2,3,4}	Ca^{3,4}	Ca	Ca^{2,3,4}	Ca^{2,3,4}	Ca^{4}
	Mg	—	Mg^{3,4}	Mg^{4}	—	—	—
	KELLY^{3}	KELLY	KELLY	KELLY^{1,2,3}	KELLY^{1,2}	KELLY^{1,2,3}	KELLY
	SAR^{3}	SAR	SAR^{1,2,3}	—	—	—	SAR^{1,2,3}

HAR = Total Hardness; ALK = Total Alkalinity; Ca = Calcium; KELLY = Kelly's ratio; SAR = SAR ratic; Mg = Magnesium; 1 - 4 = Spots 1 to 4.

As regards the quality of water for irrigation, two important parameters viz sodium absorption ratio (SAR) and Kelly's ratio were calculated. The SAR values range between 6.5 and 12.7 implying that the water belongs to S_1C_2 class of U.S. Salinity Laboratory Staff diagram (1954). This indicates that the water is, by and large, suitable for irrigation. The Kelly's ratio further supports this inference. However, the ratios of stream water at spot 1 and 2 as well as below spot 4 are slightly higher than the permissible limit during February and March, and therefore the water should be used cautiously for irrigation during this period.

Seasonally, although the ionic concentration is high during summer months than during the winter months, the contents go on decreasing downstream, due to ionic precipitation. Thus, there is a natural improvement of water quality longitudinally.

Water Quality and Human Health

In the absence of any other source of water, the local population has no choice, but to use the water of Shukadeo stream for domestic and irrigation purposes. In addition to the local population, people from nearby areas visit the shrine located between spot 2 and 3 during annual fairs and consume the "hard" water. Personal interviews with the locals and local doctors have revealed that due to excess presence of Ca and Mg, the population suffers for rheumatism and urinary infection (WHO, 1978a). There are cases which suggest formation of kidney stones, due to high concentration of Mg (Thornton, 1983). Cases of hypertension, coronary heart diseases or malfunctioning (arrhythmia) also seem to be related to the high concentration of sodium and magnesium (WHO, 1979) on the stream water.

The present analysis has shown beyond doubt that the stream water between spot 1 and 2 is not safe for human consumption, particularly during summer months. Paradoxically, the present site for water withdrawal for drinking is located between spot 1 and 2 and this has further aggravated the problem. The requirements and quality characteristics of water for livestock are fundamentally the same for human consumption. However, animals can drink (tolerate) water with moderately high dissolved solids (Matthess, 1982). Therefore, the stream water at the present site is safe for livestock.

Water Quality and Irrigation

The ion rich waters of the stream are being stored in two small percolation tanks constructed near Hivathad village (Fig. 22.1), for the past few years. As a result of this, the area is experiencing problems like water logging and salinization. During drier months, precipitation of salts is distinctly observed. As a result of this, the limited agricultural land is being converted into wasteland. And it appears that it will not be very long, before the entire fertile zone becomes a culturable waste.

Conclusion and Recommendations

The geo-chemical analysis of the Sukadeo stream water collected during post- and pre-monsoon season exhibits a distinct downstream as well as at-a-station variation in ionic concentration. The analysis further reveals that the chemical properties of the stream water are dominated by alkaline earths (Ca + Mg) and weak acids (HCO_3 + CO_3).

The data show that there is a natural improvement in water-quality downstream indicating that the stream water is safer for human consumption, below station 4. By and large, the stream water may be classified as safe for livestock and good to permissible for irrigation.

The concentration of chloride (Cl—) is low in the stream water seasonally as well as longitudinally. It can therefore, be concluded that human impact is minimum and that it has not reached alarming levels. This can be attributed to the location of a shrine near site 3 which discourages and inhibits the local population from using the stream water for other than drinking purposes.

On the basis of the above discussion following recommendations are made:

(1) Water for human consumption should be used from below spot 4, particularly between February and May and during fairs.

(2) If the local population insists that the water should be withdrawn from the presently existing spot, than the stream water can be softened with lime, which raises the pH and precipitates Ca. At higher pH magnesium hydroxide is also

precipitated and so the quality of water can be improved for drinking (WHO, 1979).

(3) Water for irrigation, should be used only after March, when the SAR values are below the permissible units, and not during January and March.

(4) Water should be alternately released from the two percolation tanks once every three or four years to flush out the saline precipitates.

Acknowledgement

The authors are grateful to Mr. B.D. Tarate and Mr. R.L. Shelke for their assistance on the sample collection as well as in analysis.

REFERENCES

Angino, E.E. (1983), Geochemistry and water quality in Thornton (ed), *Applied Environmental Geochemistry,* pp. 171-98.

Drever, J.J. (1982), *The Geochemistry of Natural Waters,* Prentice-Hall Inc.

Handa, B.K., Tamta, S.P. and Goel, D.K. (1986), Metamorphism of natural waters in the crust of weathering in the Sina-Man Sub-basins : A chemical model, in K.B. Powar and S.S. Thigale (eds), *Hydrogeology of Volcanic Terrains*, pp. 105-20.

H.M.S.O., (1972), *Analysis of Raw, Potable and Waste Waters.* Her Majesty's Stationary Office, London.

Indian Standards Institution, (1974), Indian Standard Tolerance limits for inland surface waters subject to pollution, *Indian Standards,* 2296-1974.

Matthes, G. and J.C. Harvey, (1982), *The Properties of Groundwater.* John Wiley and Sons, New York, pp. 406.

M.E.R.I. (1982), Study of Water Quality of Rivers and Reservoirs in Maharashtra State, Nasik.

Pawar, N.J. (1985), *Hydrogeology of Pune Metropolis with special reference to Chemistry of surface and groundwaters.* Unpublished Ph.D. Thesis, University of Poona. 182 pp.

Phadtare, P.N. (1986) ; Hydrogeology of the Deccan Basalts at the Sina-Man Basins, Maharashtra; in (Eds. K.B. Powar and S.S. Thigale), (eds), *Hydrogeology of Volcanic Terrains.*

Piper, A.M. (1953), A graphic procedure in geochemical interpretation of water analysis. *U.S. Geological Survey, Groundwater Note 12.*

Sivagnanam, N. and K. Kumarswamy (1983), Groundwater Quality in the Vaippar basin, South India. *The Indian Geographical Journal,* Vol. 58, No. 2, pp. 127-42.

Subramanyam, V. (1987), Environmental geochemistry of Indian River basins. *Geological Society*, Vol. 29, No. 2.

Tarate, B., (1988), Waterfall Tufa and palaeo environment Khanapur District Sangli, M.Sc. dissertation submitted to University of Poona.

Thornton, I. (1983), *Applied Environmental Geochemistry,* Academic Press, Inc. London, pp. 501.

U.S. Salinity Laboratory Staff, (1954), Diagnosis and improvement of Saline and alkali soils : *U.S. Department of Agriculture, Hand Book 60.*

World Health Organisation (1972), *International Standards for Drinking Water.* Ed. 3. WHO, Geneva.

World Health Organisation (1979), *Sodium chlorides and conductivity in drinking water.* Report on a WHO Working Group, The Hague.

23

ASSESSMENT OF KHARIF RICE PRODUCTION USING CLIMATIC MODELS IN THE SEMI-ARID NELLORE DISTRICT OF ANDHRA PRADESH

A.R. SUBRAMANIAM and P. SANJEEVA RAO

In major parts of India, insufficient and intense rainfall is the major constraint to agricultural production, as most crops depend on rainfall to meet their water requirements. As such, drought is the single most important factor affecting regional as well as national food security and the conditions and stability of the resource from which food is derived. The cost of drought in an agricultural region can be measured in many ways including the loss of human life and domestic animals, reduced agricultural production etc.

An agroclimatic model is the foundation for assessing climatic impact on agriculture. The agroclimatic crop condition index models which are based on historic data, have been used to assess crop conditions and to obtain relative crop yield information. The models are based on an agro-climatic variable such as cumulative rainfall, potential evapotranspiration (PET), soil moisture and others which are directly associated with the year to year variations in crop yield.

Steyaert *et al* (1981) proposed a climatic impact assessment programme for early warning food shortages on the basis of four agroclimatic models. Two of these models, namely generalised monsoon index (GMI) and yield moisture index (YMI) were tested for qualitative assessment of rice crop in India (Sarkar, 1985) and elsewhere (Oranut, 1982). These studies indicates the applicability of these simple models for macro-regional planning. In the present study, an attempt has been made to assess the kharif rice production of a micro-region, namely the Nellore revenue district of coastal

Andhra Pradesh, which has semi-arid type of climate.

Among the nine coastal districts, Nellore occupies an important place for rice production of superior quality under rainfed as well as irrigated conditions. The rainfall during the southwest monsoon season amounts to 31 per cent and about 50 per cent during north-east monsoon, of the annual rainfall (952 mm). About 25 per cent of the total geographical area (32.52 lakh acres) is under cultivation. Kavali, Kovur, Buchireddipalem, Nellore and Indukurpet blocks occupy more than 75 per cent of cropped area and are under rice crop during July to October, called the kharif rice.

Material and Methods

The available meteorological data during 1901 to 1984, and kharif rice yield information from 1967 to 1983 were obtained from the publications of Indian Meteorological Department and seasonal and crop reports of Government of Andhra Pradesh. The potential evapotranspiration (PET) and water balances were computed (Thornthwaite and Mather, 1955). The following two agroclimatic models proposed by Achutuni *et al* (1982) were utilised to demarcate different intensities of drought and hence crop production.

(a) Generalized monsoon index (GMI)

The generalized monsoon index is defined as,

$$GMI = \sum_{i=i}^{4} W_i P_i$$

where W_i is the weighing factor and P_i is the actual monthly rainfall in the ith month. Here, 1, 2, 3, 4 indicate southwest monsoon months (June-September). By considering the normal rainfall, the weighing factor utilised in the present study are 0.182 (June), 0.267 (July), 0.260 (August) and 0.291 (September).

(b) Yield moisture index (YMI):

The yield moisture index is defined as,

$$YMI_j = \sum_{i=1}^{4} P_i K_{cij}$$

where YMI_j is the yield moisture index for the jth crop (e.g. rice, maize etc.); P_i is the rainfall during i^{th} crop stage (e.g. i=1 for planting, i=2 for vegetative, i=3 for reproductive/flowering, i=4 for maturity) and K_{cij} is the appropriate crop coefficient for the jth crop and the ith crop stage.

This model uses crop coefficients to weigh rainfall during the growing season according to relative water requirements. The different phytophases of kharif rice crop are sowing, transplanting, tillering, elongation, flowering, maturity and harvesting. At Nellore district, rice is transplanted during July. The vegetative period coincides with August while flowering coincides with September and maturity in October. The crop coefficient values utilized are 1.1 for transplanting, 1.8 for vegetative stage, 1.25 for flowering and 1.43 for maturity based on the lysimeter data available for Nellore district (Sarkar, 1985).

The percentiles are worked out to assess the year to year drought intensity. The intensity of drought is defined as severe (0-20 percentile), moderate (20-30 percentile), and mild (30-40 percentile). The episodic events noticed during the period 1960 to 1984 are used to evaluate how each should be interpreted and the potential for use as an operational technique to monitor general agricultural conditions of a district.

Discussions and Results

Dry spells with large soil moisture deficits are normal feature of the dry climates (Subramaniam and Sanjeeva Rao, 1988). Such dry spells should not be referred to as drought unless they reach an unusual intensity or are abnormally long for the particular area. Drought is generally understood as 'a period of dryness due to lack of rain'. However, agricultural drought may be expressed in terms of which growing plants have been adversely affected by an abnormal moisture deficiency. It occurs when soil moisture and rainfall are inadequate for growth and development of crops for optimum production.

For a precise understanding of the agricultural drought problems of a crop in a region; one must know the growth periods and its climatic requirements. For example, the onset and advancement of monsoon are important for starting sowing operations of the kharif crops. Further, it is expected that the kharif rice yield will depend heavily on the onset of monsoon, its duration, amount and the year to year fluctuation depends on the variability of rainfall. Rice requires a growing period of 120 to 150 days, temperature between 15 and 38 °C and prevailing temperatures of 20 °C at night and 30 °C during day. Rice also requires abundant water and long duration of sunshine.

The water balance conditions of a significant drought year at Nellore region is illustrated in Fig. 23.1. Climatically the year 1972 represents a moderate drought during kharif season and no drought during rabi season. The crop yields reported during the kharif and rabi seasons for rice, bajra, groundnut, cotton, jowar and tobacco are below normal. The receipt of 1,300 mm annual rainfall indicates above normal conditions. However, it can be noted that the accumulated actual evapotranspiration (AET) line is above the accumulated PET/2 line only after October, i.e., during the rabi season. The corresponding water balance condition indicated the water deficit condition throughout the kharif season, hence all the crops grown in the region experienced a severe agricultural drought and hence the low crop yield and productivity. However, during October to December, the rainfall received over the region is more than the consumptive use of all the crops. The low yields reported may be due to the water logging conditions owing to excess water (472 mm) which may cause physiological damage to crops.

In order to monitor general agricultural conditions, the time series of the GMI and YMI are being plotted and the records of episodic events have been then used to identify the weather impact years and the critical thresholds of the model associated; for example, normal crops, moderate drought impact on crops (slightly below normal yields), drought impact on crops (drastically reduced yields or crop failure). The model values for such years will be considered as a threshold value for each category.

The results of the GMI indicates 31 drought years, of which 16 severe, 8 moderate, and 7 mild (Fig. 23.2). The frequency of severe drought is more during 1930-43 (8 years). This model indicates, the year 1982 which is a drought year is earmarked as a *moderate*, but the year 1984 represents a *normal*.

The analysis of the YMI indicates 30 drought years, in which 15 are severe, 8 moderate and 7 mild (Fig. 23.3). However, YMI represents the year 1982 as a *normal* year for kharif rice production at Nellore. But the year 1984 indicates a *severe* drought, which is not reflected in GMI.

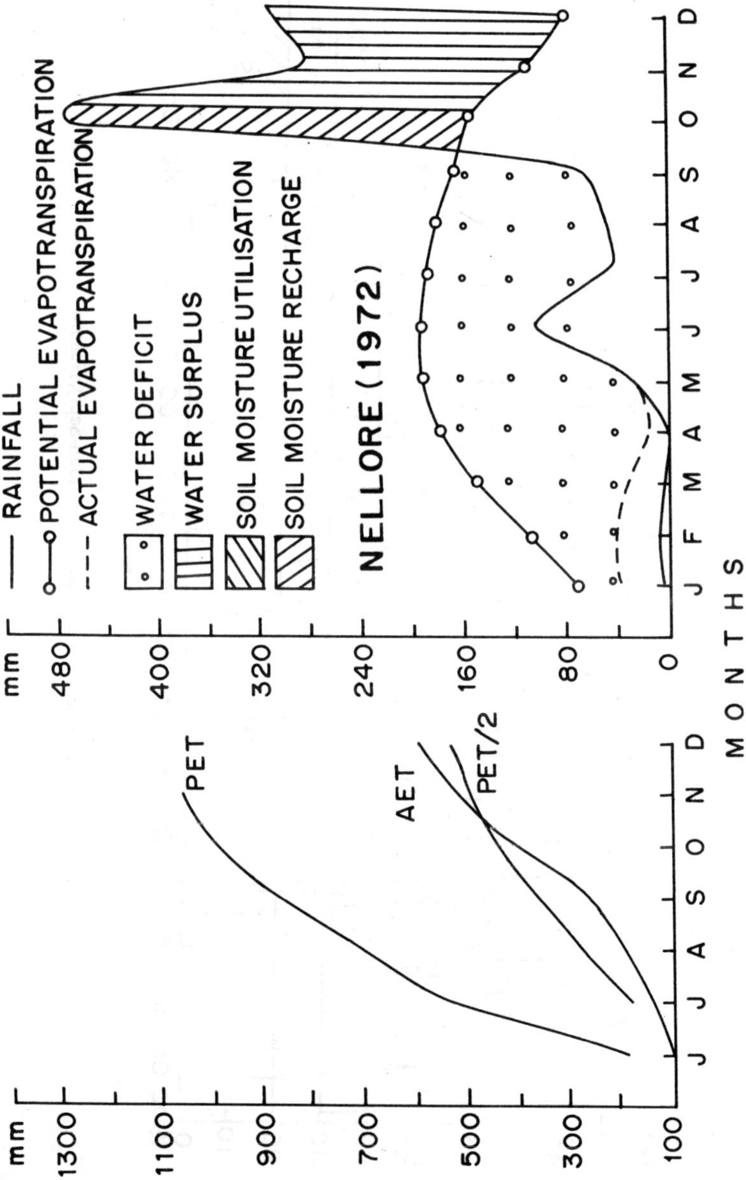

FIG. 23.1: Accumulated PET, PET/2, AET and water balance for crops during agricultural drought

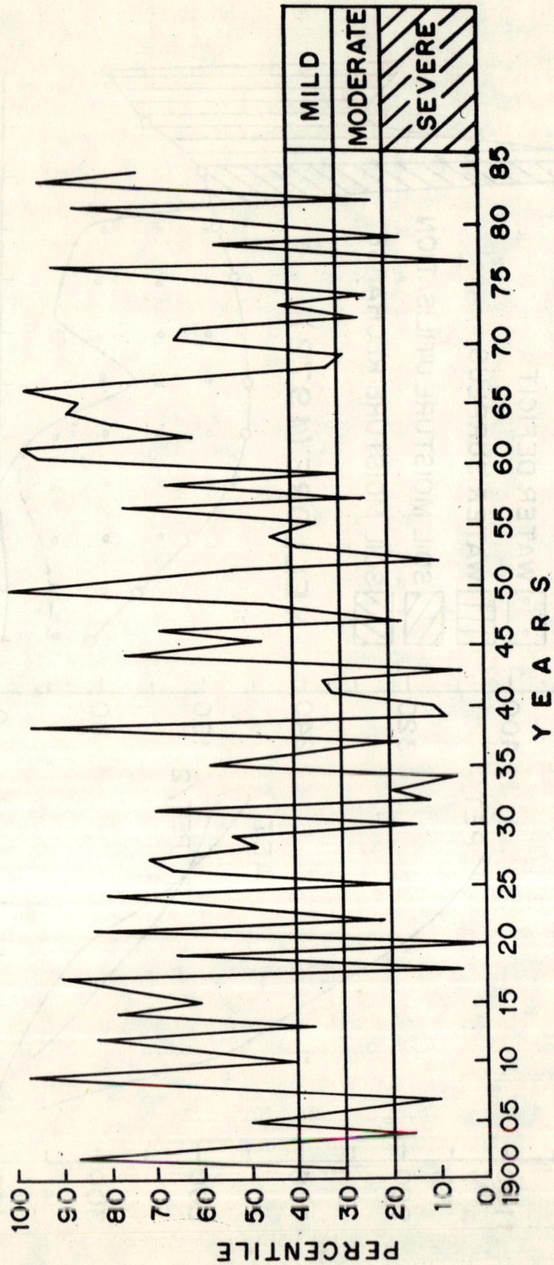

FIG. 23.2: Generalised Monsoon Index (Jun-Sept), Nellore

A comparison of both the agroclimatic models indicates that, in some years such as 1979 both the models showed a *severe* drought as evidenced by the episodic data; and in some years such as 1977 the GMI indicates a *severe* drought, because of the prolonged dry spell, whereas normal kharif rice production is indicated by YMI. This indicates the importance of considering crop growth stages into account in assessing the crop production. The percentile ranks of GMI and YMI have shown a gradual change, either increasing or decreasing, with the advancement of time. The recent extreme drought years, namely 1967, 1974, 1979 and 1984 are reflected well in the YMI, whereas droughts are reflected in GMI only for 1979.

In order to verify the drought assessment with the agroclimatic models, the kharif rice yields are quantified and summarised in (Table 23.1).

TABLE 23.1: Comparison of yield with GMI and YMI at Nellore during 1967-1983 (except '69)

Yield in % of normal	No. of years	Drought Intensity	No. of Years	
			GMI	YMI
90	10	Nil	7	6
		Mild	0	2
		Moderate	1	1
		Severe	2	1
80-90	1	Nil	0	0
		Mild	0	0
		Moderate	1	0
		Severe	0	1
80	5	Nil	3	2
		Mild	1	1
		Moderate	1	1
		Severe	0	1

If the reported crop yields are greater than 90 per cent of the normal, the GMI indicates two *severe* and one moderate drought conditions whereas YMI indicates two mild, one moderate and one *severe* drought conditions. If the reported yield is 80-90 per cent of the normal; one moderate drought with GMI and one severe drought with YMI is noticed. If the yield is less than 80 per cent of the normal one

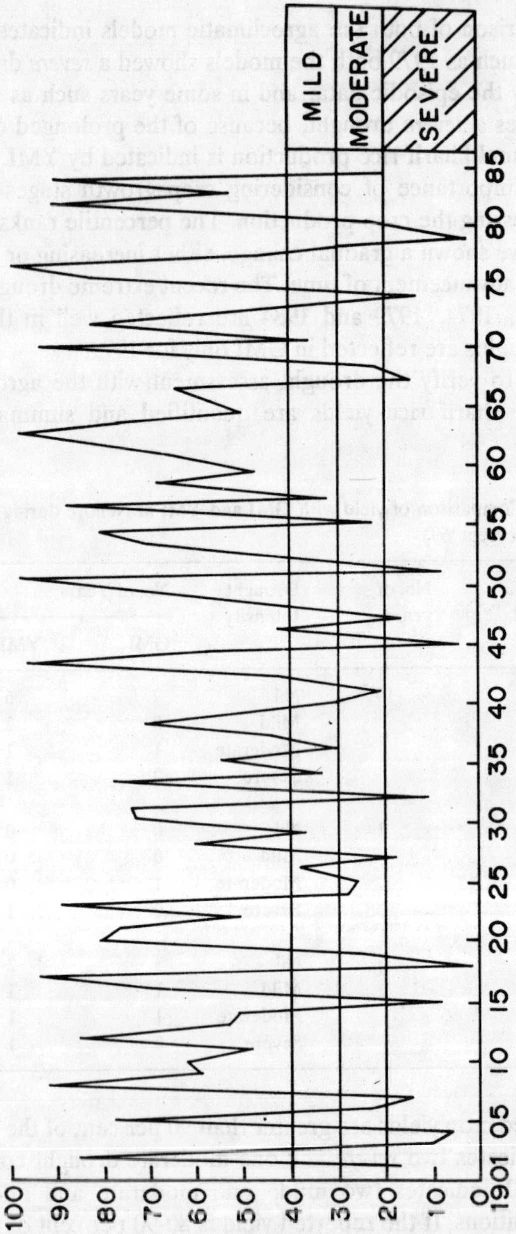

FIG. 23.3: Yield Moisture Index (July-October), Nellore, Kharif Rice

mild, one moderate and one *severe* drought with GMI; where YMI indicated one mild, one moderate and one *severe* drought conditions. Thus, agricultural drought assessment with YMI is better than GMI.

Conclusion

The water balance elements clearly brought out the necessity of good hydrological budgeting to achieve the targeted production under the prevailing agroclimatic conditions. The kharif rice yield assessment based on the climate and yield information, primarily focusses on potential brought impact as related to short and long term economic planning needs, from a micro-region to a macro-region.

If the crop failure is assessed early in the growing season, users of the information can some time recommend alternative cropping more suitable for existing and expected weather conditions. This type of early warning of potential crop failure due to drought can be provided 30-60 days before crop harvest. This may represent a 3-6 months lead time. In the short term, weekly or decadal assessment reports on climatic impact can be used by economic analysts, agricultural forecasters and extension officials to develop strategies to mitigate drought impact. These agroclimatic models also offer potential long-term economic benefits to agricultural planning and rural development, e.g. minimizing climatic risk in land and water resource management.

The models are mainly useful in assessing drought impact on dryland crops, but not in the irrigated areas (flooding impact), pests and diseases or wind damage some times associated with cyclonic storms. An early indication of sub-normal or above average crop conditions for a particular date and also for a region can be readily determined from LANDSAT imagery which then complements and verifies the results of simple crop-weather models as discussed above. Other weather related phenomena such as soil conditions, nutrient deficiencies, lodging, plant diseases and insect infections are also reflects in the imagery. Hence, it is suggested that this technique may be used, for better assessment of agricultural production.

REFERENCES

Achutuni, V.R., Steyaert, L.T. and Sakamoto, C.M., (1982), Agroclimatic assessment methods for drought/food shortages in south and southeast Asia — Test and evaluation, Final report to AID, *USFDA*, Washington D.C. by NOAA (AISC) and Univ. of Missouri, Columbia.

Oranut, P., (1982), Proposed decision making, Economic planning and rural development in Thailand. Met. Dept. Bangkok, p. 161.

Sarkar, R.P., (1985), Early assessment of yield using agroclimatic indices. Paper presented at the WMO/FAO IMD Workshop on 'Application of operational techniques in Agrometeorology', Pune, India.

Steyaert, L.T., Rao, Achutuni and Atmas, Todorov, (1981), Agroclimatic assessment methods for drought/food shortages in south and southeast Asia — proposed early warning programme, Final Report of AID, USFDA, Washington by NOAA.

Subramaniam, A.R. and Sanjeeva Rao, P., (1988), Dry spell sequences over south coastal Andhra, *Mausam* (in Press)

Thornthwaite, C.W. and Mather, J.R., (1955), The water balance, *Climatology*, 8(1).

24

ENVIRONMENTAL POLLUTION AS A CONSEQUENCE OF URBAN DEVELOPMENT WITH SPECIAL REFERENCE TO WATER POLLUTION IN INDIA

M.D. NANDESHWAR, S.A. ABBASI, P.C. NIPANEY and R. SONI

Population, when unchecked, increases in a geometric proportion, while subsistences increases only in an arithmetic proportion (Malthus, 1798). In consideration with space and time, the present situation of human growth is alarming. Specially, India is a country with a very dense and large population, next only to the Peoples Republic of China. The growth rate of India's population has also been far higher than most of the other countries of the world and particularly the growth rate of the urban population of India is still higher, causing urban development crisis of unprecedented magnitude and complexity. There is a correspondingly intensive and consistent demand on the utilisation of the countries total annually available water (1900 billion m^3) resulting in problems of water pollution whose magnitude and complexity is comparable to the basic problem of population boon (Khoshoo, 1983). In the present study, it will be difficult to give a comprehensive review of the state-of-art but an attempt is made to provide glimpses of some of the problems and the efforts of combating them.

Pollution Pressure in Bombay

Increasingly urbanisation is serving as an index of frustration and as an indicator of the potential explosiveness in urban slums. One can conclude that any increment in urban population will lead to a further deterioration in the quality of urban life, as the gap between services available and those needed widen rapidly. Table 24.1 gives the population increase in Greater Bombay through the years. While Table 24.2 depicts the vehicular activity in Bombay as compared with the whole Maharashtra state and India. Table 24.3 provides an idea of the major industries in Bombay.

TABLE 24.1: Population in Greater Bombay, 1901-1981

Year	Population	Decadal increase	Decadal per cent increase
1901	812912	—	
1911	1018388	+205476	25.27
1921	1244934	+226546	22.24
1931	1268306	+23372	1.87
1941	1686127	+417821	32.94
1951	2966902	+1280775	75.95
1961	4152056	+1185154	39.94
1971	5970575	+1818519	43.79
1981	8227332	+2256757	37.79

Source: Census of India, 1981, Series 1, Paper 2 of 1981, Provisional Population Totals, Rural-Urban Distribution.

TABLE 24.2: Vehicular Activity

Type	Greater Bombay (1975)	Maharashtra (March 1977)	India (March 1977)
Motor Cycle	47,164	1,82,553	12,16,701
Motor Cars	1,16,791	1,69,835	6,27,808
Taxies	19,751	25,096	77,808
Trucks	31,364	71,847	3,54,852
Others	5,562	71,640	6,60,270
Total	2,20,632	5,19,971	29,37,439

Source: Jussawalla, D.J. (1980) 'Man Vs Environment – Fact or Fancy' Proc. of Sem. on Mag. of En., Bombay.

One can only expect gross pollution with this kind of large population, industries and vehicles. The predominant wind directions and industrial regions of Bombay are indicated in Fig. 24.1. Winds in Bombay mostly blow from southwest in monsoon season and from northwest during other seasons, carrying air pollutants towards land. This clearly refutes the normal claim that air pollution problem of Bombay is not serious. The proximity of Bombay to sea is not helping in the removal of air pollutants into sea. In fact, due to the wind direction towards land, higher concentrations of pollution levels in New Bombay have been measured. The level of sulphur dioxide (SO_2) in New Bombay as measured by NEERI is 108 $\mu g/m^3$ as compared to 61 $\mu g/m^3$ for old Bombay (Jussawalla, 1980). A study carried out by India Meteorological Department concludes that the potential does exist in Bombay during October to March for over-stagnation of pollutants once every third day during night and early morning hours. This suggests the higher chances of occurrence of air pollution episodes in Bombay (Desai, 1980). When we consider pollution from all sources as shown in Table 24.4 we find that SO_2 is contributed mainly from industries whereas in the case of particulates both industries and other sources contribute more or less equally. The major sources of air pollution in Bombay are the industries and the transport.

TABLE 24.3: Major Industries in Greater Bombay, 1973

Type	Number
Textile & Silk	155
Chemicals	71
Pharmaceutical	28
Engineering & Foundry	28
Rubber	27
Oil	14
Glass	11
Paints	9
Plastics	8
Food	7
Miscellaneous	59
Total	417

Status of Water Pollution

Pollution due to Domestic and Municipal Wastes

In India only the urban population, which constitutes approximately 20 per cent of the overall population of 700 million, has facilities of domestic waste disposal and treatment to some extent. Of these, at

TABLE 24.4: Pollutants from Different Sources – Greater Bombay

Source	SO_2		PM		NO		CO		HC	
	tns/day	%	tns/day	%	tns/day	%	tns/day	%	tns/day	%
Domestic	2.7	0.8	19.0	20.0	3.3	3.6	7.0	2.0	2.0	2.4
Commercial and Institutional	1.7	0.6	2.2	2.8	1.0	1.1	–	–	–	–
Industrial	236.0	68.8	14.0	15.0	35.7	36.4	111.0	29.0	41.6	51.4
Power Generation	62.0	18.0	28.0	30.0	30.6	2.0	0.5	–	–	–
Transportation	40.6	11.8	30.0	32.0	27.7	28.0	263.0	68.5	37.4	46.2
Total	343.0	100.0	93.20	100.0	97.7	100.0	382.6	100.0	81.0	100.0

Source: Jussawala, 1980, 'Man Vs Environment – Fact or Fancy' Proceeding of the Seminar on Management of Environment, Bombay.

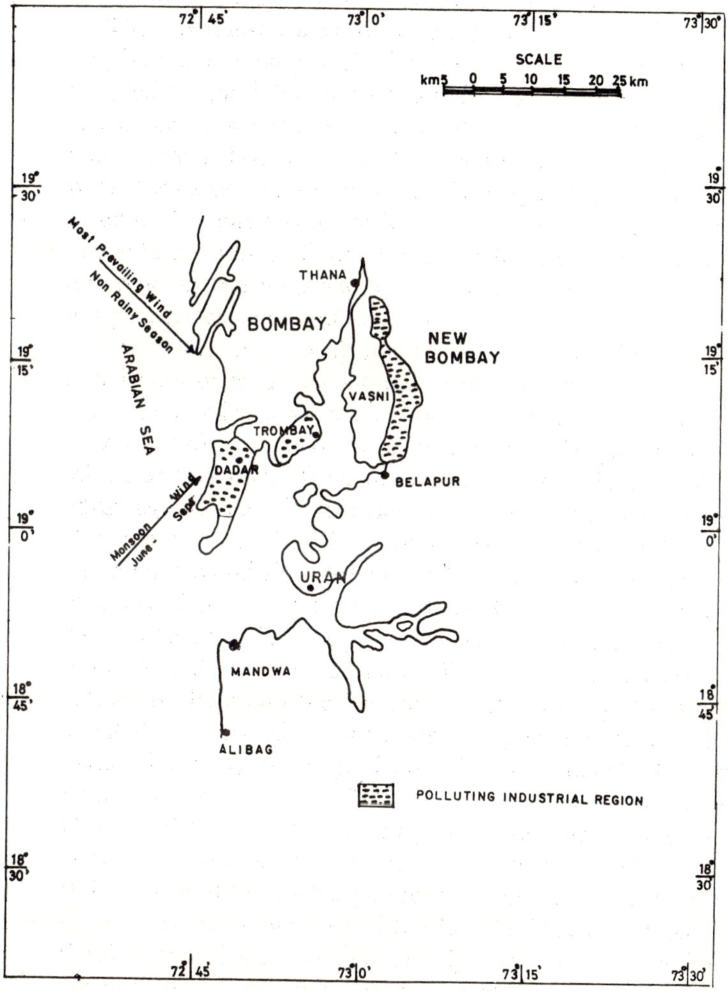

FIG. 24.1: Bombay: Industrial Polluting Region.

present, only 15 Class I cities (above 0.1 million) out of a total of 142 are fully sewered, while this facility exists in only 7 of the 190 Class II cities (0.5 - 0.1 million). Moreover, 55 Class I cities and 35 Class II towns have partial sewerage. Partial treatment is available in 27 Class I cities and 12 Class II towns. Only 8 Class I cities and 3 Class II towns have full sewerage and full treatment facilities, whereas 72 Class I cities and 147 Class II towns have neither sewer nor treatment facilities. The rural population which is about 5 times the urban population is provided with negligible environmental sanitation.

Such gross inadequacy in sewerage treatment systems in major human settlements is the principle cause of water pollution in India. Although sewerage system is inadequate in Class II towns but it is comparatively lower than Class I cities.

Pollution due to Industrial Sources

There are about 2,700 large and medium industries, out of which about 1,700 units contribution to pollution is significant. Already 353 large and medium industries are operating industrial wastewater treatment plants; 174 industrial effluent treatment plants are under construction and 190 units are in the planning stage, thus covering 717 industrial units out of 1,700 large and medium industries (Agrawal, 1982).

On the other hand, the large and medium industrial units contribute 70 per cent by volume of the total industrial wastes. The remaining 30 per cent is contributed by the small and cottage sector units. Any success in controlling 70 per cent by volume of the total industrial wastes would, however, mean only 7 to 10 per cent of the total volume of wastewater generated in the country.

At present the total volume of industrial wastewaters generated in the industrial cities of India is estimated as constituting 8-16 per cent of the total city waste. With the present rate of industrialisation, it is expected that by 2025 AD this percentage would rise to well above 50.

Pollution due to Agriculture

Of the three main water-use sectors which generate pollution, the contribution from agriculture is most difficult to assess and control because of it is locationally unspecific. Intensive agricultural practices involving high inputs of fertilisers and pesticides contribute substantially to salinisation and pollution of rivers, canals, and subsurface waters but little information is available which can enable quantification of this contribution (Sawyer & Casagranade, 1983). Several pesticides, which have been banned in other countries due to conclusive evidence of their high mammalian toxicity and low rate of biodegradation, are still used freely in India – often rather indiscriminately by private farmers. The data presented by Edwards (Table 24.5) which reveal that India has higher levels of DDT, Lindane and Aldrin in the tissues of its human population than nine other countries studied. This is alarming.

Miscellaneous Problems

There are a plathora of other water pollution problems that cut across the well defined sectors of municipal, industrial and agricultural water. Pollution of backwaters and human settlements in Kerala due to coir-retting and enormous loss in quantity and quality of water due to aquatic weeds are two examples of such problems.

Social and Other Problems caused by Water Pollution

The inadequate sewage treatment combined with inadequate treatment of raw water cause a perpetual recycling of pathogenic organisms into user population. Apart from causing epidemics and acute health problems, this situation also leads to a chronic gastroenteritic illness symptoms in very large number of individuals. Millions of man-hours are estimated to be lost due to these debilitating diseases. With the increasing public consciousness towards hazards of pollution there have been several concerted and successful public protests in recent years, specially against industrial waste discharge.

Environmental Standards with Reference to Water Pollution

At present guidelines for drinking water quality have been given by two agencies – Indian Council of Medical Research and the Indian Standards Institution (ISI). The latter agency has also given a variety of standards for allowable levels of water quality variables for different types of water uses and wastewater disposal practices. A

TABLE 24.5: Permissible Levels in Drinking Water, Irrigation & Disposal on Inland Surface Water

Metal	Discharge on inland SW 1	ISI [2] (draft 1982) Drinking water	Irrigation[3]		WHO[4] 1971	USSR[4] 1970	Japan[4] 1968	FRG[4] 1975	USEPA[4] 1975	Canadian[5] 1978
			ST	LT						
Chromium	0.1	0.05	20	5	NA	0.1	0.05	0.05	0.05	0.05
Zinc	5	5.0	10	5	5.0	1.0	0.1	2.0	NA	NA
Cadmium	2	0.01	0.5	0.005	0.01	0.01	NA	0.006	0.01	0.002
Mercury	—	0.001	—	—	0.001	0.005	0.001	0.004	0.002	0.002
Lead	0.1	0.1	20	5	0.01	0.1	0.1	0.04	0.05	0.05
Copper	1.5	1.5	5.0	0.2	0.05	0.1	10	NA	NA	1.0

1. 'Tolerance Limits for Industrial Effluents, part 1. General Limits', IS: 2490(1), 1982.
2. 'Specification for Drinking Water', Draft Indian Standard DOC: CDC, 26(8686), 1982.
3. 'Characteristics and Problems of Irrigation Return flow', US Govt. Printing Office, 1968.
4. 'Guidelines for Canadian Drinking Water Quality', Ministry of National Health and Welfare, Canada 1978.

mention of all these standards will be out of the scope of this note but a statement of ISI standards for trace metals as compared to the standards set in other countries (Table 24.5) may be of interest. Several studies by the author and co-workers, (Abbasi & Soni, 1982, 1983 a, b and Soni & Abbasi, 1981, 1982) have shown that these standards are too liberal in many cases and need thorough revision.

With regards to industries, the Central Board for Prevention and Control of Water Pollution, has set up *Minimal National Standards* (MINAS). These standards are evolved keeping in view the technological competence of the country and the economic sustainability of the industry. The annual cost of control, termed as *annual burden*, is correlated with the annual turnover of the industry to ascertain the MINAS. For most of the industries documented so far, the annual burden (capital and operation, maintenance and repair costs) of pollution control never exceeds one per cent of annual turnover to control liquid effluent to the desired level of the standards. Fertilizer and Textile Industries require two per cent of annual turnover as annual burden for the pollution control. Distilleries are the lone exception; even with five per cent of annual turnover as annual burden, the pollution control system brings the effluent quality only to first stage of desired levels. It is generally agreed that the pollution from most of the industries may be contained to the desired level of three per cent of annual turnover is put towards burden for water pollution control.

The Counter Measures

By 2000 A.D. the country is likely to use more than half of the total water available (1900 thousand million m^3) annually, which would be shared between the four primary uses. However, aquatic ecologists suggests that at no point of time more than half of the total water available annually for utilisation should be withdrawn. The total quantity of return water, which becomes wastewater after utilisation, is estimated as 286 thousand million m^3.

In India, the key to success in environmental pollution control lies in the development of labour intensive, low-cost and appropriate technologies of wastewater treatment. Waste treatment methods like anaerobic digestion yielding biogas and fertilisers, oxidation ponds (which capitalise on land availability and abundant sunshine), microbial conversion of organic wastes like pulp effluents into livestock feed etc are specially attractive.

The Centre-State Relations and Pollution Control

India being a large country, the factors that cause or assimilate pollution like climate, demography, industrialisation, municipal amenities, etc. vary widely from state to state. The extent to which a large river like Beas in north can assimilate a waste load will be entirely different from a small and sloppy river like Chaliyar can do in south. If in a particular summer, in a unique isolated time, the drinking water gets saline in Cochin, it causes state wide alarm while people in Rajasthan drink such water round the year without being even conscious of it. In such a situation, it becomes necessary that instead of having a single set of standards for the entire country at present, there are standards set by different states within a broad frame work given by the national agencies. Such standards are likely to be more practical with regards to discharge regulations and environmental management and would be compatible with the natural resources, industrial growth and economy of the different states. The standards in Japan and USA are set in this fashion and the system has worked well. The interaction between central agencies and state bodies in India may be viewed in this light.

Acknowledgements

Authors are grateful to Dr. S. Vasudev, Executive Director, CWRDM for help and guidance and to Dr. E.J. James, Head-in-charge, Surface Water Division for encouragement. Efficient and timely assistance from Smt. V. Subaida, WQE Division, Smt. V.A. Sulochana, Surface Water Division and Ms. B. Kanchana and Salim Ali School of Ecology is gratefully acknowledged.

REFERENCES

Abbasi, S.A. and Soni, R. (1982), Toxicity of Cadmium (II), Zinc (II) and Lead (II) towards Freshwater Teleost *Nuria* denricus. In the *Proceedings on National Conference on Lead, Zinc and and Cadmium at Workplace: Environment and Health Care*, New Delhi, pp 433-38.

——, (1983a), Stress induced enhancement of reproduction in earthworm *Octochaetus pattoni* Exposed to chromium (VI) and mercury (II), *International Journal of Environmental Studies* (UK), in press.

——, (1983b), Toxicity of Lower than Permissible Levels of Chromium (VI) to the Freshwater Teleost *Nuria denricus. Environmental Pollution* (UK), in press.

Agrawal, G.D. (1982), An Approach to Management of Industrial Effluents in India, Proceedings of the ENPC-III Joint Workshop on Strategy and Technology for Water Quality Management, III Bombay, 1982, pp 1-6.

Desai, V.D. (1980), Management of Environment: Problems of Bombay city, *Proceedings of the Seminar on Management of Environment*, Bombay, pp 67-70.

Edwards, C.A. (1973), *Persistent Pesticides in the Environment*, CRC Press, Inc. Ohio.

Jussawalla, D.J. (1980), Man Vs environment: Fact on fancy. *Proceedings of the Seminar on Management of Environment*, Bombay, pp 1-10.

Khoshoo, T.N. (1983), Water Quality Management in India Retrospect and Prospect, Keynote Address, *Seminar on Strategy and Technology for Water Quality Management.* III Bombay, pp 1-10.

Sawyer, A.J. and Casagranade, R.A. (1983), Urban Pest Management: A conceptual Framework, *Urban Ecology*, Vol. 7. pp. 145-57.

Soni, R. and Abbasi, S.A. (1981), Mortality and Reproduction in Earthworms *Pheretima posthuma* Exposed to Chromium (VI), *International Journal of Environmental Studies* (UK), 7, pp 147-49.

——, (1982), Impact of Mercury (II), Chromium (VI) and Zinc (II) on Some Arthropods and Protozoans – Toxicity of Zinc (II), *Proceedings of National Conference on Lead, Zinc and Cadmium at Work Place: Environment and Health Care*, New Delhi, pp 439-45.

25

GHAZIPUR : Environmental Pollution

ONKAR SINGH and K.K. SINGH

With increasing population, urbanization and industrialization, the physical character of Indian cities and towns is rapidly changing. They are becoming overcrowded and life supporting systems — the quality of fresh air and potable water are deteriorating, and open spaces are becoming scarce. It has resulted into unhygienic and insanitary living conditions causing ill-effects on human health and quality of life.

Though the problem of environmental deterioration is presently not so acute at Ghazipur town as is the case with large cities but looking into the growth trend of population due to rural-urban migration and limited amenities of the town under study this problem will get unmanageable if proper management of the environment is not made right now. Keeping this in view, an attempt has been made to analyse the level of environmental degradation in the town in spatio-temporal framework. For this purpose, a detailed study of air quality calculated on the basis of number of vehicles passing through different crossings, noise levels in peak hours in different localities and solid waste disposal in all the wards have been made. In addition, house density (houses/hectare), improper disposal of domestic waste water and quality of the river Ganga water have also been considered. On the basis of these investigations, a number of measures for the ecological planning of the town have been suggested for the betterment of the quality of urban life.

House Density

Density of houses is one of the most important indicators of an urban environment which indicates the level of crowding in a town or city. There were 6,548 residential houses at Ghazipur in 1971 but the number rose to 8,231 in 1981. Even today the existing houses are not enough to fulfil the housing needs of the town's population. It has led to construction of temporary dwellings and hutments in the inner part of the city in a haphazard way. Most of the houses in this zone reveal their dilapidated conditions. Density of houses is found to be the highest (25 houses/ha) in Barbarhana ward followed by Cheet Nath and Raiganj wards with density figures of 23 and 19 houses/ha respectively. The minimum house density (2 houses/ha) has been observed in the Degree College ward. The rest of the town reveals moderately low density of houses, varying from 4 to 10 houses/ha (Table 25.1). Further, it is evident that the outermost zone of the town covering mostly agricultural land has got the lowest density of houses while the inner parts of the town adjacent to river Ganga reveal higher densities. Though the density of houses is not very significant for assessment of environment conditions in the case of Ghazipur, but the condition of houses, improper disposal of domestic waste water and other insanitary conditions reveal that in future it will become difficult for civic authorities to handle the prevailing situation. During personal survey in the inner wards of the town, the evidences of water logging and choking of open drains causing foul smell have been noticed. All these contribute towards environmental deterioration of the town causing ill-effect on human living and comfort.

TABLE 25.1: Density of Houses

Name of Ward	Area (in ha)	No. of houses	(Density/ha)
Degree College	289	588	2.03
Gorabazar	105	545	5.19
Khazuria	136	655	4.81
Saklenabad	149	771	5.27
Shastri Nagar	125	804	6.43
Kapurpur	69	229	3.31
Raiganj	37	711	19.21
Mahajan Toli	68	609	8.95
Subhash Nagar	105	747	7.11
Cheet Nath	23	520	22.60
Barbarhana	25	624	24.96
Jamalapur	73	486	6.65
Nawabganj	94	515	5.47
Ruhimandi	75	427	5.69
Total	1,373	8,231	5.99

Source: Census Office, Lucknow

Water Pollution

The main sources of water pollution in the river Ganga are domestic and industrial waste waters which join the river through three *nallahs*. It has been estimated that about 35 lakh litres of waste water is being discharged into the river Ganga every day. In addition to this the Government Opium and Alkaloids Factory situated in the centre of the city on the bank of the Ganga also discharges toxic contaminants into the river. A few small scale industries including biscuit factory, tanneries, plastic factory, and the like also discharge their wastes. Though the data regarding concentration of various pollutants are not available in detail, the results obtained with the help of the Department of Botany, P.G. College, Ghazipur, reveal that the effluents of Opium and Alkaloids Factory are toxic in nature (Table 25.2).

It is evident from the table 25.2 that the soil of effluent receiving areas has been badly affected and possesses contents of sodium and carbonate chloride beyond acceptable levels. The polluted soil is found to have higher alkalinity (14.5%) and poor porosity and water holding capacity. The alkalinity (pH up to 9.5) is found to be related with the increasing concentration of carbonate and sodium on the soil. On the other hand, it has been observed that high nitrogen content of Opium and Alkaloid Factory's effluent if used for irrigation will increases crop production. But the discharge of effluents from the factory into the Ganga or on the soil causes pollution of water and soil respectively to a great level.

Prior to Ghazipur, up stream of river Ganga, the newly established distillery and sugar factory also discharge huge amount of waste water into the river causing pollution of its water. According to a study conducted by the Department of Botany, Banaras Hindu University the quality of Ganga water at Ghazipur has changed greatly. Table 25.2 shows physico-chemical characteristics of Ganga water at different sampling sites. It is evident from the table that the maximum alkalinity (242.25 ml/l) was observed in summer at Opium factory site, while the minimum (33.43 ml/l) at control site in winter season. It shows that higher alkalinity at opium factory is due to discharge of waste water into river from Opium and Alkaloids Factory.

TABLE 25.2: Analysis of Ganga Water at GHAZIPUR GHATS (March 1987 to February 1988)

Sites	Control site (Mid stream)			Nawapura site			Opium factory site		
Parameters	S	R	W	S	R	W	S	R	W
Temperature (°C)	28.7	29.95	20.23	28.30	30.33	20.50	28.6	30.05	20.6
Transparency (cm)	85.50	49.93	38.63	64.63	36.53	33.60	46.5	27.93	20.7
pH	7.67	7.63	7.75	7.9	7.61	7.78	8.40	7.74	8.24
Dissolved oxygen (mg/l) [DO]	6.58	4.90	7.80	4.78	4.15	6.63	3.85	3.58	6.08
Free CO_2 (mg/l)	2.86	10.80	3.9	6.7	14.70	8.15	11.10	22.63	10.85
Alkalinity (mg/l)	52.85	50.05	33.43	169.25	153.75	181.6	242.25	219.4	193.43
Acidity (mg/l)	11.64	10.70	13.98	23.10	25.98	19.60	26.48	41.80	24.78
Electrical conductivity	0.72	0.66	0.75	0.753	0.755	0.83	1.01	0.97	0.91
Biochemical oxygen Demand (BOD) (mg/l)	4.0	3.58	2.3	58.90	68.35	40.60	98.53	112.55	73.93
Chemical oxygen Demand (COD) (mg/l)	10.33	9.23	5.33	134.25	187	88.0	249.5	242.25	158.0
Hardness (mg/l)	71.0	66.25	57.0	154.3	127.5	123.8	156.8	135.3	140.8
Nitrate (mg/l)	0.044	0.086	0.045	0.49	0.868	0.548	0.665	1.26	0.750
Phosphate (mg/l)	0.49	0.60	0.66	0.40	0.57	0.77	0.49	0.72	0.903
Sulphate (mg/l)	84.33	65.23	62.45	118.5	89.5	78.00	194	150.25	106.0
Chloride (mg/l)	27.75	28.25	32.28	53.25	40.75	64.23	70.08	47.33	77.95

(Contd.)

TABLE 25.2: *Contd.*

Sites	Collector ghat site			Dadarighat site			Cheet Nath Ghat		
Parameters	S	R	W	S	R	W	S	R	W
Temperature (°C)	28.7	29.4	20.80	28.23	29.78	20.70	28.8	29.68	20
Transparency (cm)	48.03	35.95	21.33	54.90	24.85	23.58	51.1	37.0	25
pH	7.85	7.70	7.78	7.86	7.70	7.88	7.9	7.7	7.8
Dissolved oxygen (mg/l) [DO]	3.93	3.78	5.9	4.7	3.83	5.85	4.25	4.1	5.9
Free CO_2 (mg/l)	8.95	23.13	10.98	7.85	21.25	9.63	10.70	20.98	9
Alkalinity (mg/l)	203.5	195.0	184.5	204.8	180.4	123.3	191.8	184.4	170
Acidity (mg/l)	23.25	35.00	23.28	25.23	37.33	22.4	24.95	39.4	21.5
Electrical conductivity	0.88	0.86	0.83	0.86	0.82	0.77	0.83	0.79	0.77
Biochemical oxygen Demand (BOD) (mg/l)	79.73	92.7	68.15	91.38	92.00	63.38	96.88	90.6	64.2
Chemical oxygen Demand (COD) (mg/l)	193.0	215.3	128.0	226.0	212.5	128.0	245.8	205.8	128
Hardness (mg/l)	195.8	168.0	160.8	184.5	154.0	161.5	186.0	153.8	150
Nitrate (mg/l)	1.07	0.98	0.78	1.04	0.88	0.77	1.053	0.858	0.7
Phosphate (mg/l)	0.76	0.86	0.50	0.74	0.85	0.53	0.71	0.73	0.8
Sulphate (mg/l)	164.9	152.0	105.0	122.8	103.8	96.75	124.0	95.25	93.0
Chloride (mg/l)	41.25	48.33	82.25	52.98	47.88	76.73	53.3	40.5	70.93

Source: S = Summer, R = Rainy and W = Winter seasons.

The concentration of BOD has been found to be the maximum (101.25 mg/l) at Cheetnath Ghat in summer season, while minimum (4.0 mg/l) at the control site. It is further observed that at other sampling stations its concentration is higher than the prescribed standard (3.0 mg/1). The DO concentration has been observed maximum (7.80 mg/l) at control site while minimum of Opium factory (3.38 mg/l). This shows that the self purifying capacity of river Ganga water in down stream channel is less. This is due to less dilution and more contamination of pollutants in the river. Besides this, at Samsan Ghat, due to burning of dead bodies, ash and suspended material contaminate the river water. A number of dead bodies of animals are also disposed off in the river causing biological pollution.

According to a personal survey, the water quality of a big pond, Pahar-Khan-Ka-Pokhara, has also deteriorated to a greater extent. This pond was earlier used as a bathing place but the pollution of its water has resulted into decrease in utility for bathing purposes by the nearby population. Another tank situated near M.H. Inter College has been found badly polluted and almost completely covered by *Jalkumbhi* and presents an ugly look and gives bad odour. The above study reveals that due to contamination of domestic and industrial waste water, the quality of Ganga water along with the important ponds of the town has become unfit for human use. Therefore, pollution control measures should be implemented by the authorities through enforcement of legislation.

Air Pollution

Truly speaking, there has never been pure air. Foreign substances have been present in the air at all times and at all places. The term air pollution is, therefore, applied when there is an excessive concentration of foreign matters in the outdoor atmosphere which is harmful to man or his environment (WHO, 1958).

It is the commonest form of environmental pollution in a town or city. Through the problem of air pollution in Ghazipur is not as serious as in cities having more than one lakh population, but in view of the increasing fossil fuel consumption, it will become a considerable problem by the turn of this century. Though the statistical data regarding the concentration of air pollutants in Ghazipur town is lacking, an attempt has been made here to examine the intensity of air pollution on the basis of road traffic survey. The main sources of air pollution are automobile exhaust, burning of coal and other fuels for domestic and industrial purposes. In addition to this burning of municipal garbage also adds to the problem.

The volume of vehicular traffic in certain pockets such as Chowk and Lanka bus stations, National Highway (No. 29), narrow roads and mixed traffic have aggrabated the situation more critical (Fig. 25.1).

Table 25.3 shows the traffic density (vehicle/hours) at important crossings of the town. It is evident from the table that highest traffic density 408 vehicle/hours has been found at Chowk followed by Lanka bus station (377 vehicle/hour) and petrol pump (339 vehicle/hour). Further, Pirnagar crossing has been observed with minimum traffic density due to its location in the western outer zone. There are standards for estimating emissions from automobiles both for petrol and diesel driven vehicles in various modes of operation (Pandit, 1974). One petrol driven vehicle emits about 46,200 PPM of CO, 4,900 PPM of HC and 500-25 PPM of NO_x. On the basis of above mentioned emission factors, the emission of CO can be calculated at all the crossings. At Chowk crossing the concentration of CO is the maximum (19.56 gm) followed by Lanka crossing, i.e., 6.95 gm.

TABLE 25.3: Traffic Density at Selected Road Crossings of the City During Peak Hours March, 1988

Name of crossings	Traffic density (vehicles/hour)	
	Petrol vehicles	Diesel vehicles
Misir Bazar	342	35
Biseswarganj Petrol Pump	193	146
Taxi Stand	134	123
Lanka Bus Station	130	149
Lal Darwaza	247	23
Chowk	372	36
Sastri Nagar	232	70
Mughalani Chak	58	131
Khudaipura	195	33
Pirnagar	100	29

Source: Compiled from Field Survey

Table 25.4 shows the emission of CO and other important toxic pollutants at main crossings of the town. The emission of CO is not higher than the prescribed standards, but looking into future growth of population and increase in number of automobiles operating in the town will make the situation more acute. It is also found that a number of shop-keepers in Chowk localities complained of eye irritation to the author during the survey. To control automobile pollution a by-pass has been recommended so that influx of heavy vehicle can be reduced from the interior of the town.

TABLE 25.4: Total Emissions from Automobiles at Different Crossings of Ghazipur Town (in gm), March, 1988

Name of crossings	Vehicles per hr.	CO	HC	NO_x
Misir Bazar	377	17.984	2.191	0.1649
Biseswarganj Petrol Pump	339	10.256	1.395	0.0978
Taxi stand	257	7.139	0.996	0.0988
Lanka Bus Station	279	6.951	1.004	0.0585
Lal Darwaza	270	12.997	1.579	0.1112
Chowk	408	19.560	2.381	0.1675
Sastri Nagar	302	12.239	1.545	0.1045
Mughalani Chak	189	3.156	0.5293	0.0261
Khudaipura	228	10.265	1.2661	0.0878
Pir Nagar	129	5.274	0.6645	0.0450

Source: Computed after field survey

Noise Pollution

The 20th century has been described as the *Century of Noise.* Noise has become a very important *Stress factor* in the environment of man. The term noise pollution has been recently coined to signify the vast Cacophony of sounds that are being produced in the modern life, leading to health hazards (Park and Park, 1983).

That which we hear is called sound, and unwanted sound is noise. The noise is an undesirable by-product of the modern technological advancement, industrialisation and above all the population boom specially the urban population. There will be hardly any exaggeration if the present-day civilization is called a noisy civilization (Singh, Kumra and Singh, 1988).

After the air and water pollution noise problem is the greatest problem of urban life. Further, the specialists in the field have noted that noise is a slow poisoning agent like smoke and the rate of its increasing level is bound to make it comparable to the disastrous

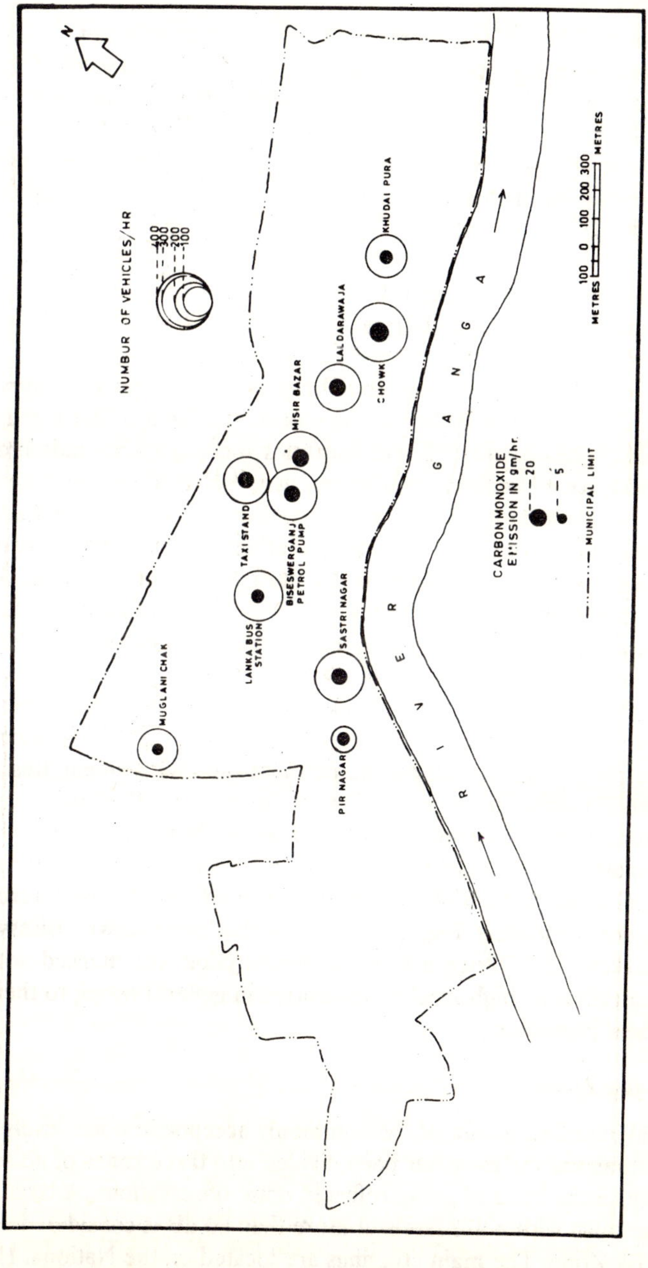

FIG. 25.1: Ghazipur Town: Traffic Density and Emission of Carbon monoxide

bombs. Though the problem of noise pollution is not as serious in towns as in big cities but looking into the future growth of the town this problem is essential to be dealt with in the present study.

A variety of relevant themes such as the nature of noise pollution, effects of noise on the human health and efficiency, noise measurement standards, noise control devices and so on have been included in the present study.

Noise Pollution Survey of Ghazipur: Approach and Methodology

In order to know the pattern of noise pollution a personal survey was conducted by the author in the month of March 1988 for measuring noise levels at 42 different localities including six roads crossings. These sound monitoring stations were selected on the criteria that they fall within the limits of Nagar Palika of Ghazipur and adjoining areas that have experienced urban influence. The maximum and minimum sound levels were recorded at these stations during peak hours from March 19 to 22, 1988 using a portable Sound Level Meter. The recorded sound levels ranges from 58 to 90 dB when observed at different crossings/localities of the town (Fig. 25.2). It was noted that the busy crossings such as Lanka Bus Station (90 dB), Mughlanichak (84 dB), Tulsisagar (86 dB), Biseswarganj (89 dB), Taxi Stand (82 dB), Railway Station (82 dB) and the Government Bus Station (80 dB) owing to their location on the National Highway reveal high sound levels. The outer areas such as Hathi Khan (57 dB), Foxganj (62 dB), etc. represent the bottom and obviously the prevailing sound level in the town exhibits a downward trend from the commercial hub to the periphery. Fig. (25.2) shows that the Chowk, Biseswarganj market, Lal Darwaza and Cheetnath Ghat are marked with concentration of high level noise centres in general owing to their functional character.

Noise Zones

Taking into account of the commonly acceptable noise levels for the human ear the town has been divided into three zones of noise intensity on the basis of above referred noise observations. A large part of the town with noise levels of more than 80 dB is considered as Noise Risk Zone. The main crossings are located on the National Highway (No. 29) from Varanasi to Gorakhpur and on the Malgodam road near the railway station and the roadways come under this zone.

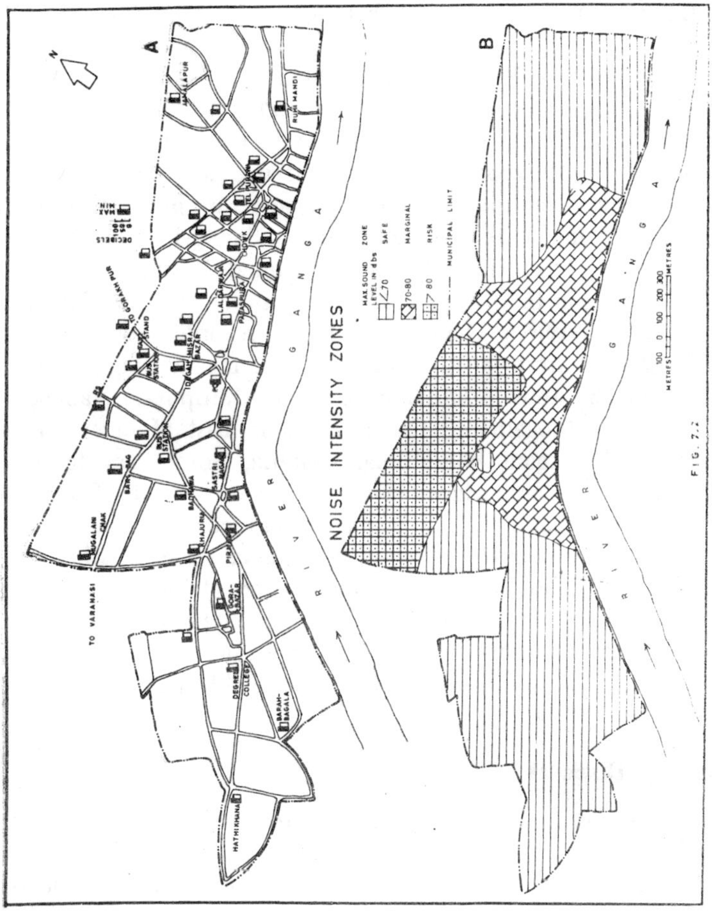

FIG. 25.2: Ghazipur Town: Sound Levels and Noise Intensity Zones

The areas having recorded noise range of 70-80 dB are designated as the marginal zone meaning thereby that these localities might be having noise levels of above 80 dB particularly at about 11 A.M. from court purposes on Kachahari road and after 11 A.M. till 4 P.M. due to marketing on Hospital road. Areas to the north of the marginal zone are regarded as Risk Zone and to the east and west areas may be called as safe zone. To identify the marginal noise rise areas, Sikandarpur, Nawapura, Subhas Nagar, Miyapura, Saiyedbarg, Mahajan Toli, Machcharhutta, Hari Sankari, Lal Darwaza, Muxalikatar, Town Hall, etc., may be named. It may be noted that people living in the marginal areas may suffer from hearing losses after long exposures. The localities having recorded less than 70 dB of noise level are included in the safe zone mostly lying at the outskirts of the town. The localities of Muftipura, Nababganj, Ruhimandi, Mogalpura, Chehalsatoon in east of the town and Mahadewa, Foxganj, Barahbagla, Khajuria, Gorabazar, etc. in the west of the town are included in the safe zone. The study of noise pollution reveals that there is an urgent need for controlling noise pollution in the town. For this purpose, at the initial stage, the widening of roads in the core town area, construction of by-pass roads and prohibition of heavy vehicles in the town during peak hours and shifting of private bus station away from the Lanka areas are urgently needed. It will not only make the people healthier but also increase their capacity to work.

Garbage Disposal

The increasing volume of garbage is becoming hazardous to the quality of urban life. Presently the Municipal Board dumps most of the garbage in lowlying areas. It is the major source of land pollution and spread of many infectious diseases such as cholera, gastro-entitis etc. The garbage characteristics indicate the class of people living there. The modern technological advancement has increased the variety of chemical materials in garbage disposed off from a town/city.

In order to know the total accumulation of solid wastes, data have been collected from the Nagar Palika, Ghazipur. Table 25.5 shows the increasing amount of garbage generation during 1982-88. This is due to increasing population trend functional characteristics, changing dietary habits of the inhabitants. Fig. 25.3 shows trhe spatial distribution of garbage disposal in the town.

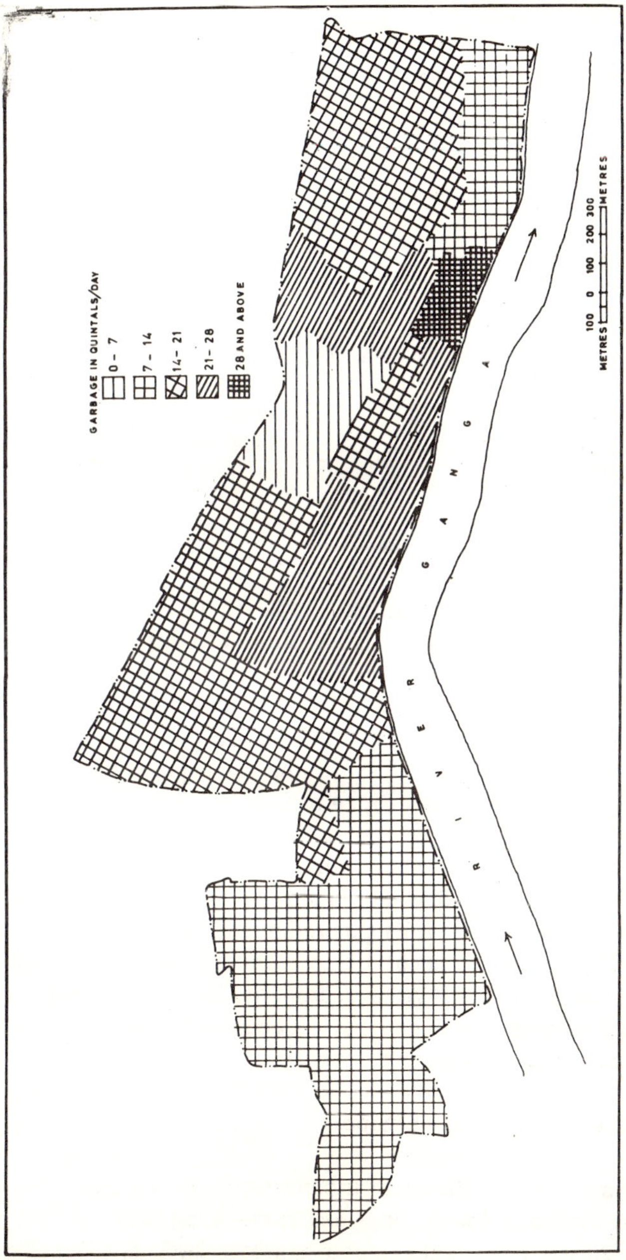

FIG. 25.3: Ghazipur Town: Garbage disposal

TABLE 25.5: Generation of Garbage at Ghazipur During 1982-88

Ward No.	Name of Ward	Garbage collection (quintal per day)
1.	Degree College	13.34
2.	Gora Bazar	13.75
3.	Khazuria	15.16
4.	Saklenabad	17.61
5.	Shastri Nagar	21.12
6.	Kapurpur	6.15
7.	Raiganj	20.17
8.	Mahajan Toli	15.67
9.	Subhash Nagar	18.25
10.	Cheet Nath	36.67
11.	Barbarhana	18.56
12.	Jamalapur	15.12
13.	Nawabganj	14.78
14.	Ruhimandi	11.35

Source: Nagar Palika, Ghazipur, 1988.

Though the Nagar Palika has employed 173 workers (59 males and 67 females) and they are being utilized to dispose off manually the garbage to the disposal sites but they are only able to collect and dispose only 60 per cent of the total garbage accumulated every day. According to Nagar Palika sources about 238 quintals per day of garbage is being generated in the town which is disposed off in low lying areas (Table 25.5). The per capita per day generation of garbage has been calculated as 0.35 Kg. Due to increasing population, change in socio-economic conditions and other related factors, the amount of garbage will increase. It will be quite difficult for the authorities to dispose off the garbage if its generation will increase with the same trend. The labour force of 173 persons is inadequate to collect, handle and dispose off garbage properly. The most suitable technique for management of garbage will be sanitary landfilling in lowlying areas adjacent to the town.

Growth and Distribution of Slums

The slums, a manifestation of urban poverty, have become the permanent feature of Indian cities and towns irrespective of their size. Nearly 20 per cent of the country's urban dwellers (33 million in

1984) reside in slums. According to an estimate Uttar Pradesh will have the largest slum population of 6.5 million by 1990 followed by Maharashtra (6.2 million), West Bengal (4.9 million) and Tamil Nadu (4.2 million). Bihar will have a slum population of 3.2 million by the end of the present decade (Singh, Kumra and Singh, 1988).

The term slum connotes an extreme condition of blight in which the housing is so unfit as to constitute a menace to the health and moral of the community (Dickinson, 1960). In other words, slum is not a building but an area having the least urban amenities, the maximum housing and population densities. The entire locality is deprived of any park or playground, the residents live in mostly unsatisfactory hygienic, social and cultural environment where the urban amenities and sanitary conditions are worst enough to cause high death and infant mortality rates. There are a few slum pockets in the town. In addition, a number of depleted buildings may be called as ahatas or slums where industrial workers, rickshaw-pullers, thela-pullers, collies and other pennyless persons are compelled to live because of their poor income. Such buildings exhibit a very poor and insanitary state of affairs.

In the town, however, the reasons for the growth of slums and *ahatas* are : (a) engulfment of rural areas on the city fringe in the process of the expansion of the town limits, and (b) deterioration and dilapidation of the residential houses lying mostly in the outer sectors of the town.

These slums and *ahatas* provide a potential take off place and have become a way of life of people of low income, viz. petty shopkeepers, domestic servants, milkmen, rickshaw-pullers and a host of others. There are no proper urban amenities in the houses — no latrines, urinals, bathrooms, proper water supply, sewage disposal and electricity. They exhibit a scene of filth with dingy lanes, overcrowded, irregular, dilapidated and ill ventilated houses having tiny rooms caused by the continuing influx of people especially from the rural areas. These localities are highly congested. According to personal surveys there are about 4,200 persons (7.0%) within 750 households living in slums of different types. It is further revealed that about 78 per cent of the workers in such localities are daily wage earners. A sample study of 100 families of Islamighat, Rajghat and Ramlila-ka-Maidan localities revealed that their income varies from Rs. 280 to 500 per month which is not sufficient even for their daily requirement. A few workers are encouraged to adopt some household industries, but still they are living below the poverty line. Most of these slum dwellers have migrated from surrounding countryside area. Further it has been found that the number of school going children is very less (about 15% of total children below 12 years of age). Due to unhygienic living conditions, they are suffering from various problems such as nutrition, health and hygiene.

Most of the slums have developed along the river in the outer zone and eastern part of the town. Both sides of the Islamighat and Rajaghat road have been covered by slums. The squatter settlements are situated along the National Highway (No.29) on the Lanka Ram Lila-Ka-Maidan. The mud-walled enclosures are roofed with straw, polythene sheets, gunny rugs, tiles, rejected asbestos sheets, etc. The majority of the population living in these settlements consists of basket makers and other such workers make their living from tiny earnings.

REFERENCES

Pandit, V.I. (1974), Automobile Exhaust: Air Pollution and its control, *Indian Jr. of Environmental Health*, Vol. 16, pp. 344-60.

Park, J.E. and Park, K. (1983). *Preventive and Social Medicine,* ninth edition, Banarsidas Bhanot Publishers, Jabalpur, pp. 173-74.

Singh, O., Kumra, V.K. and Singh J. (1988), *India's Urban Environment: Pollution, Perception and Management,* Tara Book Agency, Varanasi, p. 114.

26

NOISE POLLUTION :
Impact and Analysis

B.B. DHAR

In most of the industrial locations, and modern urban developments, *noise* is one of the most pervasive environmental problems. In a report to the President and Congress of the U.S. in mid-seventies indicated that 80-100 million people are bothered daily by environmental noise and of this 40 million are adversely affected in terms of health. Today one can imagine the likely changes. India has experienced a good industrial development and urbanization, as a result it has not been able to escape from such a hazard. In recent years, not only in industrial areas but cities like Bombay, Calcutta and Delhi have diverted the attention of all concerned towards noise hazard. Since noise by itself is a by-product of human activity, the area of exposure increases directly proportional to the population growth. The need for higher power generation, increased transport (mobility) and such other activities disturb the natural ecosystem.

The Noise Pollution

Noise pollution which are hazardous to hearing can be produced by a variety of sources, depending on the intensity and duration of exposure. Broadly they include:

- industrial activities
- construction equipments
- transportation systems, and many a times common appliances etc.

Besides this, speech communication, interference problems are common in the environment, and the community annoyance by noise generating activities, particularly in populated areas. It has been observed that population annoyed by noise on daily basis is increasing all over the world and this is resulting in social and psychological stresses — a major concern to the scientists, planners and engineers. The major related effects due to noise on human beings are as follows:

- permanent or temporary hearing loss,
- sleep interference,
- increased human annoyance,
- communication interference resulting in reduced work efficiency,
- impairment of mental and creative types of work performance, and
- possible usage of drugs like sleeping pills etc. to reduce noise stress.

Noise and vibrations have also been responsible to leave an impact on

- buildings, and other structural features in the area,
- property devaluation, and
- landuse incompatibility.

Continued exposure to noise or what we call some times *chronic noise* annoyance and distraction may lead to human errors in handling and disposal of hazardous materials, that may ultimately leave impact on land, air, water qualities, as well as while handling heavy machinery, transport equipments causing minor or major upheavals and damages. Therefore, the impact of noise may be far reaching, having a potentially significant impact on nearby area.

Here an attempt has been made to provide a simple environmental impact analysis as a result of noise produced by industrial activity and to suggest associated measures of control and the areas of R&D activity. Since the mineral industry in India has lately become very conscious of the environmental impact in general but not much is heard about industrial noise. It is hoped that the brief analysis will provide an insight as well as a thought for the planners to include this aspect as a part of the total project and not a patch work at a later stage.

The Impact Analysis

Social environment includes people (human beings) and their activities, property, groups, influence and heritage. Each and every event, be it a construction activity or a mining activity or a powerhouse, or a dam construction has its own social environment and its own surroundings. Mining is one such activity that always has its own unique environment. They are scattered all over the length and breadth of the country thereby posing a fresh set of problems each time. Due to this reason each new mining project has to go through environmental clearance at the highest level, so that the environment it is to pollute has a least impact.

Since it is not possible to highlight all the impacts here, the study will be confined only to noise impact. Industry in general, and mining in particular, has been significantly contributing to environmental pollution as a result of increasing mechanisation of mining operations, particularly in opencast mines where the induction of dozers, dumpers, heavy earth moving equipment, conveyor systems, and other associated sound emitting machineries is a must. No much technological advancement seem to have taken place in this country towards noise abatement. However, scientific studies have now been initiated in various research organisations, educational institutions and comprehensive R & D programmes have been started to analyse and to find effective control measures.

The work done by various workers in India and abroad have further proved that the noise levels produced by installed equipment in factories or mines seems to be emitting high noise levels, the frequency of which also varies significantly. Since it is not possible to go into details of the noise pollution analysis at this stage, attempt therefore has been made to highlight the impacts by the help of tables and diagrams.

Noise Sources

The major sources of harmful noise in developing countries are:

- continuous wide-band noise (wide frequency range) e.g. background noise in a mechanical workshop, etc.
- continuous narrow-band noise (sound energy mainly concentration on a small number of frequencies), e.g. circular

saws, planning machines, etc.
- impact noise (single impulses of short durations), e.g. forging, hammering;
- repetitive impact noise, e.g. rivetting, trimming, etc.;
- intermittent noise, e.g. road traffic, aircraft noise, community, activities, construction, etc.

The noise levels in some typical industries in India are summarised below (Table 26.1):

TABLE 26.1: Noise levels in some Indian Industries

Industry/Operation	Sound Pressure Level (dBA)
Detonator manufacturing and testing	94 – 140
Pharmaceutical	94 – 128
Foundry in motor manufacturing plant	104 – 120
Heavy engineering	94 – 124
Synthetic fibre manufacturing unit	90 – 117
Electronics	87 – 122
Hydel power plant	92 – 106
Road transport workshop	90 – 124
Cotton textile mills	92 – 105
Fertiliser plant	104 – 118

Source: Environmental Engineering Division, Central Labour Institute, Bombay.

Similarly, the construction equipment noise range is given in the form of a table, dividing the activities into:

- earthmoving,
- materials handling,
- stationary,
- impact equipment, and
- others.

The equipment listed in Table 26.2 covers a wide range of products commonly used in such activities, while as Fig. 26.1 gives a range for an underground coal mining machines.

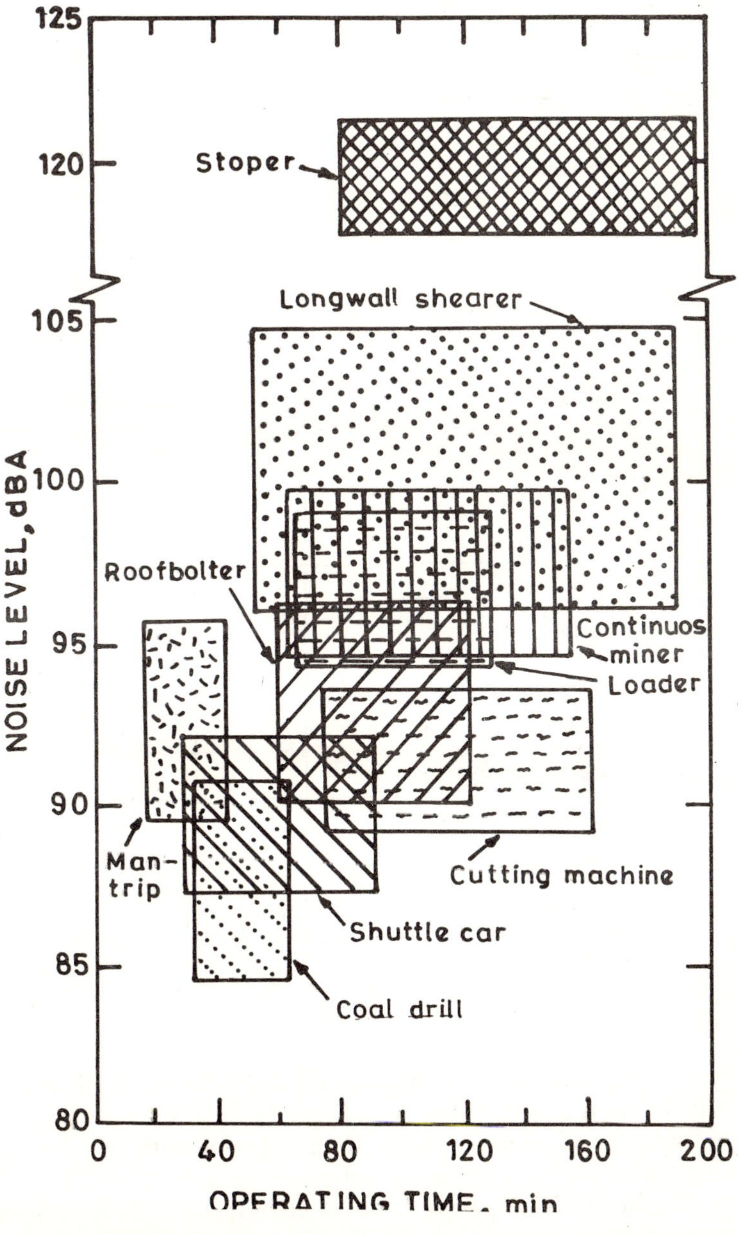

FIG. 26.1: Typical Noise Levels and Operating Times per 8-h shift of Underground Coal Mining Machines

TABLE 26.2: Construction Equipment Noise Range

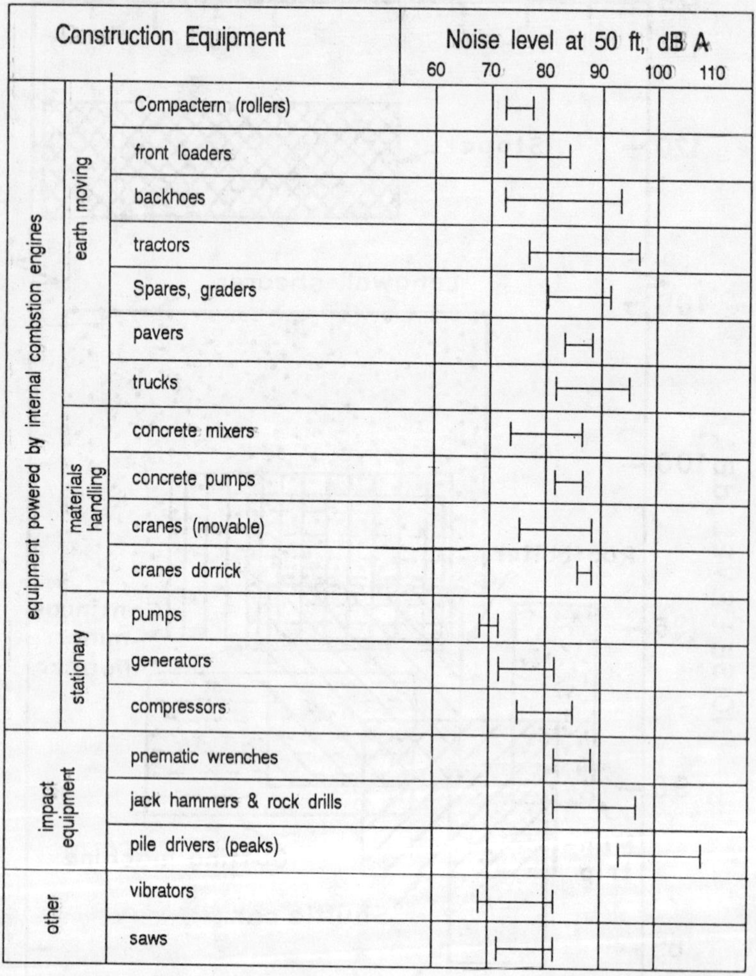

The employment pattern in noise affected industry in India as per 1978 figures is given in Table 26.3. From Table 26.3 it may be noted that mining and quarrying is the third largest industry in terms of employment next to construction only.

The cumulative exposure effect of noise for an eight-hour work each day, has been summarised below. It is a function of time and the number of exposures to varied noise levels (Table 26.4).

The percentage of hearing loss of miner's with age as per USAM report (IC-8986, 1984) is shown in Fig. 26.2.

TABLE 26.3: Employment in Some Important Noise Affected Industries

(*Figures for 1978 in lakhs*)

Industries	Persons
Cotton textile	11.09
Woollen, Silk synthetic fibres	1.72
Jute, Hemp and Mista textile	2.64
Wood and wood products, furniture & fixtures	1.33
Iron and steel	1.50
Foundries	2.13
Transport equipment and parts	3.97
Electrical machinery	2.75
Machinery, machine tools and parts (except electrical machinery)	4.72
Grain mill products	1.84
Mining and Quarrying	7.71
Construction	10.32

TABLE 26.4: Cumulative Exposure

Cumulative	No. of noise interval exposures per 8-h workday						
	1	3	7	15	35	75	150 or more
8h	90	–	–	–	–	–	–
6h	91	92	93	94	94	94	94
4h	93	94	95	96	98	99	100
2h	96	98	100	103	106	109	112
1h	99	102	105	109	114	115	–
30m	102	106	110	114	115	–	–
15m	105	110	115	–	–	–	–
8m	108	115	–	–	–	–	–
4m	111	–	–	–	–	–	–

Note: To use this table select the column headed by the number of times the noise occurs per 8-h workday, read down to the average A-weighted sound level of the noise, and locate directly to the left in the first column the total duration of dangerous noise for any 24h period.

National and International Standards

In 1976 the government of India had included *Noise-induced hearing Loss* as a notifiable disease under the factories act, 1948. Under section 89 of the Act, any medical practitioner who detects noise-induced hearing loss in a worker has to report the case to the Chief Inspector of Factories along with all other relevant information. Failure to do so is a punishable offence. A government appointed expert committee which went into the problem of compensation for occupationally-affected cases recommended that noise-induced hearing loss arising out of occupational exposure should also be included in the compensable diseases under the workmen's compensable Act (1923). The addition was in part-B of Schedule III of this Act which is enforced by the state governments. The model rules framed under the factories Act incorporate permissible levels for noise exposure and the control measures necessary to implement these.

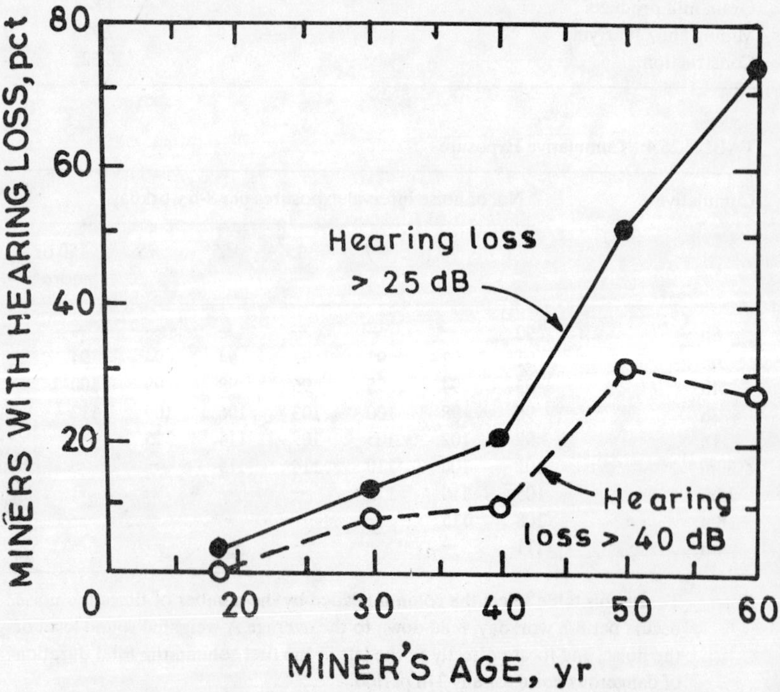

FIG. 26.2: Percentage of Hearing Loss of Miner's with Age

The comparison of international noisc-level standards of a few major countries have been summarised in Table 26.5.

Noise Control

The control programme can be divided into three main headings:

I. The criteria for the hearing conservation programme are:

(i) Conservation programme is needed in situations where the hearing loss, as measured by audiometry, is more than 10 dB at 4,000 Hz.
(ii) It is also indicated when the sound level in the working environment is more than 90 dB and there is difficulty to communicate by speech.
(iii) When it is difficult to hear a loud noise at a distance of one foot (0.3 m).
(iv) When workers find it difficult to communicate by speech.
(v) When workers experience head noises or ringing in the ears after work, and/or
(vi) When workers notice a loss of hearing persisting several hours after exposure.

II. Control programme consists of:

(i) Noise exposure analysis.
(ii) Engineering control of noise.
(iii) Protection of hearing with ear defenders where necessary.
(iv) Measurement of hearing by audiometry.

III. Reduction of noise level by engineering control method:

(i) Keeping the total noise emission from all noise sources below the dose limit in the work area.
(ii) Regulating spacing between noise sources and between sources and operations. In free-field conditions the sound level roughly varies with the square of the distance.
(iii) Enclosing the noise sources by a sound reducing (i.e. attenuating structure which prevents airborne transmission).
(iv) Reducing structure-borne transmission by isolation of the source using resilient mountings.
(v) Damping of vibrating metal structures or by replacement with materials such as wood.

TABLE 26.5: Comparison of International Noise/Level Standards

	Country	Level (Max. Permissible)	Remarks
1.	Australia (1973)	90 dB(A)	(i) 85 dB(A) in 5 years time for continuous exposure (i.e. 40 hs/week) (ii) No worker be exposed to noise level of 115 dB(A) at any time
2.	America (1972)	(i) At present 90 dB(A) for 8 hrs. is permitted; for new plants. It is 85 dB(A) for 8 hrs/day (ii) The scale however is: 80, 85, 90, 95, 100, 105, 110, 115 dB(A) for 16, 8, 4, 2, 1, 1/2, 1/4 and 1/8 hour of exposure per day respectively.	(i) Exposure to 115 dB(A) or more should not be permitted. (ii) Exposure to impact or pulse noise should not be more than 140 dB(A).
3.	Argentina	91-95, 96-100, 101-105, 106-110 & 111-115 dB(A) for 6, 4, 2, 1, 1/2 hour of exposure per day respectively.	
4.	Canada (1973)	No employee to work in our area where noise level reaches 90 dB(A) or more.	(i) However, whenever necessary (in noise authorised places of work) the exposure permitted is 87-90, 90-92, 92-95 95-97, 97-100,100-102, 102-105, 105-110,110-115 dB(A) for 8, 6, 4, 3, 2, 1 1/2, 1, 1/2, 1/4 hours of exposure per day respectively. (ii) 115 dB(A)/hour.
5.	Poland (1979)	(i) 40 dB(A): work demanding high degree of concentration. (ii) 50 dB(A): workshop & other places where high precision work is necessary. (iii) 90 dB(A): elsewhere.	
6.	U.K. (1972)	For relatively constant noise the exposure permitted is 90 dB(A)/8 hrs/day.	
7.	USSR (1971)	The exp. varies from 50 dB(A) (offices) to 85 dB(A) (in factories).	The exposure is as per work situation.
8.	Yugoslavia (1971)	Max. 90 dB(A)/8 Hrs./Day.	
9.	India (1975)	Max. 90 dB(A)/8 Hrs./Day.	DGMS Circular No. 18 of 1975.

(vi) Reducing reflected noise by use of absorbment materials on surfaces such as roof, walls and floors.

(vii) Placing attenuating screens between the operators and the sources.

(viii) Correcting imbalance and vibration by preventive maintenance.

Conclusion

In conclusion the objectives are to develop noise abatement technology in heavy industries in general and mining industry in particular. Broadly these objectives in terms of R&D could be summarised below:

— To determine major noise sources in underground and surface mines and mineral processing plants.

— To develop and evaluate both factory-installed noise controls for equipment and processing operations and noise-control field modifications to existing equipment.

— To provide more accurate noise exposure measuring techniques and instrumentation.

— To establish a procedure to evaluate the actual noise attenuation of personal hearing protectors worn by miners.

— To develop concepts and design techniques that will provide inherently quieter mining equipment and operations for the future.

— To make available to industry the technical knowledge needed to select, design and use the most effective noise-control measures.

REFERENCES

Cunniff, P.F., (1977), *Environmental Noise Pollution*, John Wiley and Sons, pp. 85-186.

Dhar, B.B. and Ratan, S., (1977), Noise Pollution, *Chemical Age of India*, Vol. 28, No. 10, October, pp. 847-52.

Dhar, B.B. and Choudhury, S.N., (1981), Noise — Occupational hazard and a source of environmental pollution, Special Science Issue, *Prajna*.

Dhar, B.B., (1986), Pollution Through Mining Industry-Fact Sheet, in *Pollution Control Handbook-1986*, Utility Publications, Hyderabad, pp. 159-67.

Dhar, B.B., (1988), Environmental Management Plan for Mining Projects — An

Update, *Jl. of Mines, Metals and Fuels*, January, pp. 13-6..

Jain, R.K., Urban, L.V. and Stacy, G.S., (1977), *Environmental Impact Analysis — A New Dimension in Decision Making*, Van Nostrand Reinhold Company, New York, pp. 54-6.

Magreal, E.B., (1975), *Environmental Noise Control*, John Wiley and Sons, pp. 48-50 and 152-55.

Mukherjee, R.N., (1986), Noise Problem Arising Out of Coal Mining Activities and Its Abatement, CMRS S&T Report, Dhanbad.

Nath Surendra, (1986), Noise Pollution, in *Pollution Control Handbook 1986*, Utility Publications Ltd., Hyderabad, pp. 49-56.

Roy, C. Bartholomae and Rober P. Parker, (1983), *Mining Machinery Noise Control Guidelines-1983*, A Bureau of Mines Handbook, United States Department of the Interior.

Singh, B., (1987), Environmental Pollution in Mines and Mining Areas and Miners Health, *Proc., Nat. Workshop on Environmental Management of Mining Operations in India* in Dhar, B.B. (ed.), BHU, February, pp. 11-40.

Webb, R.S. (1975), Noise and Extraction of Minerals, in John, M.J. (ed.), *The Institution of Mining and Metallurgy*, pp. 589-603.

27

AIR POLLUTION ANALYSIS OF VARANASI CITY

ONKAR SINGH

Air pollution may be defined as 'the presence in the gas, aerosal or particulate atmosphere of one or more contaminants or combinations thereof in such quantities and of such duration as may be, or may tend to be injurious to human, plant or animal life or property which unreasonably interferes with the comfortable enjoyment of life, or property, and the natural environment' (Pfafflin & Ziegler, 1976). The principal air pollutants responsible for damage to plants are sulphur dioxide, carbon monoxide, fluorides, smoke, soot, dust, hydrocarbons, oxides of nitrogen and aldehydes. Generally the plants are more susceptible to damage by gaseous pollutants whereas animals seem to be more susceptible to hazard either directly or indirectly by particulate pollutants (Singh *et al*, 1988). The air pollution may be classified into different categories according to its origin, state of matter or chemical composition. According to origin of the pollutants it may be grouped into two categories. : (a) primary pollutants — those emitted to the atmosphere as a result of some process; these remain for longer period in the atmosphere in chemical form in which they are emitted; and (b) secondary pollutants — those which are formed in the atmosphere as products of some reaction; this reaction generally occurs between two pollutants or between a single pollutant and a natural component of the atmosphere.

Some harmful substances are introduced to the atmosphere by natural processes but these are generally isolated and sporadic events. Of course, the massive air pollution has resulted from industrial activities. These have caused huge consumption of coal and petroleum resulting into emission of toxic air pollutants in the atmosphere. Five major categories of sources of pollution have been generally identified on the basis of the pollutant producing activity: (a) fuel combustion for transportation system, (b) fuel combustion in stationery sources, (c) industrial processes, (d) solid waste disposal, and (e) miscellaneous sources, which include forest fires, structural fires, the burning of agricultural wastes and evaporative losses from storage and marketing facilities. Among all the sources automobiles emit huge quantities of toxic gases including lead. In addition, many of the products of incomplete combustion in the presence of sunlight produce a complex mixture of pollutants called oxidants, and the atmospheric effect they generate is called *photochemical smog.* These pollutants are very typical of the atmospheres of sunny urban areas with considerable automobile traffic (Turk et al, 1978). Studies have revealed that due to increasing consumption of fossil fuels, the concentration of various toxic gases has greatly increased, resulting in global rise of the atmospheric temperature. The problem of air pollution is more acute in advanced industrial nations. In India though the situation is not the same as in developed countries but it is essential to control air pollution as it will become quite difficult to be tackled in future if not checked now.

An Outline of Air Pollution in Major Indian Cities

In India no systematic monitoring of air pollutants has been done so far to know the intensity and levels of air pollutants in major cities. For the first time in 1970-71, a short term survey was carried out by the National Environmental Engineering Research Institute (NEERI), Nagpur at four major urban centres, namely, Bombay, Calcutta, Delhi and Kanpur which revealed that the problem of air pollution exists in all the cities surveyed. Later on in 1979 an intensive survey of air pollution levels was conducted by NEERI at nine major cities of India. They monitored two major pollutants sulphur dioxide (SO_2) and suspended particulates at various sampling stations and found out the average concentrations of these for each city as a whole (Table 27.1).

Table 27.1 shows that Calcutta has the maximum and very high concentrations of SO_2 but in the city as a whole the average SO_2 concentration is not so high because there exist pockets of distinctly higher concentrations. It has been estimated that emissions weighing 1,305 tonnes including toxic substances are released into the air every day in Calcutta Metropolitan Development Area (CMDA), of which

Calcutta alone induces about 900 tonnes of toxic gases. The break up of pollutants are : particulates — 44 per cent carbon monoxide — 34 per cent, SO_2 — 9 per cent. About 90 per cent of particulate matter is emitted only from industries of Howrah industrial belt. Besides the industries, the traditional source of burning coal or charcoal in private kitchens (1,200 tonnes of soft coke every day/or over) by a large number of urban dwellers also contributes to the increasing concentration of particulates and SO_2 in the city atmosphere.

TABLE 27.1: Levels of Air Pollutants in some of the Indian Cities (1979)

City	Suspended particulates (μ/m^3)	Mean value of SO_2 ($\mu g/m$)
Calcutta	578	85
Bombay	275	83
Madras	106	16
Delhi	481	39
Kanpur	344	25
Nagpur	386	12
Hyderabad	295	26
Ahmedabad	243	71
Jaipur	379	17

Source: NEERI, Nagpur, 1979.

A recent government monitoring by the NEERI of air pollutants in Bombay revealed that air is being polluted at a rate of 1,000 tonnes of pollutants every four hours in Bombay. It is further observed that the pollutants consist of 38.44 per cent CO, 33.9 per cent SO_2 and 9.8 per cent as NO_x. In Chembur, the most dense industrial zone of the city, the residents inhaled a total of 190 tonnes of sulphur dioxide every day and many of them were suffering from cough, eye irritation, constant sneezing, bronchial disorders, chest pain and fatigue. In addition to this the exhaust gases from more than 4.2 lakh automobiles in Bombay at a rate of 0.10 ppm are also causing respiratory disorders by photochemical effects.

The problem of air pollution in Delhi is becoming more critical. A study conducted by Prof. Dave (1980-81) revealed that Motinagar industrial area had the highest concentration of suspended particulate matter (SPM) followed by Rashtrapati Bhawan, Defence Colony, Red Fort and JNU. Every square kilometre of Delhi got 328.0 kg of pollutants per day on the average. The power generation exhausts contributed 277.007 metric tonnes per day (MTPD), industrial sources 106.676 MTPD and domestic 86.957 MTPD. The automobiles contribute 466.408 MTPD, railways 48.087 MTPD and air craft 13.817 MTPD to Delhi's total air pollution (Parsai, 1986). The extent of air pollution caused by vehicles is evident from the fact that besides 4500 DTC buses, 40,000 other heavy vehicles ply every day on Delhi roads. Altogether these vehicles account for emission of over 250 tonnes of CO. In addition, these emit 40 tonnes of hydrocarbons, 30 tonnes of oxides of nitrogen and 5 tonnes of SO_2 and huge quantities of particulates including lead.

Like other cities at Ahmedabad, too, the main sources of air pollution are automobiles and industries, mostly textiles. It was found that owing to very high density of vehicular traffic in the heart of the city, concentration of lead, the most toxic air pollutant, usually remained higher than the permissible level prescribed by WHO ($2\mu g/m^3$) (Aggarwal, 1928). During summer its concentration was found to be lower than the winter.

The above study manifests that the main source of air pollution in cities are automobiles which emit huge quantities of unburnt gases and industries discharging huge quantities of smoke, dust and other particulates, sulphur dioxide and oxides of nitrogen. The increasing number of automobiles poses continuned threat to the well-being and future of the urban environment. With this purview an attempt has been made here to analyse the spatial pattern of air pollution in Varanasi city which has already crossed the million plus mark of its population size.

Air Pollution in Varanasi

Though the problem of air pollution in the city is not so alarming as that of Bombay, Calcutta, Delhi and Kanpur but its increasing number of automobiles poses serious threat to the future of the city's environment. The city of Varanasi, also popularly known as Banaras and Kashi, is said to be the oldest living city of the world. It commands unparallel importance as a combination of cultural, religious, spiritual and educational centre in the country (Fig. 27.1). In addition to its present estimated population of 1.2 million people, the city also accommodates a daily floating population of 0.2 million persons. The city is characterised by narrow roads and lanes, which further intensify the problem due to slow movement of vehicular traffic.

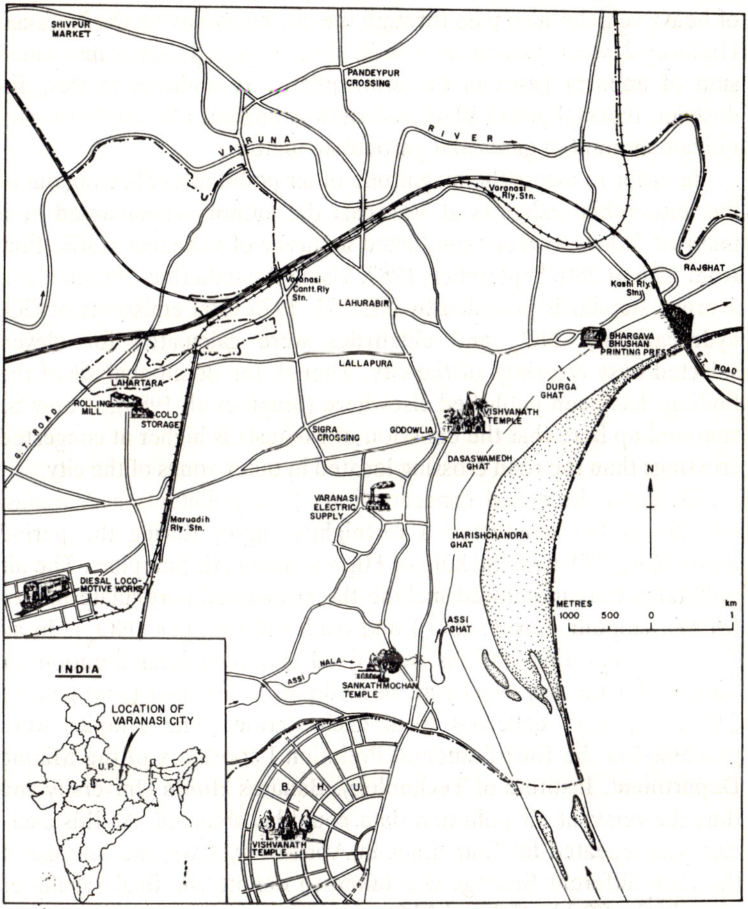

FIG. 27.1: Varanasi City

Besides a large number of automobiles plying on the roads, thousand of heavy vehicles also pass through the city every day on the National Highway-2 which runs in the middle of the city. All these cause emission of unburnt gases in the atmosphere. In addition to this, industries, thermal power plant and railway engines also discharge significant amount of gases and particulate matter.

In order to assess the magnitude of air pollutants collecting out of the automobile exhausts at Varanasi the author accompanied by a team of fellow workers conducted a survey of vehicular traffic flow from 25th to 28th September, 1987. Using the data thus obtained and average standards provided by NEERI, 1975 total emissions of CO, hydrocarbons, NO_x and aldehydes were calculated for eleven selected busy crossings of the city. Though the detail analysis of the findings has been published elsewhere (Singh et al, 1988), it may be summed up here that the emission of exhausts is higher at congested crossings than the road crossing located in outer zones of the city.

To know the actual concentration of air pollutants in Varanasi city the author conducted a fortnightly survey during the period April-May, 1988 with the help of High Volume Air Samplers. The air pollutants thus monitored include the suspended particulate matter (SPM), sulphur dioxide (SO_2) and oxides of nitrogen (NO_x). In all only 13 monitoring stations were fixed and eight hour duration air samples for the day (9.00 a.m. to 5.00 p.m.) and night (9.00 p.m. to 5.00 a.m.) were collected from each station. The samples were processed in the Environmental Pollution Laboratory of the Mining Department, Institute of Technology, Banaras Hindu University and thus the relevant air pollution data could be obtained. As this exercise was repeated for four times in April-May, 1988, the average of the four different findings was taken to denote the final results as given in the (Table 27.2 and 27.3).

Table 27.2 reveals that the presence of SPM in the ambient air of Varanasi ranges from a minimum of 189.48618 to a maximum of 1938.1556 $\mu g/m^3$. The maximum average concentration of the suspended particulate matter (1798.203 $\mu g/m^3$) was found at Golgadda crossing. This is owing to very frequent movement of heavy traffic through this site. The Chowk, Manduadih, Maidagin, Godaulia and B.H.U. Gate are other places noted for their outstanding SPM levels. All these sites are crowded with vehicles and pedestrians till late night. The localities around the clock also have got high SPM values. The concentration of SPM is not higher than the prescribed limit at most of the crossings. As the Indian cities have not been able to develop the CBD in western context, all the sampling stations virtually represent the areas of mixed functional use for which the SPM permissible limit is 500 $\mu g/m^3$. Obviously nine out of the thirteen stations have recorded SPM figures above the permissible limit. The ambient air samples collected from the B.H.U. campus, Karaundi, Circuit House and Dashashwamedh Ghat are free from SPM pollution, the reason being their advantageous locations (Fig. 27.2). The Banaras Hindu University campus is away from the main city while Karaundi is located in the outskirts. The Circuit House, though located inside the city, is not open to common people. The sampling station located at Dashashwamedh Ghat along the river Ganga is not only inaccessible to vehicular traffic but also it has got an open frontage of the river. In general the level of SPM in the atmosphere is noted to be lower during the night than during the day at Varanasi.

TABLE 27.2: Distribution of SPM in Varanasi City, April-May, 1989

Sl. No.	Station	Suspended Particulate Matter SPM ($\mu g/m^3$)		
		Day	Night	Average
1.	Karaundi (near B.H.U. campus)	456.68796	385.34125	421.0146
2.	Mining Department (B.H.U.)	326.00092	234.51815	280.25953
3.	B.H.U. Gate	1526.6358	494.01058	1010.3231
4.	Cantt. Rly. Station	594.75169	781.60545	688.17857
5.	Police Station Chowk	1938.1556	1369.6775	1653.9165
6.	Golgadda	2658.8827	937.5233	1798.203
7.	Circuit House	232.46441	428.45548	330.45994
8.	Dashashwamedh Ghat	189.48616	461.89226	325.68921
9.	Lahurabir	670.64907	769.97212	720.31059
10.	Pandeypur	800.13381	327.65023	563.89202
11.	Manduadih	1395.6368	868.39017	1132.0134
12.	Maidagin	1412.0238	682.53340	1047.2786
13.	Godaulia	1036.6015	1025.4578	1031.0296

Mean of Varanasi City : 846.35143 $\mu g/m^3$.

The ambient air samples also reveal the presence of SO_2 up to a level of 17.3419 $\mu g/m^3$. Table 27.3 reveals the spatial variation of SO_2 concentration in the city atmosphere. The concentration is found to be maximum at Lahurabir crossing (15.1816 $\mu g/m^3$) while

TABLE 27.3: Distribution of SO_2 and NO_x in Varanasi City, April-May, 1989

Sl. No.	Station	Sulphur Dioxide SO_2 ($\mu g/m^3$)			Nitrogen Oxides NO_x ($\mu g/m^3$)		
		Day	Night	Average	Day	Night	Average
1.	Karaundi	1.2138	1.8138	1.5138	—	—	—
2.	Mining Deptt.(B.H.U)	1.7133	2.1479	1.9306	—	—	—
3.	B.H.U. Gate	1.2273	1.2101	1.2187	—	—	—
4.	Cantt. Rly. Station	9.4341	11.7664	10.60025	2.1421	1.3142	1.7281
5.	Chowk Police Station	8.2628	6.3178	7.2901	2.1146	0.7842	1.4494
6.	Golgadda	14.4320	10.7132	12.5726	2.3849	1.7210	2.0529
7.	Circuit House	6.5476	9.1123	7.82995	0.6441	0.9378	0.7909
8.	Dashashwamedh	0.5341	0.6834	0.60875	—	—	—
9.	Lahurabir	17.3419	13.0214	15.1816	2.7842	2.3100	2.5471
10.	Pandeypur	13.3024	9.5313	11.41658	2.9824	1.8317	2.4070
11.	Manduadih	8.3146	4.7331	6.52385	1.6441	0.2341	0.9391
12.	Maidagin	9.3794	6.3668	7.8731	1.3201	0.3421	0.8311
13.	Godaulia	16.1348	11.4681	13.80145	2.9412	2.0014	2.4713

Mean of Varanasi city: SO_2 - 7.56628 $\mu g/m^3$
NO_x - 1.17054 $\mu g/m^3$

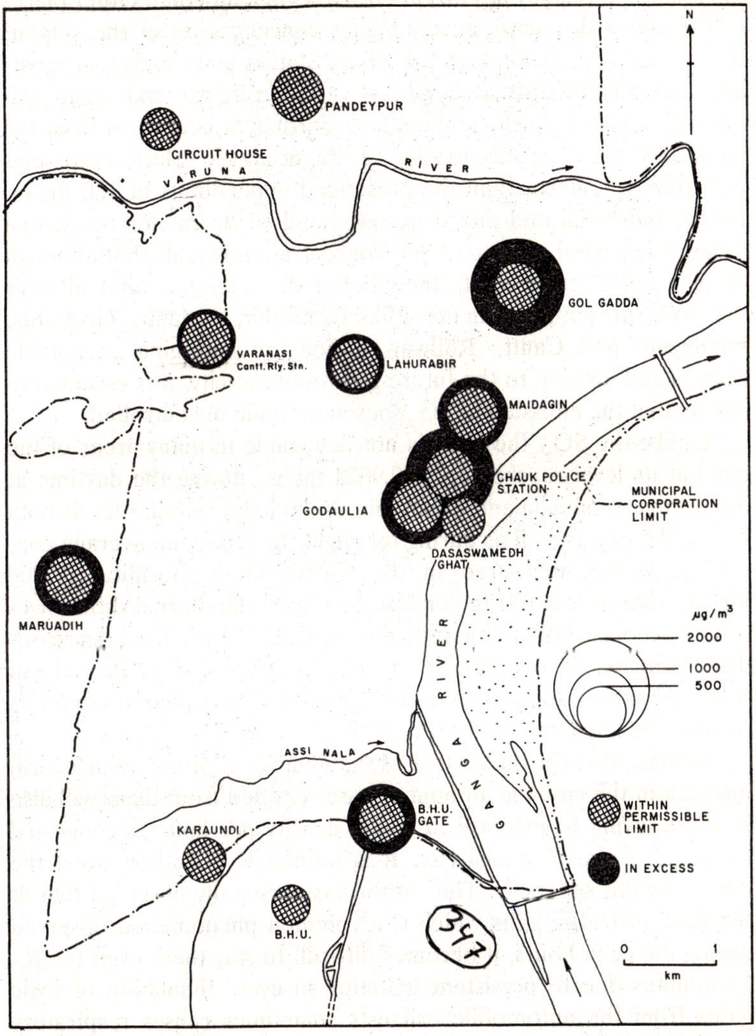

FIG. 27.2: Varanasi City: Distribution of SPM at Selected Sites, May, 1988

the minimum (0.6087 $\mu g/m^3$) at Dashashwamedh Ghat. Other places of the city with comparatively higher concentration of the sulphur dioxide are Godoualia, Cantt. Railway Station and Pandeypur crossing. The city records a significant variation in day and night concentration of SO_2 in the ambient air. Further, it is obvious from the table 27.3 that the concentration of SO_2 at all the selected crossings of Varanasi is lower than the prescribed upper limits of 120 $\mu g/m^3$ for the industrial and mixed use area and 80 $\mu g/m^3$ for residential areas. The spatial pattern of SO_2 concentration reveals that although the concentration is comfortably below the tolerance limit all over the city at present, certain areas like Lahurabir, Godaulia, Golgadda, Pandeypur and Cantt. Railway Station are emerging as notable centres and looking to the future growth of the city, it is essential to ensure that the emission of SO_2 does not remain uncontrolled.

Unlike the SO_2, the NO_X is not detectable in many areas of the city but its level reaches up to 2.9824 $\mu g/m^3$ during the daytime at Pandeypur. Godaulia and Lahurabir also maintain similar levels both during the day as well as during the night. Its minimum average concentration was measured in the Circuit House locality (0.7909 $\mu g/m^3$) due to less traffic density. As regard the diurnal differences of NO_X concentration generally it is noticed that the level decreases during nights. The emission of NO_X is quite less in the city of Varanasi and as many as four monitoring stations failed to record its presence in the ambient air. (Fig. 27.3)

Further, it is concluded that as the number of motor vehicles will increase in the city, the amount of gases emitted from them will also increase owing to increase in traffic density per hour at congested crossings of the Varanasi city. It is ultimately bound to affect the human health adversely. The author has personally observed that at the time of traffic jams which is a common phenomenon observed during the peak hours, it becomes difficult to stay there even for 10-15 minutes due to persistent irritation in eyes. Inhalation of toxic gases from the automobile exhausts sometimes causes respiratory trouble and nervousness to the citizens living along roads in the main city. These problems may be at least partly attributed to other gases also not taken up in the present survey.

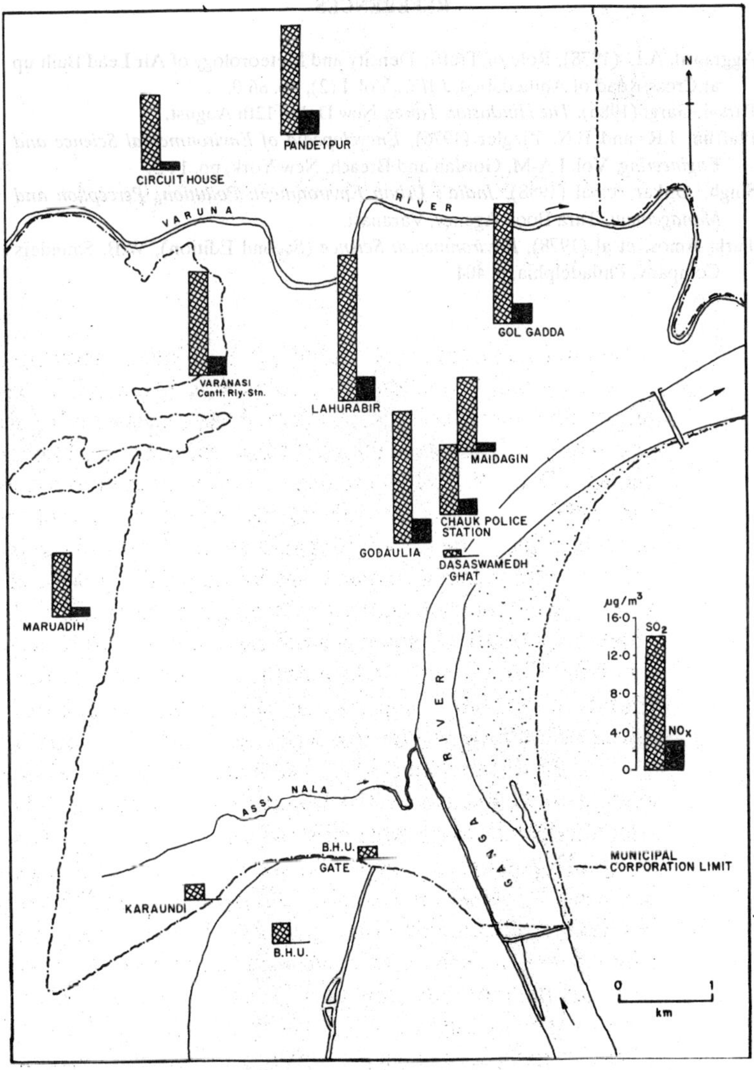

FIG. 27.3: Varanasi City: Distribution of SO_2 and NO_x at Selected Sites, May, 1988

REFERENCES

Aggrawal, A.L. (1978), Role of Traffic Density and Meteorology of Air Lead Built up at Cross Road of Ahmedabad, *JAPC,* Vol. I (2), pp. 86-9.

Parsai, Gargi (1986), *The Hindustan Times,* New Delhi, 12th August.

Pfafflin, J.R. and E.N. Ziegler (1976), *Encyclopedia of Environmental Science and Engineering,* Vol. I A-M, Gordan and Breach, New York, pp. 14-5.

Singh, Onkar, et al (1988), *India's Urban Environment: Pollution, Perception and Management,* Tara Book Agency, Varanasi.

Turk, Amos, et al (1978), *Environmental Science* (Second Edition), W.B. Saunders Company, Philadelphia, p. 404.

28

ENVIRONMENT AND CRIME :
A Case Study of Agra District

RAJENDRA PRASAD

In legal terms in the Indian context, crime is defined as any act prohibited under the Indian Penal Code (I.P.C.) or special laws committed by an individual or a group. The I.P.C. covers crimes such as murder, robbery, dacoity, theft, kidnapping, rape, riots, burglary, cheating, criminal breach of trust, counterfeiting etc. Crime is the major effects of social disequilibrium. This menace is not restricted to any nation, it is a global phenomenon. No one in society regardless of its economic, political, cultural or ethnic structure, is immune to this problem. Since the crime problem affects each and every individual and group and jeopardize their interests, it should generate intense community involvement. In the presence of excessive crime or fear of crime, people would hardly be able to live in peace and stable condition and contribute little to the progress of the society or nation. This is the reason that liberal geographers and other social scientists are now taking interest in the study of this issue by conducting researches for proper understanding and better management of the problem.

Crime is basically a human behaviour which is entirely determined by the environment around him. The basic necessities of man, namely food, clothing, shelter, health, education and recreation etc. are largely depend upon the endowment of the environmental resources. If this balance of environmental set up and human requirements is being disturbed through human or natural action, the whole human behavioural system is significantly deteriorated, which generate criminal tendencies among the people. As a consequence, the liberal mind becomes sidetracked eventually ending up on the other side of law. The balance of environmental system is mainly attributed to natural hazards (soil erosion, flood, drought, cyclone, earth-quake etc.), uncontrolled growth of population, scarcity of resources, lack of employment opportunities, fewer social amenities and facilities etc. Thus, crime is in a way a fallout of human act and or environmental problems. It can be defined as an expression of discontents which lie deeply embedded in the social system. Crime can indicate the particular environmental forces that cause it.

Now it is an accepted fact that crimes are increasing all over the world. In India, the incidence of crime is increasing at an alarming rate (Table 28.1). The total of 13,44,968 cognizable crimes under the I.P.C. were recorded in the country during 1978 against 12,67,004 recorded in 1977, showing an increase of 6.2 per cent over 1977. There is an overall increase of 56 per cent in crimes during the decade (1968-1978). The table also presents the estimated mid-year population and volume of crimes per one thousand population. All these figures indicate that crimes are increasing at a faster rate (more than double) than population.

TABLE 28.1: Incidence of Cognizable Crimes under the IPC in India

Sl. No.	Year	Estimated Population in millions	Total number of cognizable crimes	Volume of crime per thousand of population
1.	1968	511.1	862,016	1.68
2.	1969	522.5	845,167	1.61
3.	1970	534.3	955,442	1.78
4.	1971	551.2	952,581	1.72
5.	1972	563.5	984,773	1.74
6.	1973	575.9	1,077,181	1.87
7.	1974	588.3	1,192,277	2.62
8.	1975	600.8	1,160,520	1.93
9.	1976	613.3	1,093,897	1.78
10.	1977	625.8	1,267,004	2.02
11.	1978	638.4	1,344,968	2.10
Percentage change in 1978 over 1968		+24.91	+56.03	

Source: Crime in India, 1978, BPR & D, New Delhi.

Among all the states of India the problem of crime is very much serious in the administrative areas of Uttar Pradesh, Madhya Pradesh and Delhi. Due to high increase in the number of crimes in these areas, the Central Government has asked the Planning

Commission to prepare a special plan for the eradication of crime menace. The highest incidence of crimes (2,17,277) were registered in Uttar Pradesh (1978) which was 15.4 per cent of the total crimes registered in the country. In this state, the criminality is much higher in Agra district which occupies third place (1978). Certain environmental conditions are responsible for such phenomenon. Hence, this district has been taken up to examine the environmental factors associated with the causes of crime.

Environmental Set up

The study area extends between 26°44′-27°25′N latitude and 77°26′-78°32′ E longitude and covers an area of nearly 4,816 km^2. The location of Agra district is highly significant from the geographical, historical and strategic point of view. It is at the borders of Madhya Pradesh in the south and of Rajasthan in the southwest and west (Fig.28.1). The district is drained by the river Yamuna and its tributaries — Jhirna, Sirsa and Sengar in the north and Utangan and Chambal in the south. Flatland is the largest areal unit accounting for more than 3/4th of the total area of the district. The ravines along the banks of the river Yamuna, Chambal and Utangan encompass the large area of about 828.87 km^2. It is always a problem area of the district. The climate of the region may be described as semi-arid.

There are seven tahsils and 39 police circles. The human population of the district, according to the 1981 Census is 2,852,474 which gives an average density of 592 persons per km^2. The average literacy for the district is 28 per cent. There are 1,232 revenue villages spread over an area of 4,704.6 km^2 in the district including 50 uninhabited villages. The total urban area in the district is about 111.4 km^2. As many as 3.74 per cent villages of the district are connected by rail, 24.88 per cent by metalled road and 3.90 per cent by the unmetalled roads. Agriculture dominants the economy of the district, as it provides employment to nearly 52.1 per cent in the form of cultivators and agricultural labourers. Among the societies of the region, Hindus constitute 89.37 per cent, Muslims 9.01 per cent and others 1.62 per cent of the total population. The urban areas of the district are dominated by Vaishya (trader and businessman), Muslims and Scheduled Castes people. The flatlands are inhabited by Brahmins, Jats, Rajputs, Kachhis; where as the ravine areas are mainly occupied by *Nishadas, Ahirs, Gujaras, Gadarias* and some *Rajput* clans also.

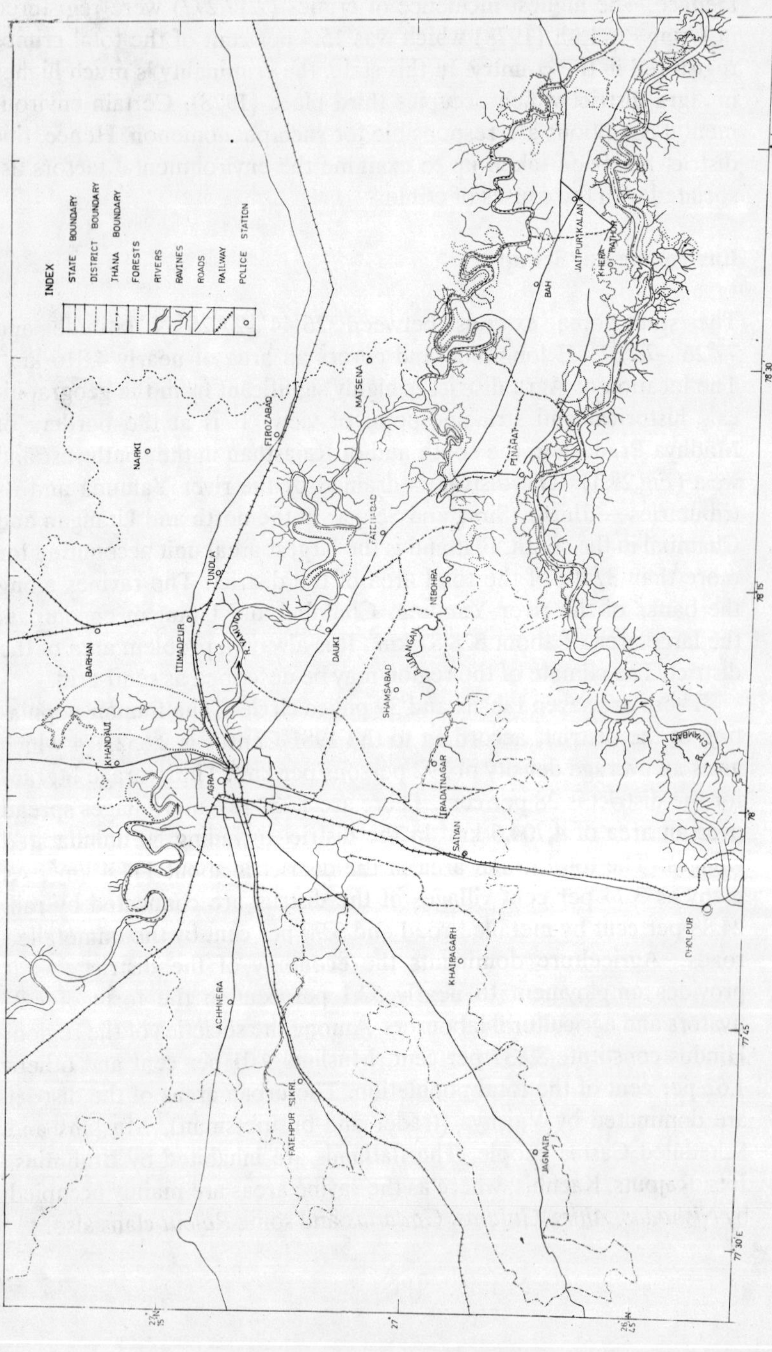

FIG. 28.1: Agra District and its Environs

Data Sources and Methodology

The present study deals with the environment and crimes, hence the statistics of crimes have been collected from the annual bulletin of *Crime in India*, BPR&D = Bureau of Police Research and Development, New Delhi. To examine the problem in depth, adequate and reliable data on crimes for Agra district have been used. Further, Police circlewise statistics on some important crimes occurring in the region under I.P.C. including dacoity, robbery, burglary, theft and kidnapping have been compiled from the records available at district police office, Agra. Based on these information, average annual incidence and the volume of important crimes in the district have been calculated and represented cartographically by Choropleth. Detailed information with regards to the criminals and environmental components, both physical and cultural, have been collected from the available records and field survey. To get a clear picture of crime in perspective of physio-cultural environment, many sources of information in the form of facts, figures and maps have been used.

Trends of Crimes in Agra District

The total number of 7,210 cognizable crimes under I.P.C., were recorded by the Police in the district during 1979 against 9,163 registered in 1977, showing a decrease of 21.33 per cent over 1977 (Table 28.2). It may be due to the much attention paid by the authority to the building up of defence against crimes. The average number of cognizable crimes committed in the district is about 8,284 per year. The volume of crimes was 0.60 per thousand of population in 1981 in the district. A comparative study of crimes committed in the district under important crime heads during the period 1977-79 shows a decrease in all heads of crimes except murder, dacoity, kidnapping and miscellaneous crimes. The highest increase of 16.22 per cent in murder, 15.35 per cent in dacoity, 8.63 per cent in kidnapping and 8.75 per cent in miscellaneous crimes was recorded during the period 1977-79. Thus, the trend of crimes in the region indicates an increasing tendency towards the serious crimes.

Distributional Pattern of Crimes

Among all the crimes under IPC, dacoity, robbery, burglary, theft and kidnapping constitute nearly 65.40 per cent of the total crimes committed in the district (Table 28.2). Therefore, in the present study, the spatial distribution pattern of these serious crimes is only analysed in detail in its environment al perspective. Police circle-wise statistics relating to these important crimes is given below.

TABLE 28.2 : Incidence of Cognizable Crime in Agra district

Sl. No.	Head of Crime	Years 1977	1978	1979	Average No.of crimes per year	Average %	Percentage variation
1.	Murder	111	112	129	117	1.42	+16.22
2.	Culpable Homo-cide not amoun-ting to murder	85	50	30	55	0.66	−64.71
3.	Rape	22	19	18	20	0.24	−18.18
4.	Kidnapping and Abduction	139	137	151	142	1.72	+ 8.63
5.	Dacoity	241	271	278	263	3.18	+15.35
6.	Robbery	345	287	206	279	3.36	−40.29
7.	Burglary	1699	1390	965	1351	16.30	−43.20
8.	Theft	3821	3447	2875	3381	40.84	−24.76
9.	Riots	443	405	294	381	4.59	−33.63
10.	Criminal breach of Trust	224	218	179	207	2.49	−20.09
11.	Cheating	136	130	121	129	1.56	−11.03
12.	Counterfeiting	2	3	1	2	0.02	−50.00
13.	Miscellaneous	1897	2010	1963	1957	23.62	+ 8.75
	Total	9,165	8,479	7,210	8,284	100.00	−21.33

Source: Crime in India, 1977, 1978, 1979, B P R and D, New Delhi.

(a) *Dacoity*

Among all the crimes, dacoity is one of the most serious crime in the region. The crime of this nature has been varying from year to year (Tables 28.2 and 28.3). The average number of dacoities (1977-81) in the district was about 198 per year. The majority of the dacoities as much as 99.1 per cent occur in the rural environment, specially around the ravines of Chambal, Yamuna and Utangan rivers. For centuries these ravines have been infested with gangs of dacoits and robbers, who have succeeded to inflict panic and terror among the villagers. The maximum number of dacoity cases were registered in Jaitpur (10) and Matsena (9) police circles during 1981. The Police circles of urban areas of Agra city registered only 0.9 per cent of the total dacoity cases of the district. More police patrols acts as a deterrent to this type of crime in the urban environment.

(b) *Robbery*

Most of the cases of robbery occur on the roads and paths and involve forceful possession of cash, articles etc. The incidences of robbery cases have decreased by 16.5 per cent in 1981 in comparison to 1980. The average number of robberies (1977-81) in the district was 223 per year. Rural police circles of alone account for 51.6 per cent of the total robbery cases registered in the district during 1981.

(c) *Burglary*

The incidences of burglary cases in the district decreased by 43.20 per cent from 1977 to 79 and again declined by 10 per cent in 1981 over 1980. The volume of such crimes is higher in urban areas, such as Agra city which accounts for 58.3 per cent of the total burglary cases registered in the district during 1981. The average number of such cases committed in the district (1977-81) is about 992 per year.

(d) *Theft*

The crime under theft are decreasing in the district (Tables 28.2 and 28.3). The average number of thefts per year for the district as a whole was about 992 (1977-81). The highest number of theft cases during 1981 were recorded in urban environment specially in Hari Parbat (95), Sadar Bazar (79) and Firozabad north (71) Police circles of the district. Recently some people belonging to Muslim and Ghoshi Caste are engaged in criminal activities like theft etc. This has increased the incidence of such crime.

(e) *Kidnapping*

Among all the aforesaid crimes occurred in the district, kidnapping is the safest profession of the dacoits of the ravines of Chambal and Yamuna. They get most of their money from kidnapping. The victims of kidnapping are usually sons of rich farmers or businessmen who are taken as captive at gun paint when they go to relieve themselves at night. Some times smaller dacoit gangs kidnap persons and then sell them to bigger gangs who collect the ransom. More people choose to pay up and in nearly all the cases the kidnapped victims are returned unhurt. In the district the number of crimes under this head were recorded 29 and 24 in the year of 1980 and 1981 respectively. The highest number of kidnapping cases during 1981 took place in Pinahat (6) followed by Jaitpur (3) and Matsena (3) police circles, which accounts for 50 per cent of the total of this head registered in the district. In urban areas of Agra city there was only one incidence of kidnapping registered in Tajganj police circle.

There is uneven rate of crimes in the region. Fluctuation in certain crimes is the interesting feature of the district. There are certain crime-infested areas in the district. On the basis of the volume of some important crimes, three types of crime-infested areas may be identified. In area distribution severe and moderate crime infested zones are larger in extent than low crime infested zone (Fig. 28.2).

1. *Severely Crime Infested Zone*

Severely crime infested area is recognised where the volume of crime per thousand of population is above 0.71. This zone includes 11 police circles of the district and mainly occur along the river systems of Yamuna and Chambal. As many as 8 police circles of this belt are located in rural environment whose large area is occupied by the gullies and ravines. The difficulty of terrain, the poverty of soils and poor means of communication made this tract a breeding ground for innumerable crimes. This is one of the most problematic zone of the district where dacoity, robbery, theft, kidnapping and other heinous crimes are occuring every day. Matsena Police circle of this belt records the highest (1.13) volume of crime.

Three police circles of the district, namely Tajganj, Sadar Bazar and New Agra of urban environment are also included in severely crime-infested zone. These areas are urban-rural fringes in character and economically prosperous. The nearby forest areas of Chhalesar and Swami Ghat villages are the best hideouts for the criminals from where they operate in these areas. The police personnel are scarcely deployed here in comparison to the main city which is also responsible for high volume of crime.

FIG. 28.2: Agra District: Crimes

2. *Moderately Crime-infested Area*

The second zone is identified where the volume of crime per thousand of population is between 0.41 to 0.70. This belt encompasses 14 police circles of the district out of which 8 police circles are located in urban environment. The problem of the crime is aggravated in these urban areas because of migration from rural to urban area. The people either come in search of jobs or some of them take shelter to protect themselves from several charges against them. They promote anti social activities in the urban areas.

3. *Low Crime-infested Zone*

This zone records the lowest volume of crime below 0.41 per thousand of population. This belt is widely distributed in 13 police circles mainly in the western parts of the districts. Here, the problem of crime is not so acute as in the previous zones because of availability of flatland and good means of communication.

Environmental Factors Responsible for the Cause of Crime

The incidences of crimes in Agra district vary from place to place. According to the field survey, nearly 79.5 per cent criminals of the district involved in dacoity, robbery, theft, burglary and kidnapping etc. are the permanent dwellers of the ravine villages. A number of environmental factors both physical and cultural determine the origin, spread and spatial distributional pattern of crimes. These causative factors may be examined as follows.

Of all the environmental factors, topography of the region plays in important role in the origin and spread of crimes. Due to heavy soil erosion, nearly 828.87 km^2 of agricultural land of the district has been transformed into ravines along the river systems of Yamuna, Chambal and Utangan. As a consequence, the inhabitants of the ravine environment have turned to the criminal way of living : dacoity, plundering, looting, kidnapping, robbing etc. to sustain their livelihood. They are pushed into this profession by harsh realities of the rugged topography, where almost nothing grows. Economic hardships are so compelling that they do not find any alternative modes of survival. Thus, the ravines serve a two fold purpose. On the one hand, they germinate the seed of crimes in the frustrated youths, and, on the other hand, they provide them shelter in their hideouts from the clutches of law. Thus, the ravines are the dwelling places of the criminals from where they operate their profession in its surroundings. It is common saying that ravines and dacoits will go together.

The scarcity of land resource is also responsible for the occurrence of crime, especially in the ravine areas. The average cultivable land per family is not even 1/4 *bigha* in many of the ravine villages. It is clear that cultivable land in ravine villages is so small that the dispute or quarrel over the land is quite common, which results a shoot-out. Malkhan Singh of Chambal ravines became a very dreadly dacoit in 1973 after he shot a man because of some dispute over the land.

The people of the ravines of Yamuna and Chambal live in an environment of physical isolation and natural freedom. They are by nature free, freedom loving and brave. If there are lightest sense of subjugation of disrespect from some particular villagers then they leave the village and flee to the ravines with guns to teach them a lesson. Many dacoits of this area have started off in this way. Man Singh, the legendary dacoit of the region formed a gang due to this reason and refused to bow even before the mighty British Government.

Social environment of the region, especially of the ravine areas, is also adding to the high volume of crimes. In this belt, the people call dacoits — *bagi* (rebel). It is a long tradition to the region that if a person kills or injures the another, the relatives of the victim party will have to take revenge. If some one wants to forget the old vendetta, the villagers won't let him to do so. The son has to take revenge to his father's and brother has to take revenge for his brother's death. In this way unending sequence of crimes are committed. Mishiriakapur (Matsena Police Circle), Pura Bhagarena (Pinahat P.C.) and many other villages of the district have been deserted on account of such tradition.

The inhuman injustice of the caste system is also causing more incidences of crime. Due to this reason humiliated pauper flee to the ravines and join the most powerful dacoits of the time. Many liberal minded people of the region like — Vikram Singh, Vidhya Ram, Kashi Ram etc. have started off in this way and had avenged their humiliation.

The people of the area pay high respect to the women. This is the reason that the incidence of rape is only 0.24 per cent of the total cognizable crimes of the district. If some one misbehaves with any woman, the people of the region become out laws to protect their family honour. Pan Singh Tomar — a military man and a former shot put champion of India who once represented the country in the Asian Games, has lately became a big dacoit of Chambal Valley on account of this reason. Pan Singh shot down a person in his village who has allegedly molested his daughter and fled to the Chambal ravines.

Incredible number of fire arms in the region has its inevitable effect on the incidences of crime. It is revealed from the field investigation that in ravine villages, there is a gun per family and the number of unlicensed guns are more then licensed ones. The importance of firearms is evident from the way, dacoit gang shares its loot. The ratio is like this : A member who has a 12 bore gun gets one part, one who has a 8315 semi-automatic rifle gets two parts and one who has an automatic weapon or a L.M.G. gets three parts. The gang leader invariably makes a net profit since he has the maximum number of guns. A good way of making money from the fallow dacoits who have no guns, is to hire guns out to them in return for half the share the gun brings.

The ravine tracts of Bah, Jaitpur, Pinahat, Pindhoura, Fatehabad, Nibohara, Matsena and Tundla Police Circles of the district have extremely low accessibility of roads and railway (7.5 km and 0.7 km per 100 km^2 respectively). Thus, one of the strongest environmental factors that has been responsible for frustrating the efforts of the government to control the crime problem is inadequacy of the transportational and communication facility. Due to this reason, information regarding the activities, plan and programmes of crime — does reach the authority quite late, who can chase them only up to a road point on the ravine fringes.

Pressure of population in rural areas is also playing an important role in increasing the incidences of crime. During the preceeding censuses (1961-81) the rural population of the district rose by 45.70 per cent. Due to very high growth of population, the people of rural areas, especially of ravine villages, where the resources are very scanty, are not getting sufficient employment to get their livelihood. As a consequence they are turned towards criminal way of living.

Caste structure is also responsible for the high volume of crime. The ravine areas of the district are mainly occupied by the latent criminality castes like *Nishadas, Gujars, Ahirs* and *Rajputs*. As a result these tracts have been the breeding ground of innumerable crimes. These criminals are also often helped by the members of their own castes and at times even the people have been known to seek their assistance.

Low literacy (average 28 per cent) is also acting as an important force to make banditry and committing crimes. In many parts of the region, literacy is too low (17 per cent). Due to this, if some one commits any sort of crime, the illiterate people of the region hope to get justice only at gun point, violating legal procedures. It has become the law of the area, specially of the ravineous land.

Besides, the political environment i.e. government policy of arranging surrender of some most notorious dacoits of the region is also generating the seeds of crimes among the local youth. This is the common saying in the region, "it is better to kill ten men than one. If you kill one man, you are hanged as a murderer, but if you kill ten men or more than that, you become a dacoit. You can then surrender and become a hero. Then government will provide you all the facilities". The surrender made it seen to the people that a dacoit is an important person who is respected even by the government. As a result, after every few years, the region produces a new crop of bandits and the cycle of crimes begins all over again.

Conclusion

It is evident from the foregoing discussions that environment has played an important role in origin and spread of crimes in Agra district. It is also clear that every crime is the result of inter play of a number of environmental factors. Though our national government is very much aware of the crime problem and its consequences and various measures have been tried for the eradication of this menace but the situation of the problem is not yet satisfactory. For the control and influencing check of this problem, certain remedies may be proposed. The implementation of these remedies may prove very useful to cope with the problem of crimes.

The policy of surrender of dacoits can solve the menace of dacoity and other crimes to a great extent. with this best intentions, Jai Prakash Narain and Gandhian workers like Subba Rao initiated the surrender of dacoits firstly in 1972. The level of crime came down for a while. But in the past few years, new gangs have replaced the old ones. The Gandhian workers failed to bring about any genuine change of heart among most of the dacoits.

TABLE 28.3 : Distributional Pattern of Crimes under important crime heads in Agra district (1980-81)

Sl. No.	Name of Police Circles	Dacoity		Robbery		Theft		Burglary		Kidnapping		Annual Average of all crime heads	Estimated volume of crime (per thousand of population
	Agra city	1980	1981	1980	1981	1980	1981	1980	1981	1980	1981		
	1	2	3	4	5	6	7	8	9	10	11	12	13
1.	Kotwali	1	—	6	3	22	35	20	20	—	—	53.5	0.43
2.	Rakabganj	—	—	7	7	52	54	18	20	—	—	73.0	0.67
3.	Shahganj	1	1	2	5	53	60	30	18	—	—	84.5	0.62
4.	Madan Mohan Gate	—	—	—	—	3	—	2	—	—	—	2.5	0.17
5.	Mantola	—	—	—	—	—	—	—	—	—	—	—	Nil
6.	Sadar Bazar	2	—	9	16	71	70	32	40	—	—	124.5	1.10
7.	Tajganj	3	—	8	3	39	43	22	27	1	1	73.50	1.01
8.	Chhatta	1	—	12	9	22	38	9	20	—	—	55.5	0.60
9.	New Agra	1	—	8	3	57	22	22	17	—	—	65.0	0.91
10.	Etmauddaula	—	—	—	—	16	—	4	—	—	—	10.0	0.30
11.	Hari Parbat	1	—	11	10	127	95	52	46	—	—	171.0	0.57
12.	Lohamandi	—	—	3	5	54	51	22	16	—	—	75.5	0.68
13.	Sikandara	1	—	2	—	14	—	8	—	—	—	12.5	0.45
14.	Jagadish Pura	—	—	1	—	8	—	6	—	—	—	7.5	0.32
15.	Naiki Mandi	—	—	2	—	6	—	2	—	—	—	5.0	0.19

(Contd.)

TABLE 28.3: *(Contd.)*

1	2	3	4	5	6	7	8	9	10	11	12	13
Agra Rural												
16. Firozabad South	2	—	1	7	27	29	15	13	—	—	47.0	0.63
17. Firozabad North	1	—	10	17	55	71	27	22	—	2	101.5	0.67
18. Naraki	9	8	3	2	25	17	14	7	—	1	43.0	0.54
19. Matsena	4	9	2	1	18	21	11	9	7	3	42.5	1.13
20. Etmad pur	2	6	5	1	18	27	8	6	—	—	36.5	0.75
21. Tundla	6	9	12	9	45	51	26	17	—	—	87.5	0.91
22. Barhan	3	8	7	1	38	33	12	12	—	—	57.0	0.63
23. Khandauli	3	2	2	2	15	15	7	7	—	—	26.5	0.40
24. Bah	5	3	2	6	18	16	12	8	3	1	36.0	0.59
25. Pinahat	8	6	2	—	19	17	11	8	—	6	38.5	0.95
26. Jaitpur	2	10	4	3	7	16	3	7	—	3	27.5	0.96
27. Pidhaura	4	6	—	—	8	5	5	3	3	2	18.0	0.90
28. Khera Rathaur	—	2	—	—	2	—	1	—	2	1	4.0	0.43
29. Fatehabad	14	2	9	2	36	21	10	13	2	1	55.0	0.93
30. Shamsabad	3	1	6	2	28	15	17	9	1	—	41.0	0.55
31. Dauki	2	1	—	1	8	5	4	3	1	—	12.5	0.19
32. Nibohara	3	5	2	2	13	12	5	5	4	—	25.5	0.73
33. Kheragarh	3	5	2	1	7	15	3	8	3	1	23.0	0.29
34. Saiyan	—	1	3	1	9	12	6	3	—	—	17.5	0.34
35. Iradat nagar	—	1	1	2	8	7	4	2	1	1	13.5	0.60
36. Jagner	1	2	—	—	12	17	7	5	2	1	23.5	0.43
37. Achhanera	2	7	2	3	15	39	7	10	—	—	42.5	0.31
38. Fatehpur Sikri	2	5	4	—	25	16	11	4	—	—	15.5	0.17
39. Malpura	4	6	1	2	32	29	16	8	—	—	49.0	0.49
Total	93	106	151	126	1032	981	491	413	30	24	1703.5	0.60

Source: District Police Office, Police Line, Agra.

The best remedies to solve the crime problem include the construction of various roads in thc ravine villages and ravine reclamation programme for the agricultural development in these areas. Cottage and small scale industries should be encouraged in these areas so that the frustrated section of the society can seek their livelihood. The number of police stations should be increased and more police patrols in crime infested areas must be deployed. The co-operation of the public is also essential in the prevention and control of crime problem.

REFERENCES

Harris, K.D. (1978), The geography of crime — A political critique, in *Fundamentals of Human Geography, —A Reader,* Harper & Row Ltd., London, pp. 325-6.

Phillips, P.D. (1978), Radical Theory, Relevance and the Geography of Crime in *Fundamentals of Human Geography — A Reader,* Harper & Row Ltd., London p. 327-29.

Khan M.G. Krishna, K.B. (1982), An analysis of crime and population in India during 1966-76, *Indian Journal of Criminology,* Vol. X, 1.

Khan, M.G. & Singh, D.R. (1980), Social differentiation in victimization by dacoit gangs in Chambal valley, *Indian Journal of Criminology,* Vol IV, 1.

Prasad, R., (1976), Ravine settlement of Lower Chambal, Unpublished *Ph.D. Thesis* in Geography, Agra University, Agra.

Prasad, R. (1981), Soil erosion in the lower catchment of Chambal River, *Geographical Review of India,* Vol. 43, No. 4.

Esha Basanti Joshi (1965), *Uttar Pradesh District Gazetteers, Agra,* Government of U.P., Lucknow.

Chandan Mitra (1985), Scourge of the Chambal, *The Statesman,* Calcutta, April, 14.

Bureau of Police Research and Development (1979), *Crimes in India,* Ministry of Home Affairs, New Delhi.

29

REDEFINING STRATEGIES IN FOREST MANAGEMENT USING REMOTE SENSING DATA

P.S. ROY and B.K. RANGANATH

Large scale deforestation could promote a net carbon dioxide flux to the atmosphere, producing higher surface temperature and associated global environmental effects besides causing various localised environmental hazards (Sagan *et al.*, 1979). Growing concerns in deforestation particularly in tropics have been expressed by various scientists. It has been emphasized that the present mapping and monitoring methodology is insufficient to cater the need of forest manager and environmentalists. There is necessity to obtain timely and reliable information concerning the forest resource base. Uses of aerial photography in forestry are well established (Hilderbrandt, 1986). Relatively, recent developments in remote sensing technology have indicated that if they are judiciously combined with ground based studies it is possible to carry out detailed forest inventories and monitoring of natural vegetation cover (Tucker *et al.*, 1985, Botkin *et al.*, 1984). The information content available in multispectral data can form common database for integrated resource inventory which has become very much necessary in modern forestry practices. The data can be manually interpreted or digitally processed providing fast access to required information.

Potential of remote sensing for forest mapping depends on the capacity of sensors to discriminate between the spectral reflectance of different forest types. In general, spectral variations are produced by the forested surface, the positions of the sun and sensor with respect to the surface, and by the atmosphere (Maxwell, 1976; Duggin, 1985). Forest types typically must have distinctive spectral reflectance patterns to be discriminated using satellite data. Variation in scene illumination related to atmospheric reflectance and changing position of the sun and sensor with respect to the forested surface, reduce spectral reflectance pattern and noise. These undesirable variations contribute in part to the signature extension problem and misclassification due to under sampling.

Broad vegetation type stratification is reported by using Landsat multispectral scanner (MSS) and Thematic Mapper (TM) digital and analogue data. Their application in India, however, is passing through very cautions phase. Studies indicate that the technology can provide valuable information with respect to forest resource (Roy *et al.*, 1985; Unni *et al.*, 1985; Jadhav *et al.*, 1987). National Remote Sensing Agency (1983) for the first time used Landsat MSS false colour composite (FCC) at national level to produce forest cover map of India using image form of data of periods 1972-75 and 1980-81. The effort was aimed on mapping natural tree clad area and to develop methodology for using satellite images for national forest monitory system.

Satellite Multispectral Data and Specifications for Forestry Studies

Data Forms

Satellite multispectral data which is available in analogue and digital form presently for vegetation studies are given in Table 29.1. The analogue representation of remote sensing data is of pictorial nature and is used for human interpretation. The digital form is handled with the help of computer. The two forms of data are in interchangeable state. Although Landsat MSS data showed the application areas of remote sensing in forestry but second generation Landsat series are better suited for forest studies as far as operational applications are concerned.

Data Specification

The repetitive coverage provided by satellite data helps in selecting right kind of data for forest studies. Following aspects need to be considered for selecting right kind of data:

– phenological state of vegetation

- phytogeographical region
- landuse practice
- sun elevation and azimuth
- cloud and snow coverage.

TABLE 29.1: Satellite Multispectral Data Characteristics and their Possible Application in Vegetation Mapping

Satellite	Sensor	Resolution		
		Temporal	Spatial	Spectral (Microns)
1	2	3	4	5
Landsat - 1,2,3,4 & 5 (American)	Multispectral Scanner (MSS)	16-18 days	79mX79m	0.5-0.6 0.6-0.7 0.7-0.8 0.8-1.1
Landsat- 4 & 5 (American)	Thematic Mapper (TM)	16 days	30mX30m	0.45-0.52 0.52-0.60 0.63-0.69 0.76-0.90 1.55-1.75 10.40-12.50 2.08-2.35
SPOT-1	Multispectral mode	26 days	20mX20m	0.50-0.59 0.61-0.68 0.79-0.89
(French)	Panchromatic mode	1 day	10mX10m	0.50-0.73
NOAA Series Meteorological Satellite (American)	Advanced Very High Resolution Radiometer (AVHRR)	Twice daily	1.1 km^2	0.55-0.68 0.725-1.10 3.55-3.93 10.33-11.30 11.50-12.50
IRS-1	LISS I		73mX73m	0.45-0.52 0.52-0.60 0.63-0.69 0.73-0.90
(Indian)	LISS II		36.5mX36.5m	0.45-0.52 0.52-0.60 0.63-0.69 0.73-0.90

Proper season of the image is largely a function of bioclimatic region under study (Table 29.2).

TABLE 29.2: Proper Season for Studying Natural Forest Cover in Different Bioclimatic Regions and Vegetation Zones

Sl. No.	Region/Vegetation Zone/Geographical Set-up	Proper Season
1.	Humid & moist evergreen and semi-evergreen vegetation in Western & Eastern Ghats	January-February
2.	Humid & moist evergreen and semi-evergreen vegetation of North-Eastern Region and Andaman & Nicobar Islands	February-March
3.	Tropical moist deciduous vegetation of Northern and Central India	December-January
4.	Temperate evergreen vegetation of Western Himalayas	March-May
5.	Temperate, sub-alpine, alpine evergreen & deciduous vegetation of Jammu & Kashmir	September-October
6.	Arid and semi-arid dry deciduous and scrub vegetation	October-December
7.	Tropical coastal mangrove vegetation	February-March

Classification System of Vegetation and its Applicability in Remote Sensing

Forest vegetation description has been made on the basis of using various properties, viz. physiognomic, structural, functional and compositional (Forsberg, 1967). On the other hand, vegetation has also been classified based on height classes and distinction based on woody tissue (UNESCO, 1969). The most acceptable classification of Indian vegetation todate is Champion and Seth Classification (1968). This classification is a climatic classification giving a clear picture of successional status of vegetation, ecological status and floristic composition. In remote sensing following three observations are important and should be used while classifying and mapping vegetation:

- certain vegetation boundaries are easy to draw because they are virtually self evident but other boundaries too can be drawn rather arbitrarily between two centres of structural or species compositional changes — continuum problem in vegetation.

- integrating vegetation mapping units across a larger terrain — classification continuum problem.
- definition of plant community in a particular geographical scale — geographical scale problem.

Remote sensing images provide certain basic elements of vegetation sensed:

- size (i.e. scale),
- shape,
- shadow,
- tone or colour,
- texture (rough or smooth),
- repetition pattern,
- site (i.e. the indirect indication of what is there),
- association (objects associated with vegetation cover),
- resolution (which puts a limit on what can be recognised on the image).

Using the above elements available in remote sensing data vegetation can be classified theoretically with all the criteria described by Forsberg (1967), i.e. physiognomical, structural, functional and compositional. Functional classification, however, can only be achieved through manual interpretation. However, many studies carried out using satellite remote sensing digital data indicate that a rather non-technical classification approach is adopted by combining different criteria represented in the form of spectral properties.

Hierarchical classification suggested by Howard and Schade (1982) has found wider acceptance for vegetation mapping using remote sensing data. The classification uses following parameters:

- vegetal physiognomy
- climatic zones
- plant formation
- plant sub-formations
- induced vegetation
- regrowth
- succession
- floristic features

After reviewing available literature and testing remote sensing data it can be concluded that the most appropriate method of classifying vegetation using remotely sensed data is (a) to identify the plant formation in terms of the macro-climatic zone in which they occur, (b) to describe from remote sensing data for each plant formation in terms of its physiognomy, including lifeforms (but not floristics), (c) to divide the plant formations into subformations according to the physiognomy of the subdominant vegetation which includes multistage sampling and stereoscopic examination of pairs of aerial photographs, and (d) to ensure that the method is open ended so that — if required attention can be given to induced vegetation, the floristics of the plant communities, dominant timber species, plant succession, regrowth etc.

Information Extraction Procedures

Manual Mapping

Delineation of forest vegetation is one of the most challenging application of space remote sensing data. During visual interpretation most of the information can be extracted from tone and colour exclusively. However, texture, site, shape are also used as standard parameter in identification of vegetation. The interpretation comprises at least three mental acts which may or may not be performed simultaneously: (a) measurement of objects on the image, (b) identification of objects imaged, and (c) appropriate use of this information in the solution of the problem at hand through mental comprehension. First two acts are relatively simple and can be converted into keys but the third act is complex and required mastery of basic information and both synthetic and analystic mode of thought about ground vegetation cover and classification system.

Computer Aided Processing

Enhancements

There are well defined image enhancement techniques which greatly increases the amount of information which can be visually interpreted from the image data by improving the apparent contrast between features in the scene. The enhancement techniques commonly applied in forestry are shown in Table 29.3.

TABLE 29.3: Image Enhancement Techniques and its Applications in Forest Mapping

Sl. No.	Enhancement Techniques	Function Achieved
1.	Density Slicing	Colour representation of land cover features in individual band data.
2.	Addition	Reduction of number of image and judicious combination in the form of false colour composition allows vegetation to appear different (like, different intensity of red).
3.	Substraction	Detects change in vegetation at two particular time periods.
4.	Ratioing	Enhances the subtle contrast between different vegetation types. Provides idea about green foliage density and quantity of vegetation. Removes atmospheric affects and impact of partial hill shadows.
5.	Contrast stretching	Increases the separation between two vegetation type boundaries allowing higher delineation accuracy.
6.	Textural analysis	Idea about forest density and heterogeneity
7.	Temporal analysis	Allows to incorporate phenological features for interpretation. Change detection of vegetation at two time periods.
8.	Principal component	Reduction of number of image and judicious combination of different transformations sometimes enhances certain vegetation related features.

Vegetation indexes developed from remote sensing data compresses available information content of various parts of electromagnetic region. Ratio based vegetation indexes developed so far are presented in Table 29.4.

TABLE 29.4: Spectral Vegetation Indices

Sl. No.	Vegetation Index	Computation	Author
1.	Simple Vegetation Index (SVI)	IR/R	Curran & Milton, 1968
2.	Vegetation Index (VI)	IR/R + Other Wavelengths	Curran, 1980
3.	Normalised Vegetation Index (NVI)	IR-R/IR+R	Curran, 1982
4.	Transformed Vegetation Index (TVI)	$\frac{\sqrt{IR-R}}{IR+R} + 0.5$	Rouse *et al.* 1973
5.	Perpendicular Vegetation Index (PVI)	$\sqrt{(R_{soil} - R_{veg})2 + (IR_{Soil} - IR_{veg})2}$	Richardson & Wiegand, 1977

Classification

Picture elements (pixel) of digital database are available on computer compatible tapes. The grey values of each pixel in multispectral domain are classified into vegetation types based on spectral signatures. The classification is achieved through dedicated image processing systems which allows man machine interaction environment. During the digital classification one must train the classifier to execute the algorithm by ascertaining values of the population means and covariance matrix elements. The training procedure is either supervised or unsupervised. In both the approaches one can encounter situation in which the forest vegetation classes can not be separated spectrally. Under supervised training it is difficult to know how many classes are separable using spectral signatures. With interactive analysis, a variety of unsupervised clustering can be employed. Many classes of high interest in forestry are differentiated only by subtle spectral differences, whereas some classes that are easily separable spectrally are of low analyst interest. As a result supervised approach which is performed by assigning spectral signature can prevent the needless discrimination of low interest categories.

Temporal Classification

Temporal classification exploits the usefulness of time related features as another elements for interpretation. Temporal analysis uses two basic time interpretive functions in forestry applications:

(a) Time related phenological features as additional discriminant, and
(b) Time related change in forest cover features.

In temporal image analysis it is prerequisite to achieve geometrical/spatial registration of different time data. Thus the temporal analysis can be used to monitor significant level of change in the forest vegetation canopy versus unaltered areas. Once change areas are identified, finer resolution systems and ground observation could be used to provide details necessary for making management decisions. There are mainly two procedures, viz., differencing and ratioing are available to transfer the data for change detection using multi date satellite data.

Multispectral Classification in Digital Terrain Model Domain

Forest vegetation is largely controlled by site factors like, altitude, slope, aspect and location. These parameters if included in multispectral domain improve the forest vegetation classification. The spatial information with respect to altitude, slope, aspect and location can be digitised and used as classification domain. They have proved very useful in classifying conifer forests. However, its operational application is limited in India due to absence of digital terrain model (DTM).

Results of Case Studies

Forest Cover Type Mapping Using Aerial Photographs

Aerial photographs (Black/White panchromatic) on 1: 25,000 scale of Godavari Basin catchment area has been interpreted to stratify forest types with intensive ground truth data. The area processes predominantly teak (*Tectona grandis*) bearing forest and includes hectic plantation activities. The study area has been classified into three terrain classes: (a) flat/relatively flat terrain with slopes below 10 per cent, (b) undulating/undulating terrain with slopes between 10 and 30 per cent and (c) hilly/hilly terrain with slopes more than 30 per cent. The type identification was mainly on the basis of occurrence of teak. Thus the class *'teak'* means forest with 30 per cent and above teak; *mixed teak* means stands with teak percentage between 10 and 30 per cent, and *miscellaneous* means stands less than 10 per cent teak. However, in some area where *Hardwickia binata* dominated, two more types were identified such as *Hardwickia binata mixed* (between 10 and 30%). Wherever, bamboo occurred in pure form and also within the miscellaneous category, it was also identified with appropriate symbols. Plantations were identified with high accuracy. Four density and two height classes were identified on the basis of average density and height respectively (Fig. 29.1)

Forest Cover Mapping Using Satellite Images

Manual interpretation of individual images covering forested tract of Central India has been carried out on 1:1 million scale to map forests (natural tree clad areas) from Landsat MSS FCC of 1980-81 time frame. The classes mapped are : closed forests (>40% tree density), open/degraded forest (<40% and >10% tree density) and non-forest areas (<10% tree density, shrubland, grassland fallow and agriculture land). The reduced map on 1:7 million showing spatial distribution of forest cover in Central India is presented as Fig. 29.2. Interpretation is primarily based on tonal characteristics of images and their calibration based on ground truth.

Meteorological satellites like NOAA (AVHRR on board) are also capable of providing information about forest cover. The obtained information is valuable for monitoring:

- seasonal dynamics of vegetation
- green foliage activity
- large scale deforestation/vegetation canopy alterations.

To investigate the above NOAA-AVHRR data over the same area covered by Landsat MSS scenes has been analysed to identify forest cover. Normalised vegetation index (IR-R/IR+R) has been generated subsequent to geometric and panoramic correction.

The forest cover maps generated by visual interpretation of Landsat MSS FCC and Normalised vegetation index generated from NOAA-AVHRR data shows matching of boundaries of closed forest, open/degraded forest and non-forest areas.

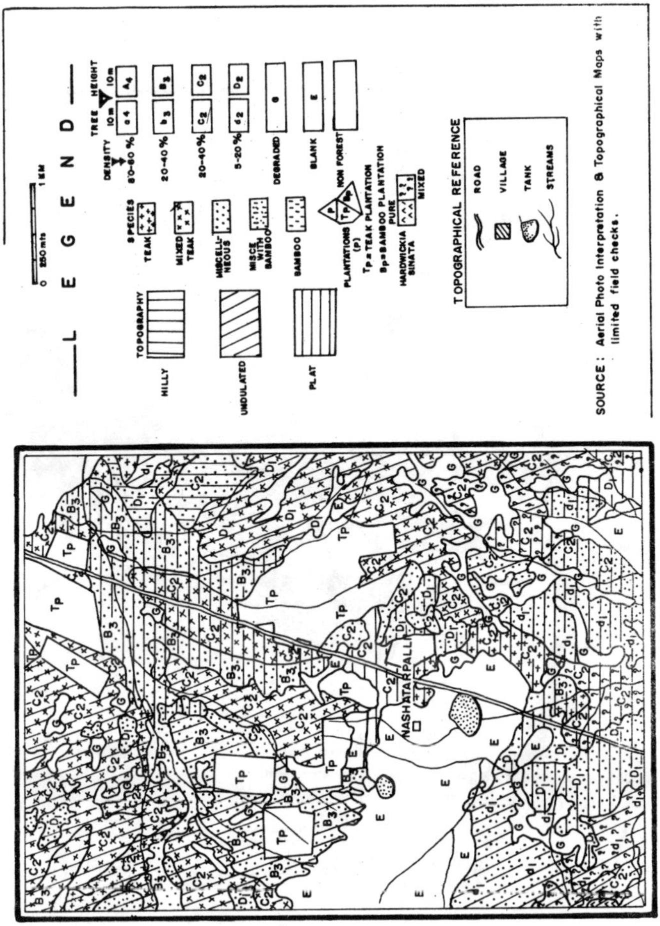

FIG. 29.1: Forest Type Map : Part of Godavari Basin Area

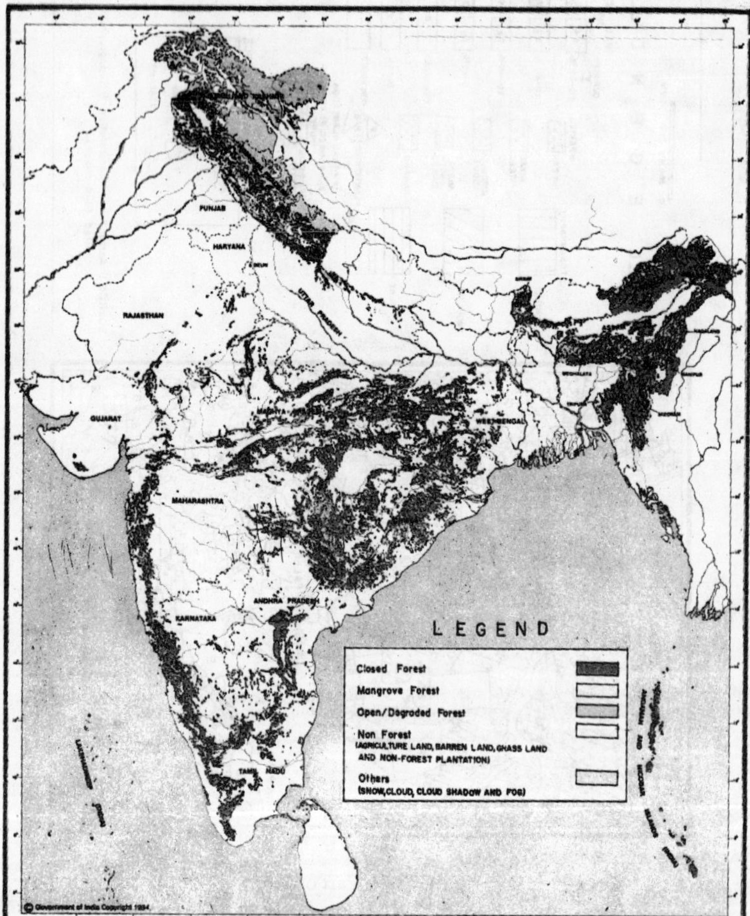

FIG. 29.2: Forest Cover Map of India, 1980-82

Forest Condition Mapping Using Satellite Images

Bastar district of Madhya Pradesh consists of largely forested landscape. These forests have, however, faced serious problems due to increase of influx of population. Inhabitant tribes also practise shifting cultivation in hilly tracts. Landsat MSS false colour composite has been used to assess spatial record of physical condition of forest resource of the district using images of 1973-75 period. The classes mapped are as below (Fig. 29.3).

Class	Class description
– Closed forest	Forests having >40% density
– Closed forest affected by shifting cultivation and encroachment	Forest having > 40% density and mosaiced by openings
– Low density forest	Forest having <40% & >10% density
– Degraded forest	Woodlands having <10% density
– Grass/Scrubland	Areas under shrub & grasses
– Riparian vegetation/Crop	Vegetation along the river courses and croplands
– Non-forest	Fallow, permanent fallow and barren areas

The map highlights the areas having serious human impact due to encroachment and undergoing practice of shifting cultivation.

Vegetation Type Mapping using Satellite Digital Data

The spectral variation of the forest/vegetation cover type canopy mainly allows mapping using remote sensing techniques. These variations can effectively be separated and mapped using digital classification techniques. Doon valley and surroundings in Uttar Pradesh are bounded by Shiwaliks in south and Himalayas in north. Shiwaliks have dense sal forest (*Shorea robusta*) in the northern aspects and moist deciduous to dry deciduous vegetation in the southern aspects. Shiwaliks of north of Doon Valley contains good sal forest. However, higher elevation are covered by Oak (*Quercus incana*) and conifers (mainly *Pinus roxburghii* and *Cedrus deodara*). Doon valley has faced serious environmental degradation in recent years due to heavy influx of population and limestone mining in Mussoorie syncline.

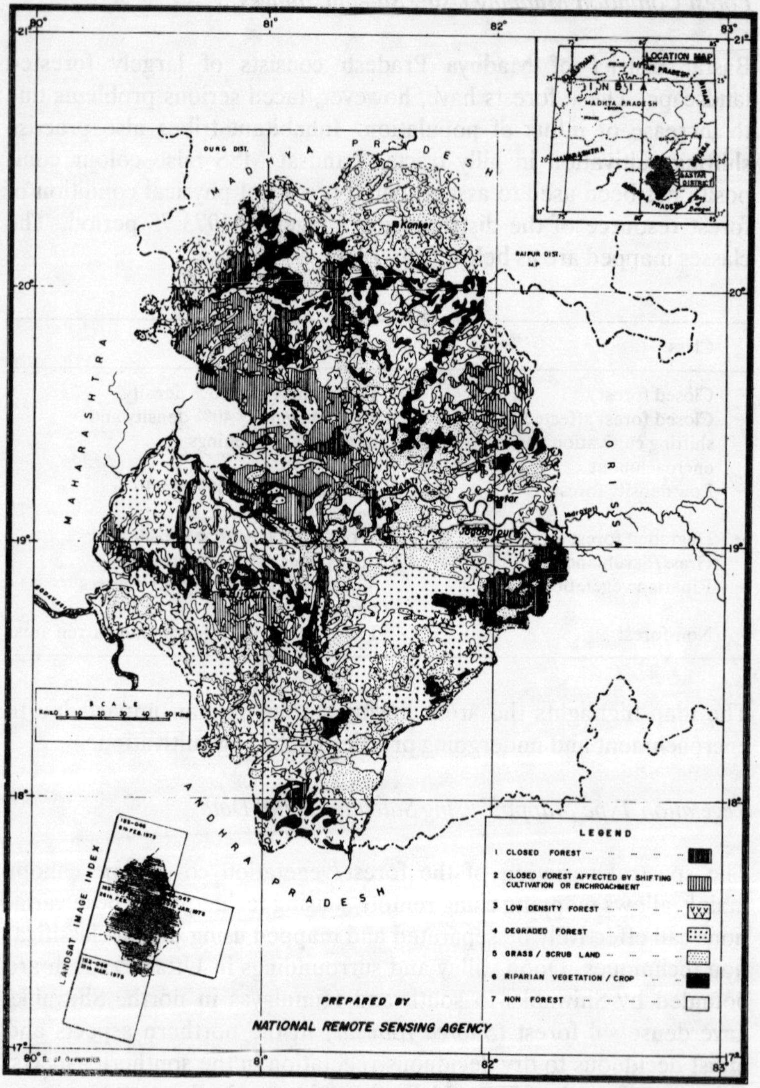

FIG. 29.3: Forest Map of Bastar District, M.P. Based on Visual Interpretation of Landsat Data

Landsat MSS digital data of November 1980 has been used to analyse the vegetation types of Doon valley and surroundings using VIPS-32 software on Regional Remote Sensing Service Centre (RRSSC), Dehra Dun computer facility. Supervised classification of Landsat MSS data (all four bands) has provided stratification of following vegetation types.

Class	Class description
– Sal forest	– Forest having pure stands of *Shorea robusta* (Sal)
. Sal mature dense	. Sal forest >60% density and >40 cm diameter class
. Sal young dense	. Sal forest >60% density and <40 cm diameter class
. Sal low density/lopped forest	. Sal forest <60% density but >20% density and heavily lopped canopies
– Pine forest	– Forest having pure stands of *Pinus roxburghi* (Pine)
– Oak forest	– Forest having pure stands of *Quercus incana* (Oak)
– Riverine forest	– Forests located along riverine system dominated by *Acacia catechu, Delbergia sissoo*
– Miscellaneous deciduous forest	– Forest comprising of mixed species of moist and dry deciduous nature
– Degraded forest	– Forests <20% density and with visible erosional pattern
– Forest Plantation	– Plantation of forest tree species with homogenous age group
– Cropped area	– Agriculture/cultivated area
– Shrub/Bushlands	– Areas covered by bushes, shrubs and grasses
– Fallow/Barren	– Non-vegetated area
– River bed	– Sandy and rocky area along river courses
– Settlement	– Habitation areas
– Water	– Water courses of river, tanks, reservoirs
– Hill shadow	– Shadowed areas

A part of Doon valley — Timli Forest and surrounding was analysed using Landsat TM digital data of March 3, 1987. The supervised classification was performed using VIPS-32 software on RRSSC computer facility using ground data as *training set.* Classes having more than 85 per cent class separability were used as training sets. Final classification was performed using maximum likelihood algorithm. The vegetation types identified are as below:

Class	Class description
— Sal	— Forests having pure stands of *Shorea robusta* (Sal)
. Sal dense mature	— Sal forest >60% density and > 40 cm diameter
. Sal dense young	— Sal forest > 60% density and <40 cm diameter
. Sal dense young (with heavy understorey)	— Sal forest >60% density and <40 cm diameter and heavy undergrowth
. Sal low density	— Sal forest <60% density but >20% density
. Sal Mixed	Sal mixed with miscellaneous species. Percentage of Sal is <30%
— Miscellaneous deciduous forest	— Forest comprise of mixed species of moist and dry deciduous nature
— Pine Mixed	— Forest on ridge of Shiwaliks having <40% density of *Pinus roxburghii*
— Riverine	— Forests located along riverine system dominated by *Acacia catechu* and *Delbergia sissoo.*
— Degraded forest	— Forests having <20% density and with visible erosional pattern
— Woodlands	— Orchards and tree clad area in agriculture landscapes
— Shrubland	— Areas covered by shrubs
— Grasslands	— Areas covered by grasses
— Tea Garden	— Tea garden having shade trees of Campher and Silver Oak
— Agriculture	— Cropped and current fallow areas, barren
— Non-forest	— Non-vegetated areas and permanent fallows

Discussion

Aerial photography is an established mapping methodology and is being used elsewhere on an operational basis for forest cover type mapping structural and volumetric analysis. Information which can be obtained from aerial photographs are presented in (Table 29.5). Most commonly used aerial photography in tropical forest inventory is black/white panchromatic photography. The data generated from aerial photointerpretation has obtained wider acceptance due to very reliability. However, Table 29.5 clearly indicates that information is directly function of scale of the photography. Therefore, aerial photography tasks have not remained cost effective for certain applications like forest cover/cover type mapping and regional vegetation mapping. Satellite remote sensing technology with varied available platforms/sensors can provide reliable information with respect to forest cover, condition and vegetation type mapping. Since the data is regular available at regular interval, it can be used effectively for monitoring the forest resources.

The vegetation spectral response in multispectral domain is unique (Knipling, 1970). This can be separated visually on images or classified using digital form of data. Images can be greatly improved by digital enhancement techniques for visual interpretation. The boundaries between vegetation cover and other land cover classes can be greatly enhanced using linear contrast strecting (Roy, 1986). The subtle differences between vegetation types can be enhanced effectively by combining information of red and infrared spectral region in the form of ratio (Yool *et al*, 1986).

The spectral properties of forest are largely controlled by dominant crown foliage type. Digital classification of multispectral data has been used widely in India to classify forest types in various regions of the country using landsat MSS data. (Unni *et al.*, 1985 and Roy *et al.*, 1985 & Roy *et al.*, 1986). The studies indicate that only broad vegetation types can be classified. Certain monospecies formations growing in gregarious form can also be identified. The present study carried out in Doon Valley using Landsat MSS data also indicate that identified vegetation types occur in single species formation only. Further stratification in sal forest areas could be possible due to pure spectral signature variation of sal canopies at different density/maturity. The results obtained from Landsat TM data analysis, however, indicate that the TM digital data contains more information spectrally for discrimination of vegetation types. This is probably due to availability of data in the region of middle infrared which is known to contain valuable information with respect to canopy moisture variation of various types.

TABLE 29.5: Forest Related Information Extraction Capabilities Using Aerial Photographs

Information	Type of Photography	Recommended Scale of Photography
Forest Cover	B/W	S(1:90,000-1:50,000)
Clear Cut/Shifting Cultivation	B/W	S(1:90,000-1:50,000)
Forest Cover Change	B/W (Temporal)	S(1:90,000-1:50,000)
Forest Type	B/W or CIR	M(1:50,000-1:25,000)
Terrain Classes	B/W	S/M(1:90,000-1:25,000)
Accessibility	B/W	M/L(1:50,000-1:10,000)
Site Suitability	B/W	M/L(1:50,000-1:10,000)
Erosion Status	B/W	M/L(1:50,000-1:10,000)
Stand Volume	B/W	M/L(1:50,000-1:10,000)
Biomass and Production	B/W	L(1:15,000-1:10,000)
Stand Height	B/W	M/L(1:50,000-1:10,000)
Stand Density	B/W	M/L(1:50,000-1:10,000)
Tree Height	B/W	L(1:10,000-1:5,000)
Forest Fire	CIR	M/L(1:50,000-1:10,000)
Forest Disease/Pollution Impact	CIR or B/W Infrared	M/L(1:50,000-1:10,000)
Wildlife Habitat Analysis	B/W or CIR	L(1:15,000-1:8,000)
Wildlife Census	B/W, CIR or Thermal	L(1:10,000-1:1,000)

B/W = Black and white aerial photography
CIR = Colour infrared aerial photography
S = Small scale
M = Medium Scale
L = Large scale

Forest classification in rugged mountainous terrain is complicated due to absence of spectrally homogenous classes caused by partial and complete shadow pattern. However, it can be reasonably countered by selecting data during higher sun elevation. Roy (1987) observed that addition of normalised vegetation index in normal spectral data of Landsat MSS reduces the misclassification due to influence of partial hill shadow.

Impact of environmental variable, like, soil, understorey, regrowth and tree density variation complicate the spectral separability (Karaska *et al.*, 1986). These influences are however, reduced in low resolution satellite data like Landsat MSS. While designing the ground truth collection, proper care needs to be taken to collect ground data of different vegetation types and environmental variable to know the cause of changes in spectral properties. Thus, total spectral variation of particular cover type can be separated in multispectral domain.

Seasonal impact of spectral separability and mapping accuracy is also quite significant (Roy, 1986). Therefore, it is important to account phenological behaviours of vegetation types while classifying using remote sensing data. Stratification and mapping using remote sensing data provides an excellent basis for ecological studies in the regional context. The vegetation zones and their interdependence can be evaluated in logical manner. The sampling procedure for production/growing stock/biomass can be based on the stratification obtained from remote sensing, enabling more reliable information.

Besides, environmental studies on regional basis needs analysis through an integrated approach. Mapping units which are available on spatial domain can be suggested for evaluation based in their inter-relationship and interaction. The spatial domain data can be handled through Geographic Information System (GIS) for landscape analysis and designing/management strategies. The ancillary ground based data, like, population, livestock, land holding, climatic condition, water resource availability etc can be mixed with resource spatial data, felicitating management decisions.

In the light of above, it is emphasised that multiplatform remote sensing technology provides viable answer to improved data collection, inventory and monitoring possibilities of forest resources in the country. The decision making in forest management can be made scientific and rational by using the above technology. There is only need to make present surveying and mapping procedures open ended to take help from this new technology.

Acknowledgements

The authors are grateful to Prof. B.L. Deekshatulu, Director, National Remote Sensing Agency (NRSA), Hyderabad and Prof. S.K. Bhan, Head, Indian Institute of Remote Sensing (NRSA), Dehra Dun for their constant encouragement and providing facilities to carry out the work reported in this publication.

REFERENCES

Botkin, D.B. and Estes, J.E., Mc Donald, R.M. and Wilson, M.V. (1984), Studying the earth's vegetation from space, *Bioscience,* 34, 508.

Champion, H.G. and Seth, S.K., (1962), *A Revised Survey of Forest Types of India,* Government Publications, New Delhi.

Curran, P.J., (1980) Multispectral remote sensing of vegetation amount, *Progress in physical geography,* 4, 315-41.

Curran, P.J., (1982), Multispectral photographic remote sensing of green vegetation biomass and productivity, *Photogram Eng. & Remote Sensing.,* Vol. 48.(2), 243-50.

Curran, P.J. and Milton, E.J., (1983), The relationship between the chlorophyll concentration, LAI and bi-directional reflectance of a simple vegetation canopy. *Int. J. Remote Sensing,* Vol. 4. (2), 247-55.

Duggin, M.J., (1985), Factors limiting the discrimination and quantification of terrestrial features using remotely sensed radiance, *Int. J. Remote Sensing,* 6(3).

Forsberg, R.F., (1967), A classification of vegetation for general purposes in G.F. Peterken (ed) *Guide to Check Sheet for IBP Areas,* IBP Handbook No. 4, 73-102, Blackwell Scientific Publications, Oxford, Edinburgh, 133.

Fontaine, R.G., (1978), Inventory and Survey: International activities, Tropical Forest Ecosystems: A state of knowledge report prepared by UNESCO/UNEP/FAO. *UNESCO Natural Resources Research,* Vol. XIV, 17-32, 683.

Hilderbrandt, G., (1986), Potential and limitations of space remote sensing for forest inventory and mapping in report *Deutsch Geselschaft Fuer Luft-und Raumfahrt,* Bonn, FRG pp. 165-85.

Howard, J.A. and Shade, J., (1982), Towards a standardized hierarchical classification of vegetation for remote sensing, Food and Agriculture Organization of the United Nations, Rome, RSC. Series II (Bulletin).

Jadhav, R.N., Srivastava, V.K., Kandya, A.K., Sarat Babu, G.U., Kimothi, M.M., (1987), Large scale forest type mapping using satellite data, International Astronautical Congress (38th Brightoon), England, October, 10-17, 5.

Karaska, M.A., Walsh, S.J. and Butler, D.R., (1986), Impact of environmental variables on spectral signatures acquired by the Landsat Thematic Mapper. *Int. J. of Remote Sensing.* Vol. 7 (12), 1653-67.

Knipling, E.B., (1970), Physical and physiological basis for the reflectance of visible and near-infrared radiation from vegetation, *Remote Sensing of Environment,* 1:155-59.

Maxwell, E.L., (1976), Multivariate system analysis of multispectral imagery. *Photogram. Engg. Remote Sensing,* 43, 1173.

NRSA, (1983), Nationwide mapping of Forest and non-forest areas using Landsat False Colour Composites for the periods 1972-75 and 1980-82, Project Report. Vol. I. p. 34.

Richardson, A.J. and Wiegang, C.L., (1977), Distinguishing vegetation from soil background, *Photogram. Eng. Remote Sensing*, 43, 1541.

Rouse, J.W., Haas, R.H., Schell, J.A. and Deering, D.W., (1973), Monitoring vegetation systems in the great plains with ERTS, Proceedings of the Third ERTS Symposium, NASA Special Publication No. 351, Scientific and Tech. Inform. Br., WA. D.C. pp 309-17.

Roy, P.S., Kaul, R.N., Sharma, M.R. and Garbyal, S.G., (1985), Forest type stratification and delineation of shifting cultivation areas in the eastern part of Arunachal Pradesh using Landsat MSS data, *Int. J. Remote Sensing*, Vol. 6. (3 & 4), 411-18.

Roy, P.S., (1986), Satellite remote sensing data interpretation using visual and computer aided techniques for wildlife habitat analysis in D.S. Kamat and H.S. Panwar (eds). Proc. Seminar-cum-workshop Wildlife Habitat Evaiuation using Remote Sensing Techniques.

Roy, P.S., Saxena, K.G. and Pant, D.N., (1986), Analysis of vegetation types using remote sensing techniques for wildlife habitat evaluation in Kanha National Park in D.S. Kamat and H.S. Panwar (eds). Proc. Seminar-cum-workshop. Wildlife Habitat Evaluation using Remote Sensing Techniques, October 22-23, 83-116.

Roy, P.S., (1987), Montaine vegetation stratification through digital processing of Landsat MSS data, Geocarto, *Int. Jour. Remote Sensing*, Vol. 2(1), 19-26.

Tucker, C.J., Townshend, J.R.G. and Golf, T.E., (1985), African land cover classification using satellite data, *Science*, N.T. 227, 369.

Unni. N.V.M., Roy, P.S. and Parthasarthy, V., (1985), Evaluation of Landsat and airborne multispectral data and aerial photographs for mapping forest features and phenomena in a part of the Godavari basin, *Int. J. Remote Sensing*, Vol. 6 (3 & 4), 419-31.

Yool. S.R., Star, J.L., Estes, J.E., Botkin, D.B., Eckhardt, D.W. and Frank, D.W. (1986), Performance analysis of image processing algorithms for classification of natural vegetation in the mountains of southern California, *Int. J. Remote Sensing*, Vol.7, (5), 683-702.

30

DEVELOPMENT OF AN ENVIRONMENTAL INFORMATION SYSTEM

PRITHVISH NAG and H. PRASAD

It appears that there has been a persisting concern for environment all over. Institutions, commissions, committees, statutory bodies, university departments, societies and voluntary organisations have been formed for environmental assessment and improvement. Hazards and impending catastrophe have been considered as a major area of interest. The phase of doomwatchers of 1960's is not over. This is followed by another phase of accentric eco-activists. There are other fields where environmental scientists can contribute in a positive way (Caldwell, 1984; Dutt, 1980; Dutt *et al,* 1987; Nag, 1983). Otherwise with all institutions interested in environmental-related issues would not have left the scope of myths in environmental management and assessment (Holling, 1978, pp 2-5).

Myths of Environmental Management

1. The central goal for designing is to produce policies and developments that result in stable social, economic, and environmental behaviour.
2. Development programmes are fixed sets of actions that will not involve extensive modifications, revision, or additional investment after development occurs.
3. Policies should be designed on the basis of economic and social goals, with environmental concerns added subsequently as constraints during a review process.
4. Environmental concerns can be dealt with appropriately only by changing institutional constraints.

Myths of Environmental Assessment

1. Environmental assessment should consider all possible impacts of the proposed development.
2. Each new assessment is unique. There are a few relevant background principles, information, or even comparable past cases.
3. Comprehensive "state of the system" surveys (species lists, soil conditions, and the like), are a necessary step in environmental assessment.
4. Detailed descriptive studies of the present condition of system parts can be integrated by systems to provide overall understanding and predictions of system impacts.
5. Any good scientific study contributes to better decision making.
6. Physical boundaries based on watershed areas or political jurisdictions can provide sensible limits for impact investigations.
7. Systems analysis will allow effective selection of the past alternative from several proposed plans and programmes.
8. Ecological evaluations and impact assessments aim to eliminate uncertainty regarding the consequences of proposed developments.

From the above descriptions, it is apparent that what ever methods have been adopted so far is not on a continuous basis. Assessments are time bound, projects bound and purpose specific. Hence these assessments are concerned either with pre- or post-action consequences. A system of continuous monitoring of information on environment is required (Dutt, 1982). Some attempts have been made in this regard with the help of geographical information system (GIS) technology which can be a possible solution to remove some of the myths.

Global Concern for an Environmental Information System

Probably, the first systematic attempts to conceptualise the idea of GIS was held in Canada (Coppock and Anderson, 1987), following this the First International Advanced Study Symposium on Topological Data Structures for Geographic Information Systems was organised in Dedham, Massachusetts in October 1977 (Dutton, 1978).

There was an exchange of ideas, experiences and expertise on different aspects of GIS technology. Since then, this technology has developed into a multifaceted field of study. Persons engaged in handling data structures, remote sensing, computer graphics, mapping and surveys have been expanding their activities in order to take advantage of this technology for processing of data, comparison, cross-checking and to have a comprehensive view of a thematic feature. This technology is boosted recently by bringing out an International Journal on GIS from Taylor and Francis Ltd., London, New York and Philadelphia (Coppock and Anderson, 1987).

Meanwhile the study of environment also developed into another multidiscipline having links with all sorts of disciplines, such as geography, botany, zoology, geology and the like. In 1976, the International Geographical Union (IGU) and the International Cartographic Association (ICA) appointed a joint Working Group on Environmental Maps and Atlases (WGEMA). This working group is making sincere attempts to built up a data base related to environmental aspects, and is very active (Bickmore, 1984; Dutt, 1986).

A serious attempt has been made in this direction following a symposium on 'Global Change' in Ottawa in 1984 (US Committee on IGBP, 1986). One of the project emerged from this symposium was on the "Thematic Maps of the World" under the auspices of the IGU but in close collaboration with the ICA and other constituent units of the International Council of Scientific Unions (ICSU). The aim is to assist in monitoring changes in the geosphere and biosphere. Apart from the preparing basic maps of geology, soils, vegetation, climate and the like, topics such as the frequency of serious natural hazards, the extent of past glaciations and aspects of agricultural production or human health and welfare were considered for mapping. On the other hand, WGEMA was entrusted with the preparation of a digital world data base for environmental sciences. This was also a follow up action of Ottawa symposium under the aegis of ICSU which was concerned with proposing a serious international study of global change in the terrestrial environment and the life that inhabits it.

Knowledge about the up-to-date of the sets of basic points, lines and areas, is *sine qua non* for organising the evidence about global change. For decades work of this kind has necessarily assumed the existence of printed base maps. Such base maps not only characterise topography but at the same time provide the underlying geographic scaffolding or indexing for the specialised information. The desire to see inter-relatedness of patterns in a geographical sense is now capable of being handled efficiently within computers so as to provide fast results in both statistical or map form; the techniques concerned are broadly known as GIS. Further, recording both geographical patterns and their statistical attributes in digital forms enable the data sets — the central information in the system — to be continuously updated without invoking the somewhat protracted procedures of classical cartography. The cartographic data base or GIS now becomes a primary objective and the derived map or statistics becomes secondary to the data base and capable of being presented in CRT form or as a printed document. The data base becomes a living, growing entity serving the particular geographical needs of environmental scientists (Bickmore, 1984).

Some more attempts have been made to process the environmental-related information by digital system. UNEP/FAO took initiative to digitise world soil map at 1:5 m. This base provides a basis for comparison with other set of information on environment. "Some basic elements of world soil classification, like slope, might benefit from direct links with a global file of relief and is able to provide information in direction and angle of slope, even if only to a resolution of $10km^2$ initially" (Bickmore, 1984). Further, UNEP has a programme to develop a Global Resource Information Database (GRID) for environmental monitoring. The maps prepared by the International Union of Geological Sciences (IUGS) at 1:5 m can be used as a base map for processing environmental information. CIA has a programme to prepare a world data base. IGN, Paris is planning to digitise their 1:5 m world series maps. IGU/ICA WGEMA decided to design a study of a world digital database for environmental sciences (WDDES). Private companies, international bodies and government sponsored agencies in different countries are trying to develop information system suited for their requirements. Processing information on environment becomes an implicit or explicit policy of their activities (Marble *et al,* 1984; Rhind *et al,* 1986; Shelton, 1979).

Development of Information System in India

Some attempts have been made by different organisations in the country to develop information system for different purposes (Nag, 1987). The National Natural Resource Management System (NNRMS) of the Indian Space Research Organisation appointed several task forces for this purposes. These task forces were required to study the existing system of generating information, data processing and management aspects, highlighting their strength and weakness, and identifying the specific elements amenable to techniques of remote sensing. Onc of the task force is on National Resource Information System (NRIS). With similar objectives, the Department of Science and Technology (DST) sponsored a project known as National Resource Data Management System (NRDMS). The purpose of this pilot project was to evolve an appropriate methodology for the collection, collation, storage, analysis and dissemination of information in a specific region in its totality. Further, Survey of India is trying to develop an information system for their own use. Similarly, the Department of Environment is acquiring hardware and software for having an information system for updating their forest maps. The National Wasteland Development Board of the same Department is trying to get GIS software through the World Bank for monitoring the wasteland in the country.

GIS Technology

GIS represents a rapidly developing field lying at the intersection of many disciplines — among them cartography, computing, geography, photogrammetry, remote sensing, statistics, surveying and other disciplines concerned with handling and analysing spatially-referenced data (Coppock and Anderson, 1987). The application of this integrated system has been realised by planners, resource managers, oil explorers, environmentalists, property assessors and the like. This technology helps us in integrating complementary and diverse information, and consequently allows comparison and cross-checking. The conventional and non-conventional information to be processed is voluminous which can be handled by a computer-based system. GIS technology has a definite application in environmental studies, and monitoring and management of the environment. Static and dynamic information of the environment has to be processed from different sources. The design of GIS technology also aims to bring together spatial statistics, map data and image data into a single system where the environmental scientist can address his spatial problems without concerning himself with difficulties of data format differences (Jackson and Mason, 1986). The potentialities of remote sensing in developing GIS are enormous. Digital systems are now available which can interpret remotely sensed images based on spatial analysis of pixels. The human factors is likely to influence the interpretation.

At the initial stage satellite remote sensing and GIS were evolving separately. The integration of both the systems will be of great advantage to environmental assessment and management. The potential timeliness and digital format of satellite remotely sensed data are two of its greatest attractions as a source of data for existing GIS, none of which has yet addressed the 'time dimension' which repetitive satellite-sensed data could supply (Young, 1986). A GIS for environmental management has to be more dynamic, i.e. the data sets should be quickly available and processed. Table 30.1 provides the details of some of the sensors and their possible role in generating information on environment. The data repeatedly available through different sensors and in various hands provides a scope to monitor environmental changes (Nag and Dutta, 1986). It is now being realised that by feeding into the image analysis system additional environmental data sets, one could achieve more effective extraction from remotely sensed data (Jackson and Mason, 1986).

A GIS framework is shown in Fig. 30.1. The inputs can be from several sources : (a) data from maps, (b) statistics, (c) field data, (d) digital data bases, (e) satellite remote sensing, and (f) air photos. These inputs will form the Data Base Management System (DBMS). The main functions of the DBMS are: (a) collection input and correction, (b) storage and retrieval, (c) manipulation and analysis, and (d) output and reporting. DBMS can also be related with other DBMS or external statistical packages. In order to make it user-specific, communication with the users are required. The intelligent interaction between the thematic specialists users and software engineer is required for transforming a DBMS to GIS. The output can be in the form of (a) maps, (b) reports, (c) photographic products, (d) data or statistics, and (e) models.

Hardware and Software Requirements

Hardware combinations from several companies or their vendors are now available. Hardware options are based on the capacity of the central processor, such as minicomputer, micro-processor, super minicomputer and large mainframe. Other units are attached to the central processing units (CPU). A desirable GIS should have following peripherals:

- Vector digitiser for input
- Laser scanner for input
- Digital stereoplotter for input

- Flat-bed or drum plotter for output
- Visual display for output
- Line printer for output
- Disk drive for facilities
- Winchester drive for facilities
- Dunn camera for output
- Graphic interactive system for manipulations
- Work stations for input, output and manipulations.

TABLE 30.1: Sensors, Bands and Environmental Information

Sensor/band	wavelength (μm)	environmental information
Landsat (TM)		
1	0.45-0.52	– Water penetration, soil, vegetation and deciduous/coniferous flora discrimination
2	0.52-0.60	– The measurement of visible green peaks of vegetation for vigor assessment.
3	0.63-0.69	– The chlorophyll absorption band which can be used for vegetation discrimination.
4	0.76-0.90	– Determining biomass content and for delineating water bodies.
5	1.55-1.75	– Vegetation/soil moisture content and snow/cloud differentiation.
6	10.40-12.50	– The thermal inferred band which is useful in vegetation stress analysis, soil moisture determination and thermal mapping.
7	2.08-2.35	– Discrimination between rock types and hydro thermal mapping.
IRS		
1	0.45-0.52	– Sensitivity to sedimentation, discrimination of conifers.
2	0.52-0.59	– Green reflectance of healthy vegetation.
3	0.62-0.68	– Sensitivity to chlorophyll absorption of vegetation.
4	0.77-0.86	– Sensitivity to green biomass.

Source: Bulletins of the National Natural Resource Management System, Indian Space Research Organisation, Bangalore.

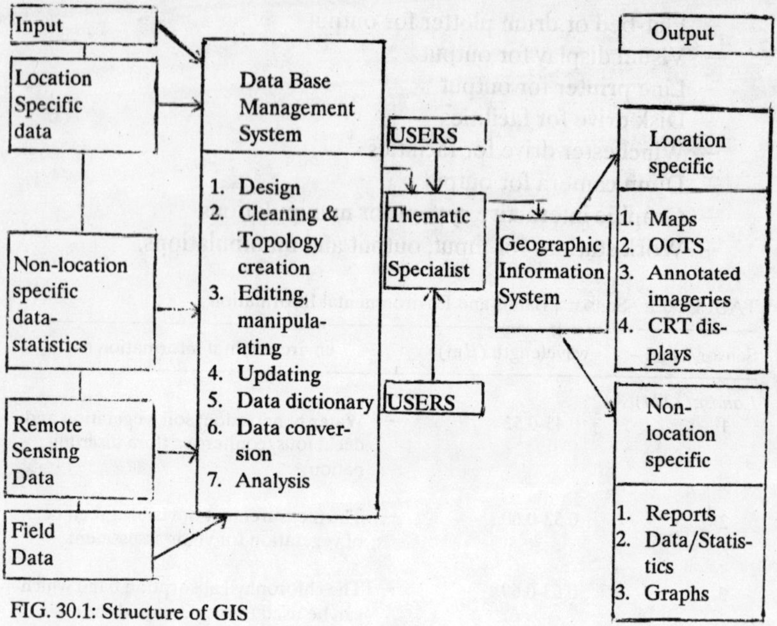

FIG. 30.1: Structure of GIS

The software should have following functions (Tomlinson and Boyle, 1981; Luscombe, 1985; Rase, 1987) :

Basic system capabilities :

- Digitizing
- Edgematching
- Polygonization
- Labeling
- Reformatting digital data from other systems
- Database creation and management
- Edit and display
- Plotting
- Updating
- Browsing

Data manipulation and analysis functions:

- Create lists and reports
- Reclassify attributes
- Dissolve lines — merge attributes

- Line thinning — weeding
- Scale change
- Transforming — rubber sheer stretching
- Projection change
- Generate
- Subdivide area
- Search
- Windowing
- Measure
- Calculate
- Compare data sets — overlays
- Construct
- Perspective views.

Standard GIS software package normally has the above functions. The details of a general purpose information system is shown in Table 30.2.

Special Requirements for Environmental Information System

The information on environment can be divided into three types:

(a) Static relief, climate, soil etc.
(b) Dynamic cyclone, pollution etc.
(c) Socio-economic Housing, health etc.

Information is now available on almost all conceivable issues. The static features can be digitised and can be used repeatedly for a longer time. But, the dynamic changes in the environment requires constant monitoring. The data input can be from remote sensing sensors or from fieldwork or laboratory analysis of the field data. The data structure of these types of inputs are to be understood for digital processing of data. Information on the environment can be in the form of points, grids, lines or polygons. In order to estimate the hardware required and the capacity of the CPU, the volume of the information processed is to be calculated. Obviously, the data sets used here are the larger and voluminous in number than the general purpose GIS. The hardware and software facilities for overlaying information sets should be large in number so that monitoring and interrelationships could be possible. This will also allow to study environmental impacts.

TABLE 30.2: Functions of A General Purpose GIS Package

Data capture/Editing Updation	Data storage/ Base management	Data query Retrieval	Data manipulation Analysis	Data display Outputs
Manual cell encoding	Multiple Map sheets	Browsing	Overlay (Intersection, union)	Hard copy maps, Statistical tables
Map digitisation		Location	Reclassifying polygons	
Raster image	Multilayer, Multi user	queries	— dissolve	CRT displays
Co-ordinate editing		— point	— merge	— Graphics
- Interactive	Journaling/Recovery	— Rectangle	Proximity (shortest route,	— Attributes
- Batch	Data Dictionary	— Circle	neighbour), Contouring/	Intermediate files
		— Irregular polygon	/interpolation	
Edge Matching	Data base redefinition	—Corridor	Slope/Aspect	
		Attribute	Visibility	
Splinter Removal	Data-types	queries	Buffer (point, line, polygon)	
	— Binary			
Line-coordinate thinning	— Characters	Attribute/	Scale/Projection Change	
	— Dates	location		
Dropline		Multiple	Distortion elimination	
Polygonethinning		Mapsheets		
Vector to Gird	Data Model		Co-ordinate rotation/Translation	Generation of
Gird to vector	— Relational		Point-in-polygon (number,	Latitude/longitude latices
Polygonisation	— Hierarchy		Yes/No)	Labelling
	— Networks			
Topology creation			Distance (line, curve)	— Font choice
Symbol library/menu			Area, Perimeter	— Placement choice
Text label data			Volume	— Orientation choice
Editing cartographic elements for a window on CRT			Centroid	Cartographic symbols generation
			Angle direction	— For points
			Space-attribute-time	— Lines
Interactive entry of points, lines & polygons			correlation (rate or change)	—Regions

Source: Space Application Centre, Ahmedabad.

The third type of information sets are based on periodical collection of data by identified government agencies, such as the Registrar-General of India and the Wildlife Department. These data sets on environment are based on the administrative setup of the country. Hence to use this information, the administrative structure of the country is to be digitised. The information is not always choroplethic in nature. There are point based information as well (Nag, 1984a & 1984b).

An example of coastal monitoring of the environment has been taken here. The data input can be in the following forms : (a) shoreline, (b) suspended sediment pattern map, (c) landform, (d) wetlands, and (e) other bathymetric details. The data can be captured from (a) maps via digitizer or scanner, (b) remote sensing imagery, and (c) auto stereoplotter. The information should be analysed by overlaying, or calculating distances. The output can be in the form of maps, statistics, or a hard copy of the preferred overlayed information. Similarly, the information on pollution can be processed after knowing the carriers and barriers of the specific pollutants, the physical and climatic conditions required and the pattern of decrease or increase in the intensity of the pollutant. The source of the pollutant can be a point like factory, a line like river, or a region like urban area. The information may require laboratory analysis as well.

Another important aspect in developing an information system for environmental monitoring and management is the scale of operation. The issues involved are time-specific, region-specific and problem-specific in general (Nag and Dutta, 1986). Hence, no common scale would be applicable for all the purposes. The scale is theme related. Perhaps, for mapping wildlife 1:2 Million map would be suitable. Forest mapping is preferred at 1:1M scale, while urban pollution at about 1:20,000 scale or larger. The information system should be possibly scale free. This system should have a scope for enlargement so that the priority or target areas could be related to the overall system. The scale of operation is linked to the nature of data available. Similar data sets are not necessarily available or desirable for the whole country (Smith *et al,* 1987; Wiggins, 1987).

The third aspect is the selection of hardware and software. As mentioned earlier, these are linked with the volume and nature of data to be processed. If the information system is based on remote sensing data primarily, hardware and software for image processing is required. These information sets which are becoming more and more refined and multiple in number are to be geometrically corrected and then superimposed. The processing of information of the fast changing environment is to be done quickly. Hence, such techniques have to be developed.

Conclusion

An information system can help in removing some of the myths in environmental management and assessment, because the system allows constant monitoring of the environment. The pre- and post-development consequences can be analysed. It will provide a method for an integrated assessment of the impacts of a development process or a hazard. Simulation studies and environmental modelling would be possible. But, some inherent problems can not be removed totally, such as subjectivity in interpretation of the information, or the proposed model for assessment. The system will allow several alternatives for model building. A suitable one has to be worked out.

It appears that only big institutions and large finances can support an operational information system (Dutt and Nag, 1987; Dutt *et al,* 1987). Whether in a developing country like India should have the luxury of having such an expensive system ? where do a university department (such as geography) stands in developing this thrust area? These questions are likely to come before going for such a system. Probably a humble beginning can be made in university department by having a micro-processor based information system which have some facilities for image processing and digital mapping. This will initiate computer culture in the department. Probably after processing information on environment and conducting a series of studies, the exact requirements of software and hardware can be assessed (Godchild *et al,* 1987). The new peripherals could be identified which can be hooked up with the mainframe computer at the university centre. Time has come when geographers will have to make a valuable contribution in this field. Traditionally geographers have been doing such exercises but not with high precision. This applied field of geography should not be left for development to other disciplines. Geography is an integration science and today the information processing for environmental assessment provides a greatest challenge to this discipline.

REFERENCES

Bickmore, David (1984) *Environmental Sciences : A World Digital Data Base,*

IGU/ICA Working Group on Environmental Maps and Atlases, Oxford (Typed).

Caldwell, Lyton Keith (1984), *International Environmental Policy : Emergence and Dimension,* Duke University Press, Durham, North Carolina.

Coppock, Terry & Eric Anderson (1987), Editorial review, *International Journal on Geographical Information System* (I.J.G.I.S.), 1(1), 3-11.

Dutton, Geoffery (ed) (1978), *Harvard Papers on Geographic Information Systems,* Laboratory for Computer Graphics and Spatial Analysis, Cambridge (Mass.), in 8 volumes.

Dutt, G.K. (1982), Mapping the environment, *International Journal on Environmental Studies,* 19, 25-33.

Dutt, G.K. (1986), Issues and challenges, in H.H. Singh, et al (eds), *Geography and Environmental : Issues and Challenges,* Prof S.L. Kayastha felicitation volume, Concept Publishing Company, New Delhi, 3-10.

Dutt, G.K. & P. Nag (1987), *National Atlas & Thematic Mapping Organisation in 2001 A.D.,* NATMO Occasional Monograph 2, Calcutta (in press).

Dutt, G.K., S.M. Mishra & P. Nag (eds) (1987), *Environmental Assessment and Mapping,* NATMO Monograph, Calcutta (in press).

Godchild, Michael F. & Brian R. Rizzo (1987), Performance evaluation and workload estimation for geographic information system, *I.J.G.I.S.,* 1(1), 67-76.

Holling, C.S. (ed) (1978), *Adaptive Environmental Assessment and Management,* International Series on Applied Systems Analysis, International Institute for Applied Systems Analysis, John Wiley & Sons, Chichester, New York, Brisbane & Toronto.

Jackson, M.J. & A.C. Mason (1986), The development of integrated geo-information system, *International Journal of Remote Sensing* (I.J.R.S.), 7, 723-40.

Luscombe, B. Wayne (1985), Establishment of GIS capabilities in India, The World Bank International Finance Corporation, Office Memorandum 60220 (unpublished).

Marble, D.F., K.E. Brassel, D.J. Peuquet & H. Kishimoto (eds) (1984), *Proceedings of the International Symposium on Spatial Data Handling,* Geographisches Institute, Universitat Zurich-Irchel.

Nag, P. (1983), Perception of the environment : Issues and Challenges, *National Geographical Journal of India,* XXIX (3 & 4), 217-31.

Nag, P. (ed) (1984a), *Census Mapping Survey,* IGU Commission on Population Geography, Concept Publishing Company, New Delhi.

Nag, P. (1984b), Post-census mapping, in John I. Clarke (ed), *Geography and Population : Approaches and Applications,* Pergamon Press, Oxford, 159-69.

Nag, P. (1987), A proposed base for geographic information system for India, *I.J.G.I.S.,* 1(2) forthcoming.

Nag, P. & A. Dutta (1986) Issues in environmental mapping, in H.H. Singh, et al (eds), *op. cit,* 299-307.

Rase, Wolf-Dieter (1987), The evolution of a graduated symbol software package in a changing graphics environment, *I.J.G.I.S.,* 1(1), 51-65.

Rhind, D.W., B.K. Wyatt, D. Briggs & J.C. Wiggins (1986), The creation of an environmental information system for the European community, *Nachrichten aus dem Karten und Vermessungswesen,* Series II, No. 44.

Shelton, R.L. & J.E. Estes (1979), Integration of remote sensing and geographical information systems, in *Proceedings of the 13th International Symposium on Remote*

Sensing of Environment, Environmental Research Institute of Michigan, Ann Arbor.

Smith, T.R., Sudhakar Menon, Jeffery L. Star and John E. Estes (1987), Requirements and principles for the implementation and construction of large scale geographic information system, *I.J.G.I.S.,* 1(1), 13-31.

Tomlinson, Roger F. & A., Raymond Boyle (1981), The state of development of systems for handling natural resources inventory data, *Cartographica,* 18(4), 65-95.

U.S. Committee for an International Geosphere-Biosphere Program (1986), *Global Change in the Geosphere-Biosphere — Initial Priorities for an IGBP,* National Academy Press, Washington.

Wiggins, J.C., R.P. Hartley, M.J. Higgins & R.J. Whittaku (1987), Computing aspects of a large geographic information system for European community, *I.J.G.I.S.,* 1(1), 77-87.

Young, J.A.T. (1986), Remote sensing and an experimental geographic information system for environmental monitoring, resource planning and management, *International Journal on Remote Sensing,* 7(6), 741-4.

31

BIOCIDES INDUCED ENVIRONMENT AND HUMAN HEALTH

R.S. SINGH

Use of chemicals has been increasing day by day. This is more prominent in urban areas. Building and dwelling units are now made for specific use and a specific environment is created with the help of technology, chemicals and internal arrangements, such as use of air conditioners and pesticides or architecture. All of the factors, in different combinations, generate different types of environment which affect the human health. If we look into details of temperature, proportion of metallic cover in interior decoration etc., we will find that they contribute to specific environment which harbour different insecticides and biocides and other bio organisms. Such organism are more prominent in library, computer rooms, laboratories, hotels, factories and in other places meant for specific use. They contribute to micro environment at very small scale. The investigation into the such micro environment has become important keeping in view of the fact that environment is being transformed every day with the help of technology, chemicals and other artificial processes. Even doctors and health workers are specialising on diseases which flourish in such micro environment.

Cultural heritage deteriorated by bio-organisms is an important aspect of study for the custodians working in museums, libraries and archives. Insects and fungoids infestation are a serious problem in any institution housed with cultural property. A number of chemicals, either alone or in combination are used to control these organisms which is called 'Biocides' (Allsopp and Allsopp, 1983; Allsopp and Seal, 1986). The application of biocides is made in solid, liquid or gaseous state. The process of disinfestion involves the use of an insecticide in gaseous form is called 'fumigation'.

Fumigants

In order that a chemical can be effective as a fumigant, it must be volatile enough to produce a toxic concentration and uniform distribution of vapour in an enclosed space however small it may be. Except in vacuum fumigation, molecular diffusion must be relied upon to bring about gas penetration of material to be fumigated. The rate of diffusion in space depends upon the molecular weight of the gas, size of the openings through which the gas has to pass and the absorptive capacity of the material. Density influences distribution of fumigant. Some fumigants, are heavier than air and tend to floor and must be artificially stirred with fan.

Physical state of the fumigant at ordinary temperature also has importance. Some chemicals like methyl bromide, ethylene oxide, and hydrogen cyanide (HCN) which are gaseous at or having immediately ordinary fumigation temperature. Other chemicals like carbon tetrachloride, ethylene dichloride, carbon disulphide that are liquid at ordinary temperature and solids like paradichlorobenzene are volatile enough to vaporise within a reasonable time to produce concentration necessary for fumigation in an enclosed space. Thymol and naphthalene however, require heating for volatilisation.

Temperature has considerable importance in determining the effectiveness of a fumigant. Insects are not active at temperatures below 60°F and they become more or less dormant at 50°F or below. In the dormant state it is extremely difficult to kill them with fumigants in a given space. At 75°F they are active and their susceptibility to gases is good. The effect of atmospheric moisture upon the toxic action of fumigants is much less than that of temperature.

The following fumigants at the concentrations stated below are very effective in killing insects or pests at storage within 24 hours and have no deleterious effect on museum objects like paper (cellulosic materials), ink and other record components (Talwar, 1975).

Fumigants	
1. Hydrocyanic acid gas	Concentration per 1000 cu. ft. from 1b. of sodium cyanide or 288 c.c. liquid HCN.
2. Carbon disulphide and Carbon tetrachloride (1:4)	6 lb.

3. Ethylene dichloride and Carbon Tetrachloride (3:1)	14 lb
4. Ethylene oxide and Carbon dioxide 1:9 known as Carboxide	30 lb
5. Ethylene oxide	2 lb
6. Methylene oxide	2 lb
7. Methyl formate-carbon dioxide (2:5)	28 lb

Museum or archival fumigants may be divided into three basic classes : insecticides, moldicides, and general fumigants. Insecticides are generally recommended to be applied in the area of contaminated material rather than upon the objects. These pesticides include, among others arsenic trioxide, dieldrin, lindane, napthalene, paradichlorobenzene, pyrethrum and silica aerogels.

Properties of arsenic trioxide has been used effectively in controlling termites (*Nastutitermis exictiosus H*). Only 1.7 g is sufficient to eradicate entire colonies of the subterranean species (Svent-Ivanty, 1968). Since arsenic is highly toxic and is a recognised carcinogen, its application should be done with great precautions (Sax, 1975).

Dieldrin is recommended as a specific treatment for the control of silverfish. Fumigation is accomplished through the use of paper sheet coated with a 0.3 to 0.4 per cent weight for weight deposit of the chemical. This technique has numerous advantages including the extraordinary residual life of the chemically treated sheet which is said to be as long as ten years (Svent-Ivanty, 1968). Since this chemical is rated as highly toxic when inhaled or absorbed through the skin, and therefore, dieldrin treated sheet should be prepared by chemist under controlled conditions. The concentration of this substance should not exceed the estimated threshold limit value (TLV) of 0.25 mg/m^3 of air (Sax, 1975). Lindane (1, 2, 3, 4, 5, 6-Hexachlorocyclohexane) is a noted effective fumigant and can be used against cockroaches and silverfish. Available in powder, dust or as a liquid, it is applied in building crevices and similar hiding places particularly in urban environment. Since it is known to have adverse effects when directly applied to organic material or ink, application in the liquid form should be done with brush rather than sprayer in the infested areas (Clapp, 1974). This chemical is classified as moderately toxic and its TLV in working areas, has been estimated as

0.5 mg/m^3 (Block, 1953).

Naphthalene is also a specific biocide used against most of the insect pests particularly beetles. It is applied in a sealed box by exposing the infested material to the fumes of this volatile chemical at the rate of 47 gm/ft. (Talwar, 1975). Naphthalene is a skin irritant and is moderately toxic when inhaled or absorbed through the skin hence it is hazardous to human health. The maximum TLV in working areas is 10 parts of vapour per million parts of contaminated air by volume at 25°F and 760 mm i.e. 10 ppm or 50 g/m^3 (Sax, 1975).

Paradichlorobenzene is one of the most universally recommended general insecticide and is specific for booklice and beetles. It is moderately toxic as an irritant and only slightly toxic when inhaled. Its workroom TLV is 74 ppm or 450 mg/m^3, (Block, 1953; Sax, 1975). Since it does not destroy insect eggs, treated material should be examined at least during the year after fumigation to determine if any reinfestation has occurred (Kathpalia, 1973). O-isopropoxy-phenyl-methyl-carbamate (Baygon) is a highly toxic *Carbamate* insecticide that is said to be of significant value in the control of cockroaches (Sax, 1953). The conservators recommending its use in specific dilution to an 0.5 per cent solution that can be brushed over cracks and other places of intrusion in houses and buildings (Clapp, 1974).

Pyrethrum is noted as effective against to the cockroach, cricket and silverfish. It is a very common ingredient in household pesticides. Pyrethrum-I is a viscous liquid moderately to highly toxic when inhaled and may provoke serious reactions in persons allergic to the chemical. It has a workroom TLV as 5 mg/m^3. It is applied in structural cracks and around the base of shelves in a closed urban situation.

Silica aerogel is a finely powdered micro cellular silica foam that destroys cockroaches, termites and other soft bodies insects mechanically through its action in the destruction of the fatty layer of insects body. It may be scattered in cracks and crevices in built up structures where infestation is suspected (Gallow, 1963). In addition to powder and liquid insecticides described above, several gaseous compounds have been used as insecticides. Among more common of these are aluminium phosphide (sulphos) and sulfuryl fluoride (vikane). Because of their extreme toxicity, the application of such compounds should be done with great care.

Moldicides are applied directly to the infected material or to an entire area where contamination is suspected. In suggested area disinfestants include thymol and trichloroethane and lauryl-dimethyl-carbethoxymethyl ammonium bromide. These are fumigants that should be applied only when the infected area can be successfully sealed off from inhabited parts of the institution.

Thymol chemically known as isopropylcresol in conjunction with 1,1,1, trichloroethane is suggested for fungicidal fogging. The thymol concentration of 0.5 kg. in 5 litres of trichloroethane, the compound is said to be sufficient for an area of 20,000 cubic feet (Waters, 1975). The hazard rating for thymol has been discussed; 1.1.1, trichloroethane is rated as moderately toxic when inhaled or absorbed through skins (Sax, 1953).

Thymol mentioned above, may be used through vaporization applied directly to the infected paper in 10 per cent solution made in denatured ethyl alcohol (Cunha, 1971). Because thymol is a convenient vapour form and is often used by conservators (Meddleton, 1977; Waters, 1975). Kowalik (1980) investigated the effect of thymol on fungi decomposing old papers. Crystals of thymol (melting point of the thymol is 51.5°C) were placed on the bottom of the desiccator. To generate the vapours the bottom of the desiccator was heated to a temperature of 52°C everyday for 2 hours for 14 days. Every day 70 g/m^3 of thymol is added. Kowalik found that some fungi existed on object but did not grow anymore after 6 days of thymol fumigation. On the contrary, object inoculated with good sporulating cultures of fungi showed their growth even after 14 days of treatment with thymol vapours. Fungi were existing on object probably weakened owing to lack of water and were therefore easier to kill than fresh well fruiting fungi.

Difference in opinion among conservation specialists exists as to the amount of thymol to be used and duration of exposure. Plenderleith (1971) suggests 50 gms to a cubic meter of air which should continue for a period of 14 days. Kathpalia (1971) maintains that 120 gms is the recommended dose for a cubic meter of air but claims success with 20 gms in a single hour exposure. Wardly (1971) suggests 25 gms as the proper amount of chemical to use, while Clapp (1974) suggests a four days exposures to vapours. Werner, a well known paper conservator, says, thymol has disadvantages. It can act as a very powerful solvent. Problem with the use of thymol includes the fact that it has been known to attack some glues, it dissolve inks and may leave some residuals on paper. Direct application or interleaving has been known to discolour the document (Strossberg, 1978).

The quaternary ammonium salt lauryl dimethyl carbethoxymethyl ammonium bromide (Cequartyl B.E.), Prepared in a 5 per cent solution with denatured alcohol, is recommended for use in area of fumigation for moulds in the proportion of 5 cm^3/m^3. It may be applied by using an air compressor and spray gun. The compound is said to be irritating to the mucous membranes. The conservation scientists recommend that it has been applied successfully in repository of paper objects and have indicated a problem with ink adhesion upon exposure to chemical and also suggest that it may adversely react with cellulose (Zappala, 1972).

Recently Kowalik has tested vapour fungitoxic action of paradichlorobenzene (Kowalik, 1980). This compound is known as insecticide. The books infected should be placed in the container and exposed to the vapours but not allowed to touch the crystals. According to Block and others p-dichlorobenzene is also effective in inhibiting the fungal growth in closed spaces (Block, 1952, 1968; Mouro, 1969). Kowalik has found that p-dichlorobenzene at an amount of 50 g/m^3 inhibits completely the growth of some fungi but is fungistatic only against a few fungi.

Orthophenyl phenol (Topane, Dowcide 1) has been suggested as a fungicide for paper and other organic objects. As a paper moldicide, the chemical is used to treat interleaving or wrapping paper in a 10 per cent aqueous solution (Clapp, 1974). Since its volatility is low, treated papers can be used for a period of six months or more without replacement of the chemical.

Paranitrophenol is another phenol compound occasionally recommended for use as a fungicide. This chemical may leave a greenish tinge on the paper when used in excess (Kathpalia, 1971).

Salicylanide (Shirlan) and sodium salicylanide are said to be useful as fungicide on paper. Caution must be taken in their use since they are known to have undesirable effects when applied in excessive quantities.

For old papers and documents destined for permanent preservation a mixture of thymol and mercuric chloride in ether-benzene solution, the sodium salt of pentachlorophenol in water and also salicylanide in trichloroethane and acetone solution, the sodium salt of dichlorophene lauryldimethyl-carboxydethyl ammonium bromide have been used with effectiveness (Hetherington, 1965; Baynes, 1971;

Kennedy, 1959; Plenderleith, 1974; Nordstrand, 1967; and Kowalik, 1972 and 1977). Salicylanide is effective in the case of cellulosic materials having a low moisture content but it is not maintained for a long period in the fibre when subjected to a low continual humidity (Block *et al.*, 1968). The mixture of mercuric chloride and thymol is strongly toxic and increases slightly the acidity in the paper (Kowalik, 1972). Pentachlorophenol besides being toxic, is weakly acidic and gives in light besides hydrochloride, a red product chloranilic acid and other colour components (Munakata and Kuwahara, 1969). According to Zappala (1972) lauryl dimethyl carboxymethyl ammonium bromide has not to be chemically inert and is deleterious towards cellulose.

For the immediate killing of micro-organisms but not immunizing the object, ethylene oxide seems to be the most suitable when 90 per cent carbon dioxide is present or about the same percentage of various fluorinated hydrocarbons ($C\ Cl_2$, F_2 and $C\ Cl_3\ F$), the resulting mixture can in turn be mixed with air in all proportions without passing through an inflammable state (Monro, 1969; Torlulson, 1966; Kereluk and Gammon, 1970 and 1972; Kereluk, Gammon and Lloyl, 1970). In this last case moisture is not so important as concentration. The mixture of ethylene oxide and carbon dioxide in the proportion 10:90 in a vacuum chamber is the most effective at 30-35 per cent R.H. and become progressively slower as the R.H. increases to 100 per cent (Block, 1968; Bruch 1973; Gaughran 1973 and Mullican 1968). Regarding the biostability of paper, there is a new polymeric antifungal substance i.e., polymerious guanidine compounds which is being synthesized in U.S.S.R. (Nyuksha, 1980). For example polyhexamethyleneguanidine — a water soluble polymer of the Kation type has chemical structure as:

$$NH_2 \left[\begin{array}{r} (CH_2)_6 NHC-NH \\ ,NH-HCL \end{array} \right] \begin{array}{c} H \\ n \end{array} \qquad n = 4 \pm 6$$

This substance is a transparent, hard, yellowish, odourless resin with a softening temperature of 80°-100° C. The resin is hygroscopic, water soluble, any concentration is possible and is in one per cent solution of polyhexamethyleneguanidine, has pH 8-10, stable in storage. In decomposition polymerous carbamide (urea) (NH_2-CO-NH_2) and ammonium chloride (NH_4Cl) are formed. Thus, polyhexamethleneguanidine does not have bad effect on the physical and chemical state and properties of paper during artificial aging and prolonged storage. A micro examination of paper impregnated with polyhexamethyleneguanidine shows that the substance fills the spaces between fibres evenly and covers the fibre surface completely. It has a fungicidal and fungistatic effect on paper damaging fungi.

Insecticides of Plant Origin

Plant products have been used as insecticide since ancient times in India. Their use can be still observed in the rural and tribal areas. Plant insecticide have low mammalian toxicity. Chinese were the first to discover the value of Derris (*Derris ellipptica* and *Derris Malaecenisis*) as an effective insecticide.

Chemical substances like alkaloids and isoflevoroids present in plants are the substances which act as insecticides. Alkaloids such as nicotine from tobacco (Nicotina spp), Rotenone from Derris (*Derris ellipptica*) and pyrethrins and cinerins (pyrethroids) from pyrethrum (*Chrysanthemum cineraridefolium*) are good plant insecticides. Most plant insecticides act as contact poisons being absorbed through the cuticle of the insects.

'Leguminosae' or the 'Pea family' has several plants known to have promising insecticidal values. Logwood (*Haemotoxylon campohianum*) extracts are used in controlling the Japanese beetle. Alcoholic extracts of roots of the fish-poison climber (*Millettia pachycarpa*) controls some aphids. This plant contains two insecticidal compounds known as rotenone and saponin which act as contact and stomach poisons. Rotenone is a complex isoflevoroid placed in the group 'Rotenoids' obtained from the roots of plant of genera Darnis, Lonchocarpus and Tephrosia. Roots of 'devil's shoe — string' (*Tephrosia virginiana*) contains rotenone. *Derris ellipptica* are used as the source of rotenone in pest control. Extracts of Darris roots mixed with Pyrethrum are also used as insecticidal sprays. Rotenone is not harmful to man.

Plants belonging to the 'Solanaceae' or 'potato family' are also promising insecticidal plants. The active principles in these plants are alkaloids called 'Solanaccous alkaloids'. Anabasine is obtained from the tobacco (Necotiana glauca) plant. Crude extract of its leaves can be used to control aphids. Nicotine, an alkaloid obtained from tobacco (Nicotiana spp.) is graded as contact insecticide. Nicotine fumigation kills many insect pests. In India practices like spraying of the water extracts of tobacco stems, leaves, etc. for pest control are still prevalent. Another practice widely followed in India, especially in rural areas is the application of ground tobacco for plant protection. Nicotine sulphate, a commercial compound obtained by the action of nicotine with sulphuric acid is used as a contact spray. Nicotine is highly volatile. Its conversion into the sulphate lowers its volatility. Nicotine sulphate sprays effectively control many kind of insects and pests. Nicotine dust mixed with gypsum or clay and nicotine sulphate are promising insecticides against many varieties of insects. Nicotine fumes, obtained by evaporation of liquid nicotine base over a heater or by burning nicotine, are important in house fumigation. There are several other solanaceous plants that contain promising insecticidal properties.

Pyrethrum (insect powder) is obtained from the floral parts of *Chrysanthemum cinesariaefolium*) belonging to the 'Compositae' or 'Sun flower family'. It is a safe, promising household insecticide used either in sprays or dusts. The active principles in pyrethrum are pyrethrins I and II and cinerins I and II. They are unstable compounds as they decompose when exposed to air or sunlight.

Pyrethrins are toxic contact and stomach poisons. They are non-toxic to human beings. Their action is similar to that of DDT Domestic pests such as cockroaches are effectively eradicated by application of pyrethrum sprays.

Pyrethrum is used in following ways:

(i) As Pyrethrum dust, popularly known as insect powder,
(ii) Oil extracts, and
(iii) Water soluble extracts.

Plants like Castor (*Ricinus communis*), common basil (*Oeimum basilicum*), Cockroach plant (*Haplopnyton cumicidum*), China berry (*Melia Azendarch*), Caraway (*Carum Carvi*), Coriander (*Coriandrum sativum*), Anise (*Pimpinella anisum*) and several others also have effective insecticidal properties against many domestic pests.

Conclusion

We are all familiar with the serious environmental pollution caused by application of synthetic insecticides and pesticides, like organochlorine and organophosphorus compounds. The environment can be saved from the synthetic chemical (insecticides and pesticides) pollution if more use is made of the safe and effective naturally occurring plant insecticides. Besides the use of synthetic insecticides discussed in foregone, it is worth suggestion that insecticide of plant origin would always be superior over the others because they are safe and least hazardous to human being, easily available to use and does not effect the environment and ecology.

REFERENCES

Allsopp, C. and Allsopp, D. (1983), An uptodate survey of Commercial Products used to protect material against Biodeterioration, *Int. Bio. Bull.,* 19 (314), pp. 99-146.

Allsopp, D. and Seal, K.J., (1986), *Introduction to Biodeterioration,* Edward Arnold Ltd., London.

Baynes-Cope A.D., in Walters, A.H. & Huek Vander Plas E.H. (eds.) (1971), *Biodeterioration of Materials,* Applied Science Pub. Ltd., London, Vol. 2, p. 381.

Block, S.S., (1952), Experiment in Mildew Presentation, *Engineering Progress at the University of Florida,* Vol. IV : 27-34.

Block, S.S., (1953), Humidity Requirement for Mould Growth, *Engineering Progress at the University of Florida*, p. 287.

Block, S., in Lawrence, C.A. & Block, S.S. (1968), *Disinfection Sterilization and Preservation,* Lea and Febiger, Philadelphia, p. 575.

Clapp, A.F. (1979), *Curatorial Care of Works of Arts on Paper,* 2nd Ed. Rev., Intermuseum, Laboratory Oberlin Ohio, p. 34.

Cunha, G.M., Cunha, D.C. (1971). *Conservation of Library materials : A Manual and Bibliography on the Care Repair and Restoration of Library Materials,* 2nd ed., Vol. I, p. 147.

Gallow, Fausta (1963), *Biological Agents which Damage Paper Materials in Libraries and Archives in Recent Advances in Conservation*, Butterworth, London, p. 59.

Gaughran, E.R.L. (1973), Newer Methods and Trends in Sterilization, *Development in Industrial Microbiology,* 14, pp. 57-64.

Hetherington, D.C. (1965), Mold Preventive for Book Binding, *Science*, 101, p. 223.

Kathpalia, Y.P. (1973), *Conservation and Restoration of Archival Material,* UNESCO, Paris.

Kennedy, R.A. (1959), *Conservation in Wet Tropics.* Museum Association of Middle Africa, Technical Bulletin 1.

Kereluk, K. & Gammon, R.A. (1972), The Microbiological Activity of Ethylene oxide, *Development in Industrial Microbiology*, pp.28-41.

Kereluk, K., Gammon, R.A. and Lloyl, R.S. (1970), Microbiological Aspects of Ethylene Oxide Sterilization II, Microbial Resistance to Ethylene oxide, *Applied Microbiology*, 19:1 152-6.

Kowalik, R. (1972), Some Problems of Microbiological Deterioration of Paper, *Ammali Della Scuola Speciale per Archivistic Bibliotecari dell'*, Universita di Rome, Nuova Tecuica Graphia Rome 1-2, pp. 68-80.

Kowalik, R. (1971), Some Aspect of Microbiology of Paper and Parchment in Petersen, D.F., (edit.) *Herzog Augest Bibliothek Wolfenbitteler Forschungen,* Jacobi Verlag. Breman and Wolfenbiittel Band 1, pp. 61-90.

Kowalik, R. (1980), Decomposition of Paper, *Restourator,* Vol. 4.

Meddleton, B.C. (1977), Book Preservation for the Librarian, in William, J.C. (ed.) *Preservation of Paper and Textiles of Historic and Artistic value,* Advances in Chemistry Series, 164, American Chemical Society, Washington D.C., pp. 5-23.

Mouro, H.A.U. (1969), *Manual of Fumigation for Insect Control,* F.A.O. of United Nations, Rome.

Mullican, C.L., & Hoffman, R.K. (1968), Dry Heat of Gasous Chemical Resistance of *Balillus subtitis* var niger spores. Included water soluble crystals, *Applied Microbiology,* Vol. 16 pp. 1110-3.

Munakata, K. and Kuwahara, M. (1969), Photochemical Degradation Product of Pentacholophenol, *Residue Reviews,* Vol 25.

Nordstrand, O.K. (1967), Chinese Double-leaved Books and their Restoration, *Libri,* Vol. 17 pp. 104-30.

Nyuksha, J.P. (1980), Biodeterioration and Biostability of Library Materials, *Restourator,* Vol.4, No. 1.

Plenderleith, H.J. and Werner, A.E.A. (1974), *The Conservation of Antiquity and Works of Arts,* Oxford University Press, London.

Sax, Newton I, *et al.* (1975), *Dangerous Properties of Industrial Materials*, New York Van Nostrand Reinhold Co., p. 419.

Strossberg, Richard (1978), The use of Fumigants in Archival Repositories, *The American Archivist*, Vol. 41, No. 1, pp. 25-36.

Svent-Ivanty, J.J. (1968), *Identification and Control of Insect Pest in Conservation of Cultural Property with Special reference to Tropical Conditions,* UNESCO, p. 62.

Talwar, V.V. (1975), Fumigation Techniques for sterilisation of paper, *J. of Conservation of Cultural Property in India,* Vol. 8, pp. 46-9.

Torkelson, T.R., Hoyle, H.R. and Rowe, V.K. (1966), Toxicological Hazards and properties of Commonly used Space, Structural and Certain other Fumigants, *Pest Control,* Vol 34, 7 pp. 13-4.

Wardly, D.B. (1971), *Document Repair*, Society of Archivest London, p. 16.

Waters, Peter (1975), *Procedures for the salvage of water Damaged Library Materials Washington,* Library of Congress, pp. 12-3.

Zappala, M. (1972), Some Aspects of Chemical Research in Institute Di Patologia del Libro *In Conservation on Paintings and the Graphic Arts,* Lisbon Conference, pp. 995-1000.

32

DEPLETING VASCULAR CRYPTOGAMS: A Case Study of Kumaun Himalaya

P.C. PANDE, A.K. SINGH and H.C. PANDE

Dangerous devegetation in Kumaun Himalaya (28°44′ to 30°44′ N and 78°45′ to 81°5′ E) has resulted in rapid depletion of some useful species. Every year several species are either becoming endangered or extinct. This drastically upsets ecological equilibrium achieved over a long period of time. The continuous, intensive and reckless devegetation of herbal species has become a matter of grave concern. Owing to the reckless devegetation Kumaun Himalaya is experiencing tremendous changes in the composition of native plant communities and their dependent microbial environment. The most prominent interference with the Kumaun Himalaya ecosystem is the drastic reduction of vegetal cover caused by anthropogenic changes.

The present paper has been designed to focus on the extinct and threatened pteridophytes of Kumaun Himalaya. The work is based on the informations collected and collated. Though there is ample published literature on Kumaun pteridophytes (Clarke, 1880; Beddome, 1893; Hope, 1899-1904; Duthie, 1906; Loyal and Verma, 1960; Pande, 1973; Bir and Trikha, 1974; Holttum, 1976; Saxena, 1976; Biswas and Biswas, 1976; Dixit and Das, 1979; Niranjan and Roy, 1979; Verma and Khullar, 1980; Dhir, 1980; Khullar and Gupta, 1980; Khullar and Sharma, 1980; Khullar *et al*, 1983; Khullar, 1984; Awasthi and Pande, 1984; Pande *et al*, 1984; Pande, 1985 a, b; Pande and Bir, 1985; Pande *et al*, 1985; Punetha, *et al*, 1985; Satija and Bir, 1985; Pande and Kandpal, 1986; Joshi and Pande, 1986; Pande and Pande, 1987; Khullar and Sharma, 1987; Khullar *et al*, 1987; Pangtey and Punetha, 1987), but none of the researchers have attempted to analyse the deteriorating ferns and fern-allies population of this region. Keeping in view, the importance of ferns from environmental point of view authors have attempted to describe pteridophytes of Kumaun Himalaya in the following groups categorised by the Survival Service Commission (IUCN, *Plant Red Data Book*, 1978).

Sl. No.	Name of the plants	Place of occurrence in Kumaun	Remarks
CATEGORY I : EXTINCT TAXA			
Plants reported before 1906 and not reported/collected thereafter by any worker from the region, and most probably these plants have lost their existence from the area under study.			
1.	*Archniodes aristata* (Frost) Tindale	Kumaun	Recorded only by Hope (1903). Not collected from this region thereafter.
2.	*Arthromeris repandula* Ching	Below Nainital	Recorded only by Trotter (1891) as *Pleopeltis Himalayensis* from this region (vide Satija and Bir, (1985). Not collected since then.
3.	*Asplenium rutamuraria* Linn.	Upper Kali & Dhauli valley between 1100-1300 m	Reported by Duthie (1906). It is collected in recent years.
4.	*Athyrium mackinnoni* (Hope) C.Chr.	Kumaun	This is one of the largest species of *Athyrium*, recorded only by Duthie (1906) from Kumaun without giving the exact locality of collection. Not collected thereafter from this region
5.	*Athyrium setigerum* C.Chr.	Near Jauljibi in Gori valley	Recorded only by Duthie (1906). Not collected since then.
6.	*Athyrium tenellum* Hope	Jageshwar between 1500-2000 m	Reported by Duthie (1906). Not collected since then.
7.	*Athyrium tenuifolium* D. Don	Nainital	Champion (1937) vide Dhir (1980). Not collected by subsequent collectors till date from this region.

(Contd.)

(Contd.)

Sl. No.	Name of the plants	Place of occurrence in Kumaun	Remarks
8.	*Bolbitis varience* (Wall. ex Hook. et Grev.) Schott.	Kumaun between (300-900 m height)	Only Duthie (1906) reported from this region, has not been recollected since then.
9.	*Ctenopteris subfalcata* (Bl.) Kunze	Gini, Sosa and Kalamuni ridge	Duthir (1884, 1906) reported this from this region. Not collected since then.
10.	*Pronephrium repandum* (Fee) Holtt.	British Kumaon (Kumaun and Garhwal)	Clarke (1863) and Hope (1903) reported it from British Kumaon (vide Khullar *et al.* 1983). Lateron Duthie (1906) also reported it from the same locality, shows that it was quitc abundant in nineteenth century becomes extinct after a lapse of 80 years.
11.	*Cystopteris montana* (Linn.) Bernh. in Desv.	Milam Glaciers and Dhauli valley	It was only reported by early British Botanists (Duthie, 1884; Beddome, 1892; and Hope, 1990) from this region. Not collected since then.
12.	*Dryoathyrium boryanum* (Willd.) Ching	Sarju valley (between 3000-1000 m)	Recorded only by Duthie (1906). Not collected since then. According to Dhir (1980) there is no recent records of this ferns from Western Himalayas also.
13.	*Macrothelopteris ornata* Bedd.	Gola valley around 600 m altitude (Nainital)	Only Hope (1903) reported this fern from this region and could not be recollec-ed afterwards by any sub-sequent workers since then.

(Contd.)

(Contd.)

Sl. No.	Name of the plants	Place of occurrence in Kumaun	Remarks
14.	*Microlepia hirta* (Kaulf.) Presl	Ramganga river (800 m)	Strachey (1848) vide Dhir (1980) reported this from this region. This specie is not collected afterwards.
15.	*Notholaena marantae* (L.) Desv.	Kali valley (2900-3300 m) and Pindari gorges (2500 m)	Strachey & Winterbottom, Duthie, Reid and Trotter vide Dhir (1980) reported it from this region and not collected since then.
16.	*Onychium japonicum* (Thunb) Kze	Phurkia (3300 m) Karmi, Askot, Pithoragarh and Ramganga valley	The British botanists (Strachey & Winterbottom, Hume, Duthie, Trotter and McLeod) collected this ferns from different parts of Kumaun Himalaya between the years 1846-1901. (cf. Hope, 1901). Not collected in recent years.
17.	*Pallaea hastata* (Thunb) Prantle	Kumaun, Gangas valley (1200-1500 m) and Khairna below Almora (1200-1500 m)	Strachey (1848), Reid (1884) and Duthie reported this species (cf. Dhir, 1980). Not collected since then.
18.	*Phegopteris connectiles* (Michx.) Watt.	Kumaun between (3300-3400 m)	Only Duthie (1906) recorded this ferns from Kumaun. Not collected since then.
19.	*Polystichum duthie* (Hope) Christ.	Lebong pass	Only known from two old collections from Kumaun region (i.e. Duthie, 1884 and 1886). No subsequent collection made after the year 1886.
20.	*Vandenboschia pyxidifera* (L.) Copel.	Pindari gorges (2400 m)	The only collection (Hope, 1990) of this fern from Kumaun region of Western Himalayas.

(Contd.)

(Contd.)

Sl. No.	Name of the plants	Place of occurrence in Kumaun	Remarks
21.	*Woodsia alpine* (Balton) Gray	Lidder valley (between 1100-1500 m)	Recorded only by Duthie (1893). Not collected since then.
22.	*Woodsia andersanii* (Bedd.) Christ.	Sundhardhunga glaciers (Pindar valley)	Recorded only by Beddome (1883). Not collected since then from Kumaun. According to Dixit (1983) the specie is known by only five old collections in India.
23.	*Woodsia lanosa* Hook	Kali (3600 m) and Pindar valley (3600 m)	Recorded by Duthie (1884), Hope (1903) and Strachey (1906) from this region. Not collected in recent years.

CATEGORY II : ENDANGERED TAXA

Plants reported before 1906 and rarely once or twice collected by subsequent authors.

Sl. No.	Name of the plants	Place of occurrence in Kumaun	Remarks
1.	*Asplenium nidus* Linn.	Kumaun (Hope, 1901) Askot (Duthie, 1884) Gori Ganga (Duthie, 1886) Ramganga valley (Strachey, 1848)	Recently it has been relocated by Pangtey (1987) from Sandeo near Didihat, after a lapse of 86 years. Quite rare probably only one or two individuals are growing at Sandeo.
2.	*Asplenium tenuifolium* Don	Nainital (Duthie, 1906; Champion, 1937) vide Dhir, 1980.	Pangtey (1987) relocated this species after 49 years from the same locality.
3.	*Asplenium virde* Hudson	Martoli and Pindari (between 3400-4000 m)	This species relocated after 81 years from Pindari by Dhir (1980). The senior author has also collected from the same locality in 1982.

(Contd.)

(Contd.)

Sl. No.	Name of the plants	Place of occurrence in Kumaun	Remarks
4.	*Athyrium duthie* (Bedd.) Bedd.	Darma and Byans between 4800-5500 m (Hope, 1861 and Beddome, 1892)	There are only two recent collections of this species from India i.e. (i) Dhir (1980) from Nainital (Kumaun), (ii) Smith (1955) from Sikkim.
5.	*Blechnum orientale* Linn.	Mackinnion (1904) vide Dhir (1980) reported this from Garhwal comes under British Kumaun, Duthie (1906) reported this from the same locality.	Recently Pangtey *et al* (1985) collected this species from Nolara (900 m) in Ramganga valley (Pithoragarh district).
6.	*Botrychium virginianum* (L.) Sw.	British Kumaun (Duthie, 1906)	This species is collected by Dhir (1980) from Khati and Punetha (1987) from Pithoragarh district. It is extremely rare.
7.	*Coniogramme affinis* Wall. ex Hieron.	Only Macknnon (1848) vide Dixit and Das (1979) collected from Kumaun and not collected since than till 1980	Recently Verma and Khullar (1980) reported it from Nainital. One of us (PCP) collected this species form Munsyari in 1985.
8.	*Davallodes membranulosum* (Wall. ex Hook.) Copel.	Mohargari pass, 2000-3000 m (Strachey and Winterbottom in 1946-49), below Nainital (Hope, 1984), near Askot (Duthie, 1886), Sarju valley (Inyal, 1900) (cf. Dhir, 1980).	Recently it is collected by Samant (1986) from Patwadangar (1200-1400 m) near Nainital after a lapse of 86 years from Kumaun.
9.	*Dennstaedlia scabra* (Wall. ex Hook.)	Nainital (Hope, 1861; Duthie 1906).	It has been recollected after a lapse of 80 years from Nainital by Pangtey (1987).

(Contd.)

(Contd.)

Sl. No.	Name of the plants	Place of occurrence in Kumaun	Remarks
10.	*Equisetum arvensis* Linn.	Pindari (Strachey and Winterbottom, 1846-49; Duthie, 1906).	Recently one of us (PCP) collected the same species from above Dwali near Pindari Glacier. Quite rare.
11.	*Nephrolepis falcata* (Cav.) Christ.	Nainital (Hope, 1861)	Only Niranjan and Ray (1978-79) reported this species from Pithoragarh district. Extremely rare. The species known by only two collections from Western Himalaya (Both from Kumaun; Hope, 1861 and Niranjan and Ray, 1978-79).
12.	*Peranema cytheoides* D. Don	British Kumaun (Blanford, 1910; duthie, 1906).	Recently Pande (1985) recollected it from Kapkot for the first time from Kumaun is Western Himalayas.
13.	*Polystichum duthie* (Hope C. Chr.)	Kutty valley (Hope, 1899)	It has been recollected by Dhir (1980) from Pindari after 91 years.
14.	*Polystichum lachenense* (Hook.) Bedd.	Darma, Byans and Ralam valley in upper Kali Valley (Duthie, 1906)	Also recollected by Dhir (1980) from Pindari after 74 years.
15.	*Psilotum nudum* (Linn.) P. Beauv.	Kumaun (Thomas-Thomson) Gagas river near Ranikhet (Strachey and Winterbottom, 1946-49).	Recollected after 132 years from Kumaun (Quarab-Almora).
16.	*Pteris subquinata* Wall. ex Agardh	Bageshwar (Beddome, 1892; Duthie, 1906).	Recollected by Dhir (1980) from the same locality.

(Contd.)

(Contd.)

Sl. No.	Name of the plants	Place of occurrence in Kumaun	Remarks
17.	*Stenogramma pozoi* (Lagasca) K. Iwast.	Jageshwar (Duthie, 1906)	Recently Pande *et al* (1985) collected from Jageshwar (1810 m). it is the only place in Kumaun where it is growing occasionally.

CATEGORY III : VULNERABLE AND RARE

Plants of rare occurrence which are under threat.

Sl. No.	Name of the plants	Place of occurrence in Kumaun	Remarks
1.	*Actinopteris radiata* (Sm.) Link.	Ranikhet (Pande, 1973)	Reported by Pande (1973). Not collected since then. Quite rare.
2.	*Adiantum pedatum* Linn.	Below Phurkia (near Pindari)	Phurkia is the only place in Kumaun Himalaya where it is growing in about 1 sq. m area only, along road sides.
3.	*Asplenium adiantum-nigrum* Linn.	Loharkhet	Reported by Dhir (1980) only, not collected afterwards.
4.	*Asplenium aethiopicum* (Burm. f.) Becherer	Ranikhet	Recorded only by Pande (1973). Not collected thereafter.
5.	*Asplenium sarelii* Sarelii	Didihat	Extremely rare, only collected by Punetha (1987).
6.	*Asplenium unilaterale* Lam.	Throughout Kumaun	Quite rare
7.	*Athyrium flabellulatum* (Clarke) Tardieu	Pindari (3300 m)	Recorded only by Dhir (1980). Not collected afterwards.
8.	*Athyrium hohenackerianum* (Kze.) Moore	Gangolihat and Pithoragarh	Recorded only by Niranjan and Roy (1978-79). Not collected afterwards.

(Contd.)

(Contd.)

Sl. No.	Name of the plants	Place of occurrence in Kumaun	Remarks
9.	*Athyrium proliferum* Moore	Ranikhet	Recorded only by Pande (1973). Not collected afterwards.
10.	*Botrychium ternatum* (Thunb.) Sw.	Throughout Kumaun	An extremely rare species of Kumaun Himalaya.
11.	*Ceterach officinarum* DC, in Lam.	Bharari (Almora)	Quite rare, only Dhir (1980) collected it as a new record for Kumaun hills and not collected since then.
12.	*Christella Kumaunicum* Holtt.	Kumaun	Holttum (1976) described it as a new specie from Kumaun and not collected since then.
13.	*Cyathea spinulosa* Wall. ex Hook.	Didihat	Extremely rare, only 2 or 3 individuals are found growing.
14.	*Drynaria tibetica* Ching et Wu	Naina peak (Nainital)	Recorded only by Khullar (1984), however, Ching (1983) recorded the same species from Tibet.
15.	*Dryopteris conjugata* Ching	Pithoragarh (2400 m)	Recorded only by Punetha (1986). Extremely rare.
16.	*Equisetum hymale* Linn.	Dogaon (Nainital)	Recorded only by Saxena (1976) as a new record for India. Not collected afterwards.
17.	*Equisetum ramosissimum* Desf. subsp. *ramosissimum*	Ranibagh (Nainital)	Quite rare, collected by Pande (1985)
18.	*Gymnopteris delavayi* (Bak.) Und.	Garbyan valley	Rare, once collected by Biswas & Biswas (1976).

(Contd.)

(Contd.)

Sl. No.	Name of the plants	Place of occurrence in Kumaun	Remarks
19.	*Helminthostachis zealanica* (Linn.) Hook.	Tanakpur	Quite rare, Joshi & Pande (1986) reported it as a new record for Western Himalaya.
20.	*Lepisorus kuchenensis* (Wu) Ching	Nainital	Recorded only Dhir (1980). Not collected thereafter.
21.	*Lindsaya cultrata* (Willd) Sm.	Bharari (Almora)	Reported only by Dhir (1980). Extremely rate.
22.	*Loxogramme lanceolata* (Sw.) Presl	Khati-Dwali enroute Pindari	Reported only by Dhir (1980). This was first record of its occurrence in Western Himalaya.
23.	*Lygodium circinatum* (Burm.) Sw.	Tanakpur (300 m)	Reported only by Dhir (1980) as a new record for Western Himalaya.
24.	*Mecodium javanicum* (Spr.) Copel.	Khati enroute Pindari	Recorded only by Dhir (1980) as a rare species.
25.	*Microlepis firma* Mett.	Ramganga valley	Rather scarce in Kumaun hills.
26.	*Microlepis stignosa* (Thunb.) Presl	Didihat, Lohaghat	Rather rare in Kumaun Himalaya. Recently collected by Punetha (1987).
27.	*Microsorium zipelii* (Thunb.) Ching	Jageshwar	Reported by Dhir (1980) as a new record for Western Himalaya. Extremely rare.
28.	*Monachosorum subdigitatum* (Bl.) Kuhn.	Pindari	Pane (1987) reported this as new record for Western Himalayas. It is very rare specie.

(Contd.)

(Contd.)

Sl. No.	Name of the plants	Place of occurrence in Kumaun	Remarks
29.	*Onychium fragile* Verma et Khullar	Jageshwar	Pande and Bir (1985) reported this as a new record for Kumaun. Extremely rare.
30.	*Osmunda regalis* Linn.	Pindari and Munsyari	One of the rarest species of ferns of the Kumaun Himalaya.
31.	*Phymatopteris stewartii* (Bedd.) Pic.-Ser.	Lariakanta (Nainital) Bhowali (Nainital and Dhakuri	Hope (1861) and Dhir (1980) reported this from Kumaun. We have never come across this species in the localities mentioned above. But the senior author collected it from Khati enroute Pindari. Extremely rare.
32.	*Phymatopteris strachey* (Ching) Pichi Sermolli	Dhakuri and Kilberry	Extremely rare in Kumaun Himalaya. Dhir (1980) and Verma & Khullar (1980) reported this from Kumaun.
33.	*Polypodium amoenum* Wall. & Mett. var. *Pinnatifida*	Bharari (Almora)	Recorded only by Dhir (1980) described it as a new species.
34.	*Pronephrium nudatum* (Roxb.) Chandra	Found in low altitudes	Rare in Kumaun hills.
35.	*Pyrrosia stigmosa* (Sw.) Ching	Pithoragarh	Rare in Kumaun hills.

Acknowledgement

Financial help provided by CSIR, New Delhi is gratefully acknowledged.

REFERENCES

Awasthi, D.K. and Pande, P.C. (1985), A note on the phytogeographical distribution of ferns and fern-allies of Almora (W. Himalaya), *Jour. Bomb. nat. Hist Soc.*, 81: 739-41.

Atkinson, E.T. (ed.) (1882), *The Himalayan District of North Western Provinces of-India* forming vol. X of the Gazetteer N.W.P., Allahabad. Reprinted in 1973 under the title *The Himalayan Gazetteer*, Cosmo Publications, Delhi-6. Vol. I, Part II; 642-54.

Beddome, R.H. (1892), *A handbook to the Ferns of British India, Ceylon and Malaya Peninsula with supplement*, Calcutta.

Bir, S.S. (1987), Pteridophytic flora of India; rare and endangered elements and their conservation. *Indian Fern J.*, 4(1&2): 425.

Bir, S.S. and Trikha, C.K. (1974), Taxonomic revision of Polypodiaceous genera of India VI *Lepisorus excavatus* group. *amer. Fern J.*, 64: 49-63.

Clarke, C.B. (1880), A Review of the ferns of Northern India, *Trans. Linn. Soc. Lond. Ser.* II (Bot), 1: 425-611.

Datta, A. (1983), Occurrence and distribution of some rare ferns in India in Jain, S.K. and Rao, R.R. (ed), *An Assessment of threatened plants of India*, B.S.I., Howrah, pp. 323-27.

Dhir, K.K. (1980), Ferns of North-Western Himalayas. *Bibliotheco Pteridologica*, Vaduz: *J. Cramer*, 1: 1-158.

Dixit, R.D. and Das, Anjali, (1979), The genus *Coniogramme* Fee in India, *Proc. Indian Acad. Sci.*, 88(B) II, 4: 253-68.

Dixit, R.D. (1982), Rare and interesting pteridophytes of India II, in Jain, S.K. and Rao, R.R. (ed.), *Assessment of Threatened Plants of India*, B.S.I., Howrah, pp. 328-34.

Duthie, J.F. (1906), *Catalogue of Plants of Kumaun and adjacent portion of Garhwal and Tibet* based on the collections made by Strachey and Winterbottom during the years 1846-49, London.

Holttum, R.E. (1976), The genus *Christella* Lavell sect. *Christella*, studies in the family Thelyptridaceae, *Kew Bull.*, 31(2): 293-339.

Hope, C.W. (1899-1904), Ferns of North Western India, Bishen Singh Mahendra Pal Singh, Dehra Dun, India.

Jain, S.K. and Sastry, A.R.K. (1980), Threatened Plants of India, A State of Art Report, New Delhi.

Joshi, G.C. and Pande, P.C. (1987), *Helminthostachis zeylanica* (L.) Hook. (Ophioglossaceae) – A new record for Western Himalaya, *Jour. Bomb. nat. Hist. Soc.*, 84(3): 724-25.

Khullar, S.P. (1984), The fern of Western Himalaya – a few additions, corrections and annotations, *Indian Fern J.*, 1: (1&2): 89-95.

Khullar, S.P. and Gupta, S.C. (1980), Systematics of genus *Polystichum* from the Himalayas, *Nova Hedwigia*, 32: 415–44.

Khullar, S.P. and Sharma, S.S. (1980), Systematics of Himalayan species of *Onychium*, in Bir, S.S. (ed.), *Aspects of Plant Sciences*, Vol. 3, New Delhi, pp. 63-87.

—— (1987), The Ferns of Western Himalaya excluding Uttarakhand, in Pangtey, Y.P.S. *et al.* (eds), *Western Himalaya: Environment, Problems and Development*, Gyanodaya Prakashan, Nainital, U.P., pp. 310-46.

Khullar, S.P., Sharma, S.S. and Chaudhary, Veena (1987), The Ferns of Garhwal Himalaya – a check list, in Pangtey, Y.P.S. et al. (eds.), *Western Himalaya: Environment, Problems and Development*, Gyanodaya Prakashan, Nainital, U.P., pp. 347-38.

Khullar, S.P., Sharma, S.S. and Kaur, P. (1983), The Thelyptridaceae of Western Himalaya, *Nova Hedwigia*, 38: 617-67.

Loyal, D.S. and Verma, C.K. (1960), Ferns of Nainital, J. Bomb. nat. Hist. Soc., 57: 479-90.

Niranjan, A.R.S. and Roy, S.K. (1978-79), Ecological distribution of some ferns of Central Himalayas, *J. Sci. Res.* B.H.U., 29: 227-32.

Pande, P.C. (1973), Pteridophytic flora of Ranikhet, *Indian For.*, 99: 49-52.

—— (1985 a), *Peranema Cytheoides* D. Don From Kumaun Himalaya, *J. Indian bot. Soc.*, 64: 227.

—— (1985 b), Psilotum nudum (Linn) Girsep, from Kumaun Himalaya – rediscovered, *J. Indian bot. Soc.*, 64: 214-15.

Pande, P.C. and Bir, S.S. (1985), Occurrence of *Onychium fragile* Verma et Khullar from Kumaun Himalaya I, *Bomb. nat. Hist. Soc.*, 82(2): 441.

Pande, P.C. and Kandpal, M.M. (1986), Pteridophytic flora of Didihat (Western Himalaya), *Acta Botanica Indica*, 14 (Suppl.): 115-22.

Pande, P.C. and Pande, T.D. (1987), An enumeration of ferns of Binsar (W. Himalaya), *Acta Botanica Indica*, 15: 101-03.

Pande, P.C. and Pangtey, Y.P.S. (1980), New records of fernallies from Kumaun Himalaya, *Indian J. For.*, 10(2): 151-52.

Pande, P.C., Andola, B.D. and Rawat, G.S. (1984), An enumeration of ferns of Jageshwar area in Almora district, *Himal. Res. & Dev.*, 3(2): 39-40.

Pangtey, Y.P.S. and Punetha, N. (1987), Pteridophytic flora of Kumaun Himalaya – an updated list, in Pangtey, Y.P.S *et al* (eds.), *Western Himalaya: Environment, Problems and Development*, Gyanodaya Prakashan, Nainital U.P., pp. 389-412.

Pangtey, Y.P.S, Samant, S.S. and Rawat, G.S. (1985), A note on the occurrence of *Blechnum orientale* Linn. in Kumaun Hills, *J. Econ. Tax. Bot.*, 6: 747-48.

Pangtey, Y.P.S., Kalakoti, B.S., Rawat, G.S. and Pande, P.C. (1982), Observation on the fern flora of Pindari area, *Himal. Res. and Dev.*, 1: 156-60.

Punetha, N. (1985 a), Taxonomic observation on some Athyriod ferns from the Pithoragarh district of Kumaun (North-West Himalaya), *Indian Fern J.*, 2 (1&2): 23-31.

—— (1985 b), Taxonomic observations on some species of Pteris from Pithoragarh district of Kumaun, West Himalaya, *Indian Fern Jour.*, 2(1&2) 65-72.

Punetha, N., Tewari, M. and Kaur, S. (1985), taxonomic studies on the families Loxogrammaceae and Polypodiaceae of the Pithoragarh district (North-West Himalayas), *Indian Fern. Journ.*, 2(1&2): 73-89.

Satija, C.K. and Bir, S.S. (1985), *Aspects of Plant Sciences*, Vol. VIII: *Polypodiaceious fern of India*, Today & Tomorrow's Printers and Publishers, New Delhi.

Saxena, D.K. (1976), Note on the occurrence of *Equisetum hymale* L. in India, *Curr. Sci.*, 20(8): 679.

Singh, A.K. and Rawat, D.S. (1985), Depletion of Oak Forest Threatening Springs: An Exploratory Survey, *National Geogr. Jl. of India*, Vol. 31(1): 44-48.

Verma, S.C. and Khullar, S.P. (1980), Ferns of Nainital (Western Himalaya): an updated list, *Fern Gaz.*, 12: 83-92.

INDEX